Thierry Meyer, Genserik Reniers
Engineering Risk Management

Also of interest

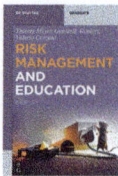

Risk Management and Education
Meyer, Reniers, Cozzani, 2025
ISBN 978-3-11-110464-5, e-ISBN (PDF) 978-3-11-110489-8,
e-ISBN (EPUB) 978-3-11-110732-5

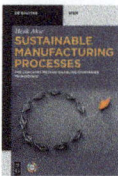

Sustainable Manufacturing Processes.
The Coaching Method Enabling Companies to Innovate
Akse, 2025
ISBN 978-3-11-138342-2, e-ISBN (PDF) 978-3-11-138366-8,
e-ISBN (EPUB) 978-3-11-138380-4

Sustainable Process Engineering
Szekely, 2024
ISBN 978-3-11-102815-6, e-ISBN (PDF) 978-3-11-102816-3,
e-ISBN (EPUB) 978-3-11-103032-6

Sustainable Products.
Life Cycle Assessment, Risk Management, Supply Chains,
Ecodesign Has, 2024
ISBN 978-3-11-131482-2, e-ISBN (PDF) 978-3-11-131546-1,
e-ISBN (EPUB) 978-3-11-131566-9

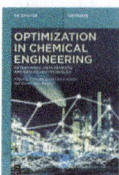

Optimization in Chemical Engineering.
Deterministic, Meta-Heuristic and Data-Driven Techniques
Gómez-Castro, Rico-Ramírez (Eds), 2025
ISBN 978-3-11-138338-5, e-ISBN (PDF) 978-3-11-138343-9,
e-ISBN (EPUB) 978-3-11-138362-0

Thierry Meyer, Genserik Reniers

Engineering Risk Management

Optimizing Operational Risk Decision-Making,
Safety and Reliability, Risk Assessment

4th, Completely Revised and Extended Edition

DE GRUYTER

Authors
MER Dr Thierry Meyer
Ecole Polytechnique Fédérale de Lausanne
EPFL-SB-ISIC-GSCP
Station 6
1015 Lausanne
Switzerland
thierry.meyer@epfl.ch

Prof. Genserik Reniers
Safety and Security Science Section
Department of Values, Technology and Innovation
Faculty of Technology, Policy and Management
Delft University of Technology
Jaffalaan 5
2628 BX, Delft
The Netherlands

ISBN 978-3-11-149058-8
e-ISBN (PDF) 978-3-11-149363-3
e-ISBN (EPUB) 978-3-11-149501-9

Library of Congress Control Number: 2025932338

Bibliographic information published by the Deutsche Nationalbibliothek
The Deutsche Nationalbibliothek lists this publication in the Deutsche Nationalbibliografie;
detailed bibliographic data are available on the Internet at http://dnb.dnb.de.

www.degruyter.com
Questions about General Product Safety Regulation:
productsafety@degruyterbrill.com

Contents

About the authors

Thierry Meyer was born in 1961 in Geneva. He obtained his MSc in chemical engineering at the Swiss Federal Institute of Technology in Lausanne (EPFL) in 1985, followed by a PhD in 1989. He joined Ciba-Geigy Inc. in 1994 as a development chemist in the pigment division, becoming acting head of development and then production manager in 1998. In 1999, he switched to the Institute of Chemical Sciences and Engineering at EPFL, heading the Polymer Reaction Unit till 2004 and since then the Chemical and Physical Safety Research Group. He teaches several courses in the field of industrial and chemical engineering, safety and risk management at bachelor, master and continuing education level. From 2005 to 2015, he was also head of occupational safety and health at EPFL for the Faculty of Basic Sciences and, since 2016, head of the EPFL Safety Competence Centre. In addition, he is the Swiss academic representative in the working party on Loss Prevention and Safety Promotion of the European Federation of Chemical Engineering.

Genserik Reniers was born in 1974 in Brussels. He obtained his MSc in chemical engineering at Vrije Universiteit Brussels and received his PhD in 2006 in applied economic sciences from the University of Antwerp, Belgium. He is full professor at the Safety and Security Science Group of the Delft University of Technology in the Netherlands, where he teaches courses related to risk analysis and risk management. At the University of Antwerp as well as at KU Leuven, both in Belgium in a part-time capacity, he is also professor lecturing among others in advanced engineering risk management. His main research interests concern the collaboration surrounding safety and security topics and socioeconomic and sociotechnical optimization within the chemical industry. He serves as an editor of the *Journal of Loss Prevention in the Process Industries* and as an associate editor of the journal *Safety Science*. He further serves as the Belgian academic representative in the working party on Loss Prevention and Safety Promotion of the European Federation of Chemical Engineering.

https://doi.org/10.1515/9783111493633-203

1 Risk management is not only a matter of financial risk

Risk continues to perplex humankind. All societies worldwide, present and past, face and have faced decisions about how to adequately confront risks. Risk is indeed a key issue affecting everyone and everything. How to effectively and efficiently manage and deal with risks has been, is and will always be a central question for policy-makers, industrialists, academics and actually for everyone (depending on the specific risk). This is because the future cannot be predicted; it is uncertain, and no one has ever been successful in forecasting it. But we are very interested in the future, and especially in possible risky decisions and how they will turn out. We all face all kinds of risks in our everyday life and going about our day-to-day business. So, would it not make sense to learn how to adequately and swiftly manage risks, so that the future becomes less obscure?

Answering this question with an engineering perspective is the heart of this book. If you went to work this morning, you faced a risk. Riding a bicycle, using public transportation, walking or driving all involved risk. Choosing to put your money in a bank, invest in stocks, or hide it under a mattress carried other forms of risk. Buying a lottery ticket introduced an element of chance – closely tied to the concept of risk. Opting for one production process over another or deciding to write a book instead of research papers were also choices involving risk. Even the choice to publish one book over another was a risk. These examples show that "risk" can take on various meanings, depending on our perspective.

The current highly competitive nature of economics might encourage firms to take on risks that could lead to more or higher profits, but at the same time also more or higher possible losses. Risks thus need to be managed with care and attention. Giving up managing such risks would or could indeed be financially destructive or even suicidal, even possibly in the short term, because once disaster has struck, it is too late to rewrite history. Some people might argue that safety is expensive. We would answer that an accident is even more expensive: the costs of a major accident are very likely to be huge in comparison with what should have been invested as prevention.

Let us take, for example, the human, ecological and financial disaster of the Deepwater Horizon drilling rig in April 2010. Are the induced costs of several billions of euros comparable to the investments that could or might have averted the catastrophe? Answering this question is difficult, a priori, because it would require assessing these uncertainties and therefore managing a myriad of risks, even the most improbable. Most decisions related to environment, health and safety are based on the concept that there exists a low level of residual risk that can be deemed as "acceptably low."

https://doi.org/10.1515/9783111493633-001

For this purpose, many companies have established their own risk acceptance or risk tolerance criteria. However, there exists many different types of risk and many methods of dealing with them, and at present, many organizations fail to do so. Taking the different types of risk into consideration, the answer to whether it would have been of benefit to the company to make all necessary risk management investments to prevent the Deepwater Horizon disaster would have been – without any doubt – "yes." And this is actually, and regretfully, the case for all human-made disastrous accidents.

In 2500 BC, the Chinese had already reduced risks associated with the boat transportation of grain by dividing and distributing their valuable loads between six boats instead of one. The ancient Egyptians (1600 BC) had identified and recognized the risks involved by the fumes released during the fusion of gold and silver [1]. Hippocrates (460–377 BC), father of modern medicine, had already established links between respiratory problems of stonemasons and their activity. Since then, the management of risks has continued to evolve:
- Pliny the Younger (first century AD) described illnesses among slaves.
- In 1472, Dr. Ellenbog of Augsburg wrote an eight-page note on the hazards of silver, mercury and lead vapors [2].
- Ailments of the lungs found in miners were described extensively in 1556 by Georg Bauer, writing under the name "Agricola" [3].
- Dating from 1667 and resulting from the great fire that destroyed a part of London, the first Fire Insurance Act was published.

Despite today's steep increase of risk and risk management knowledge and the neverending acceleration of all kinds of risk management processes, what still remains to be discovered in risk management and risk engineering is rather systemic and more complex.

As Ale [4] indicates, the essence of risk was formulated by Arnaud as early as 1662: "Fear of harm ought to be proportional not merely to the gravity of the harm, but also to the probability of the event." Hence, the essence of risk lies in the aspect of probability or uncertainty. Ale further notes that Arnaud treats probability more as a certainty than an uncertainty and as, in principle, measurable. Frank Knight even defines risk in 1921 as a "measurable uncertainty" [5]. Today, the word "risk" is used in everyday speech to describe the probability of loss, either economic or otherwise, or the likelihood of accidents of some type. It has become a common word, and is used whether the risk in question is quantifiable or not. In Chapter 2, we will further elaborate on the true definition and the description of the concept of "risk" and what constitutes it.

In response to threats to individuals, society and the environment, policymakers, regulators, practitioners from the industry and others involved in managing and controlling risks have taken a variety of approaches. Nowadays, the management of risk is a decision-making process aimed at achieving predetermined goals by reducing the number of losses of people, equipment and materials caused by accidents possibly happening while trying to achieve those goals. It is a proactive and reactive approach to accident and loss reduction. We will discuss risk management and its definition in a more general way in the next chapters, and we will define risk as having a positive and a negative side, but we will focus, in the remainder of the book, on managing risks with possibly negative consequences.

Human needs and wants for certainty can be divided into several classes. Abraham Maslow [6] discerned five fundamental types of human needs and ordered them hierarchically according to their importance in the form of a pyramid (Fig. 1.1).

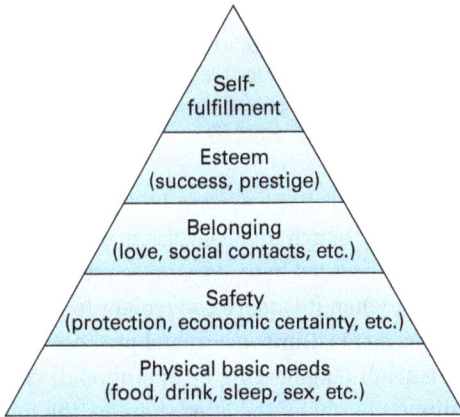

Fig. 1.1: Maslow's hierarchy of human needs.

As long as someone's basic needs at the bottom of the hierarchy are not satisfied, these needs demand attention and the other (higher) needs are more or less disregarded or – at least – they do not have the full attention they deserve. "Safety," or in other words, the striving for a decrease of uncertainty about the negative side of risks, is a very important human need, right above basic needs such as food, drink, sleep and sex. Consequently, if risks are not well managed in organizations, and people are not "safe," the organizations will not be well managed at all. They may focus upon "production" (the organizational equivalent of the physical basic needs), but the organizations will never reach a level in which they excel. Thus, engineering risk management (ERM), as discussed in this book, is essential for any organization's well-being and for its continuous improvement. Summarizing the above, following Maslow's pyramid, organizations that truly embrace safety in their culture will have employees who feel a strong belonging to the organization, there will be more creativity and profit, successfulness will increase and they will prosper.

When reading this book it is necessary to wear "engineering glasses." This means that we will look at risk management using an encompassing engineer's approach – being systemic on top of analytic:

The analytic and the systemic approaches are more complementary than opposed, yet neither one is reducible to the other. In systemic thinking – the whole is primary and the parts are secondary; in analytic thinking – the parts are primary and the whole is secondary.

The analytic approach seeks to reduce a system to its elementary elements in order to study in detail and understand the types of interaction that exist between them. By modifying one variable at a time, it tries to infer general laws that will enable us to predict the properties of a system under very different conditions. To make this prediction possible, the laws of the additivity of elementary properties must be invoked. This is the case in homogeneous systems, those composed of similar elements and having weak interactions among them. Here the laws of statistics readily apply, enabling us to understand the behavior of the multitude of disorganized complexity.

The laws of the additivity of elementary properties do not apply in highly complex systems, like the risk management, composed of a large diversity of elements linked together by strong interactions. These systems must be approached by a systemic policy. The purpose of the systemic approach is to consider a system in its totality, its complexity, and its own dynamics [7].

This book is written by engineers. This entails that it is not written by financial experts, insurers, traders, bankers or psychologists. Although financial risk management is a crucial matter of daily business, it will not be covered here. As a consequence, we will not discuss the topic of money investments, when it is more convenient to invest in bank accounts, shares or other financial placements. Our concern and purpose is to apply engineering methodologies to (nonfinancial) (engineering or operational) risk management. On the other hand, we will discuss in one of the later chapters the microeconomic essentials related to decision-making with respect to engineering risks. We believe that ERM is all about making the right decisions and allocating the budget to deal with engineering risks in the most optimal way. Such optimal decision-making requires being knowledgeable about financial aspects of risks and using microeconomics to aid in the decision-making.

What is the essential idea of (engineering) risk management? We have all heard the saying, "Give a man a fish and you feed him for a day. Teach a man how to fish, and you feed him for a lifetime." We could adapt this expression taking into account a risk management standpoint: "Put out a manager's fires, and you help him for a day. Teach a manager fire prevention, and you help him for a career." If a manager understands good risk management, he can worry about things other than firefighting.

Negative risk (or the negative aspect related to engineering risks – the positive aspect is making profits by producing products or delivering a service) could be described as the likelihood (often expressed as a frequency or probability) and magnitude of a loss, disaster or other undesirable event. In other terms: something bad *might* happen. ERM can be described as the identification, assessment and prioritization of (safety, security, environment, quality and ethics) risks, followed by the coordinated and economical application of resources to minimize, monitor and control the probability (including the exposure) and/or impact/consequences of unfortunate events. In other terms: *being smart about taking chances.*

The first step is to identify the negative risks that a company faces: a risk management strategy moves forward to evaluate those risks. The simplest formula for evaluating specific risks is to multiply (after quantifying the risk in some form) the likelihood of the risky event by the damage of the event if it would occur. In other words, consider the possibility and consequences of an unwanted event.

The keyword here is data. The best risk management specialists excel at *determining predictive data.*

The ultimate goal of risk management is to minimize risk in some area of the company relative to the opportunity being sought, given resource constraints. If the initial assessment of risk is not based on meaningful measures, the risk mitigation method, even if it could have worked, is bound to address the wrong problems. The key question to answer is: "How do we know it works?"

Several reasons could lead to the failure of risk management [8]:
– The failure to measure and validate methods as a whole or in part.
– The use of factors that are known not to be effective (many risk management methods rely on human judgment and humans misperceive and systematically underestimate risks).
– The lack of use of factors that are known to be effective (some factors are proven to be effective both in a controlled laboratory setting and in the real world, but are not used in most risk management methods).

As already mentioned, risk is often measured by the likelihood (i.e., the probability in quantitative terms) of an event and its severity. Of the two, severity is more straightforward, especially after the event. Measuring the probability is where many people encounter difficulties. All we can do is to use indirect measures of a probability, such as observing how frequently the event occurs under certain conditions or, when no information is available, make educated and expert assumptions:
– Risk has to have a component of uncertainty as well as a cost (we have uncertainty when we are unable to quantify exactly accurate the likelihood).
– Risk is generally thought of in highly partitioned subjects with very little awareness of the larger picture of risk management.
– Risk is an uncertain event or set of circumstances that, should it or they occur, will have an effect on the achievement of objectives.

Very often, we hear that experience helps in understanding risk, but:
– experience is a nonrandom, nonscientific sample of events throughout our lifetime;
– experience is memory-based, and we are very selective regarding what we chose to remember;
– what we conclude from our experience can be full of logical errors;
– unless we receive reliable feedback on past decisions, there is no reason to believe our experience will tell us much;

- no matter how much experience we accumulate, we seem to be very inconsistent in its application [8]; and
- so at least there is a need for a collective memory – the total memory of different experts with various expertise backgrounds.

There is space for improving risk management so as to adopt the attitude of modeling uncertain systems: acting as a scientist in choosing and validating models; building the community as well as the organization. Risk management is not a sole matter of managers or risk experts because it involves and affects everyone in the organization and, directly or indirectly, a lot of people who form no part of the organization.

Near-misses (being interpreted as something that "only just did not happen," but was very close to happening) tend to be much more plentiful than actual disasters and, as at least some of them are caused by the same event that caused the near-miss, we can learn something very useful from them.

The evolution of risk management has been influenced by expanding knowledge and tools as well as by the hazards that need to be addressed. Regulatory bodies, which tend to react in response to incidents, over time enacted measures to prevent recurrences. These bodies also have shaped how hazards are identified and controlled.

This book intends to provide the reader with the necessary background information, ideas, concepts, models, tools and methods that should be used in a general organized process to manage risks. The reader will discover that risk management is considered to be a process which affords assurance that:
- objectives are more likely to be achieved;
- unwanted events will not happen or are less likely to happen; and
- goals will be – or are more likely to be – achieved.

The ERM process enables the identification and evaluation of risks, helps in setting acceptable risk thresholds, ranks them, allows for the identification and mapping of controls against those risks and helps identify risk indicators that give early warning that a risk is becoming more serious or is crystallizing. Once the risk management process has been followed and controlled, a risk recording has been produced and risk owners have been identified, the risk should be monitored and reviewed. Risk management is a *never-ending process*, being more iterative than ever with the increasingly fast evolution of emerging technologies.

Every organization has to live with negative risks. They go hand in hand with the positive risks that lead to gains. Managing those negative risks in an adequate manner is therefore a vital element of good governance and management. Several types and sources of risks – either internal or external to the activity – may affect a business project, process, activity, etc. Some risks may be truly unpredictable and linked to large-scale structural causes beyond a specific activity. Others may have existed for a long time or may be foreseen to occur in the future.

As already indicated, the key idea of risk is that there is uncertainty involved. If compared with life [9], the only certainty in life is death, and the uncertainty lies in when and how death occurs. People strive to delay the final outcome of life and try to improve the quality of life in the interim. Threats to these interim objectives involve risks, some natural, some man-made, some completely beyond our control, but most of them controllable and manageable.

In summary, it is not always obvious whether something is good/right or bad/ wrong, as is often the case in movies. Some action can be bad for one person, but it can be good for another. Or something can be wrong at first sight, but it can be right in the long run. Or it can be bad for few, but good for many, etc. Deciding whether some action or process is good or bad is thus more complex than it seems on the surface. Life is not black or white, as it is so often pictured in stories and fairy tales. Life is colorful and complex. Dealing with life is complex. Risks are not only a natural part of life, they can even be compared with life. Dealing with risks is much more complex than it appears at first view: there are positive and negative sides to most risks and thus they are not black or white. Handling risks is therefore not easy or straightforward. Uncertainties in life are the very same as those related to risks. Decisions in life or in organizations are comparable with those related to risks. Making the right decisions in life (which is often very difficult because of the complex character of the decision-making process, missing information, etc.) leads to a prosperous and possibly longer life. Making the right decisions in business (which is also difficult because of the complexity of the problems at hand and the uncertainties accompanying available information) leads to sustainable profits and healthy organizations.

To make the right decisions time and time again, risks should be "engineered"; that is, they should be managed with engineering principles. This book was written to help anyone make the right decisions in life and in business, and thereby continuously improve his/her, or in case of an organization its, position.

References

[1]   Ramazzini, B. (1713). De Morbis Artificum Deatriba. In: Diseases of Workers. Wright, W.C., Editor. Chicago, IL: University of Chicago Press, 1940.
[2]   Rosen, G. (1976). A History of Public Health. New York: MD Publications.
[3]   Agricola, G. (1556). De Re Metallica T. 12th edn. Hoover, H.C., Hoover, L.H., Editors. London: The Mining Magazine, 1912.
[4]   Ale, B.J.M. (2009). Risk: An Introduction. The Concepts of Risk, Danger and Chance. Abingdon, UK: Routledge.

[5]   Knight, F.H. (1921). Risk, Uncertainty and Profit. 2005 edn. New York: Cosimo Classics.
[6]   Maslow, A.H. (1943). A theory of human motivation. Psychol. Rev. 50: 370–396.
[7]   Umpleby, S.A., Dent, E.B. (1999). The origins and purposes of several traditions in systems theory and cybernetics. Cybernet Syst. 30: 79–104.
[8]   Hubbard, D.W. (2009). The Failure of Risk Management. Hoboken, NJ: John Wiley & Sons Inc.
[9]   Rowe, W.D. (1977). An Anatomy of Risk. New York: John Wiley & Sons Inc.

2 Introduction to engineering and managing risks

2.1 Managing risks and uncertainties: an introduction

Traditionally, two broad categories of management systems can be distinguished: business management systems and risk management (RM) systems. The former are concerned with developing, deploying and executing business strategies, while the latter focus on reducing safety, health, environmental, security and ethical risks.

Business management systems specifically aim at improving the quality or business performance of an organization, through the optimization of stakeholder satisfaction, with a focus on clients – such as the ISO Standard 9001:2008 [1] – or extended to other stakeholders (e.g., employees, society and shareholders) – such as the EFQM 2010 Model for Business Excellence [2] or the ISO 9004:2009 Guidelines [3].

Some of the most popular generic examples of RM systems are the international standard for environmental management (ISO 14001:2015 [4]), the European Eco-Management and Audit Scheme (EMAS) [5], the internationally acknowledged specification for occupational safety and health (OHSAS 18001:2007 [6]) and the international standard for integrity management SA 8000 [7] and the ISO 45001:2018 Occupational Health and Safety Management Systems Requirements, which help organizations to improve employee safety, reduce workplace risks and create better, safer working conditions.

The boundaries of those two categories have seemed to fade in recent years. First, as a result of successive revisions, the increase in alignment and compatibility between the specific management systems has weakened boundaries [8]. Second, the introduction of a modular approach with a general, business-targeted, framework and several subframeworks dealing with specific issues, such as RM, as shown by the development of the EFQM Risk Management Model in 2005 has also caused distinctions to fade. Third, this evolution is illustrated by the emergence of integrated RM models, which emphasize both sides of risks: the threat of danger, loss or failure (the typical focus of RM) as well as the opportunity for increased business performance or success (the typical focus of business management). Examples of this last category are the Canadian Integrated Risk Management Framework (2001) [9] and the Australian–New Zealand standard AS/NZS 4360:2004 [10], which served as the basis for the development of the generic ISO Risk Management Standard 31000:2018 [11]. It is interesting to consider the latest insights of ISO 31000:2018 to discuss the risk concept.

A *risk* is defined by ISO 31000:2018 as "the effect of uncertainties on (achieving) objectives"; an effect is a deviation from the expected [11]. Our world can indeed not be perfectly predicted, and life and businesses are always and permanently exposed to uncertainties that have an influence on whether objectives related to a certain event or action will be reached or not. The only certainty there is about risks is that

https://doi.org/10.1515/9783111493633-002

they are characterized by uncertainty. All other features result from assumptions and interpretations.

The ISO 31000:2018 definition implies that risks (financial as well as nonfinancial, technological) are two-sided: the negative side of a risk manifests itself in the case that the outcome is negative and is not in line with the objectives related to the risk event, and the positive side of a risk is shown when the outcome turns out positive and in line with the objectives of the event, always from the perspective of the same entity or person. Remark that for one and the same risk event (following a management decision, for instance), both sides (obtaining or not obtaining the objectives) are always present for a decision-maker (Fig. 2.1). It is therefore a straightforward assumption that organizations should manage risks in a way that the negative outcomes are minimized and that the positive outcomes are maximized. This is called RM and contains, among others, a process of risk identification, analysis, evaluation, prioritization, handling and monitoring (see below) aimed at being in control of all existing risks, whether they are known or not and whether they are positive or negative.

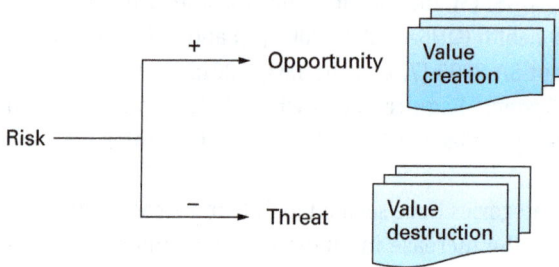

Fig. 2.1: Negative and positive sides of risk.

Continuously being concerned with the positive as well as the negative sides of risk decisions in an organization, in other words "continuous RM" lowers the risk of disruption and assesses the potential impacts of disruptions when they occur. It has been developed over time and within many sectors in order to meet diverse needs. The adoption of consistent processes within a comprehensive framework can help ensure that all types and numbers of risk are managed effectively, efficiently and coherently across an organization. In current industrial practice, RM is usually mainly, or even only, focused on the negative side of risks, and only on avoiding losses, instead of simultaneously avoiding losses and producing gains. This is mainly because companies tend to consider "risk managers" as "negative side of risk managers" for historical reasons: risk managers have been appointed in organizations mainly to satisfy legislative requirements or because of incidents and accidents that happened within firms. Hence, the only risks that needed to be managed displayed possible negative consequences. Nonetheless, to take all aspects of risks into account and to take optimal decisions, risks should ideally be viewed from a holistic viewpoint, meaning that all relevant stakeholders and experts should be involved in the RM process and that all possible knowledge and know-how should be present, to feed decisions that, besides trying to avoid losses, should try to optimize the gains as well.

The best available "classic viewpoint" of RM is one of (narrow) thinking of purely organizational RM and taking different domains within the company into account: integrating them to take decisions for maximizing the positive side of risks and minimizing the negative side. The process for taking decisions in organizations should however be more holistic than this. Risks, or attaining certain objectives, are characterized by internal as well as external uncertainties, while trying to attain the objectives. Hence, all – or as many as possible of – these uncertainties and their possible outcome(s), attaining the objectives ("positive side of risk") or not attaining the objectives ("negative side of risk"), should be anticipated, that is, identified and mapped, for every risk, and by different types of experts and stakeholders. If this can be achieved, the optimal decisions can be taken. The end goal is to use all the right people and means, at the right time, to manage all existing risks (scenarios, objectives and uncertainties) in the best possible way, or whether they are known or not. After all, a part of RM is improving our perception of reality and "thinking about the unthinkable."

There are two ways of making profits. The first way is very obvious and easy to understand by managers: there is a cause–consequence relationship between productivity/production and innovativeness, on the one hand, and profits, on the other. When the former goes up/down, the latter also increases/decreases. The second way is much more difficult to understand by many managers: that is, the relationship between safety and security on one side, and making profits on the other. Safety and security are all about avoiding losses, and avoided losses can be seen as hypothetical profits for an organization. Losses that you never have are costs that you do not have to make, hence (virtual) profits. For achieving sustainable profitability in an organization, both investments in productivity and innovation as well as engineering RM (ERM) (safety and security) investments are therefore needed. Let us use the metaphor of a football tournament: the attackers can be seen as the production people and the defenders as the safety/security managers. Football matches can be won by excellent attackers. Football tournaments (winning match after match) can only be won if the defenders are also excellent, and losses are kept to an absolute minimum. Managers should always remember that "unsafety is patient."

To manage uncertainties efficiently, a composite of three building blocks (knowledge and know-how, stakeholders and expertise and mindset) can be conceptualized, as illustrated in Fig. 2.2.

For each of the building blocks needed by an organization to efficiently perform "uncertainty management," some recommendations can be suggested. To deal with the knowledge and know-how needs, cognition should be present in the organization about past, present and (e.g., scenario-based) future data and information, and about risk-taking, risk-averse and risk-neutral items. Furthermore, the organization should

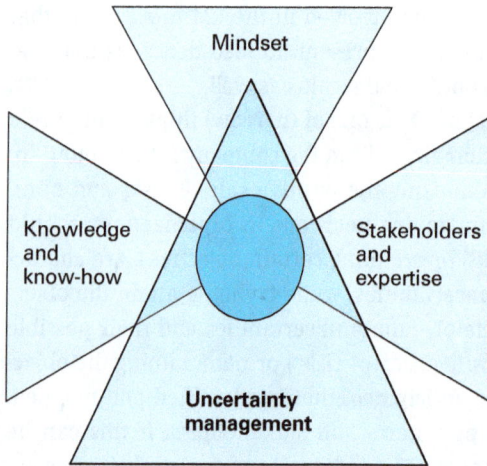

Fig. 2.2: Required capabilities for adequate uncertainty management.

collect information on laws and regulations, rules and guidelines and best practices and ideas and also on the procedural, technological and people domains. To address the stakeholders and expertise building block, involvement should be considered from different organizations, authorities and academia as well as from other stakeholders (clients, personnel, pressure groups, media, surrounding communities, etc.), and from different types of disciplines (engineers, medical people, sociologists, risk experts, psychologists, etc.), and used where deemed interesting. The mindset building block indicates that, besides existing principles, some additional principles should be followed for adequate uncertainty management: circular, nonlinear and long-term thinking; law of iterated expectations; scenario building and the precautionary principle; and operational, tactical and strategic thinking. All these requirements should be approached with an open mind.

RM can actually be compared with mathematics: both disciplines are commonly regarded as "auxiliary science" domains, helping other "true sciences" to get everything right. In the case of mathematics, true sciences indicate physics, chemistry, biochemistry and the like: without mathematics, these domains would not be able to make exact predictions, conclusions, recommendations, etc. The same holds for risk/uncertainty management: without this discipline, applied physics, industrial chemistry, applied biochemistry, etc. will yield suboptimal results, and the achievement of objectives will be hampered. Hence, mathematics is needed for correct laws in physics, chemistry, etc., and RM is required for optimized applications in physics, chemistry, etc. Figure 2.3 illustrates this line of thought.

Bearing Fig. 2.3 in mind, although the management of the negative side of risks has proven to be extremely important over the past decades to save lives and/or to avoid illnesses and injuries in all kinds of organizations, major and minor accidents as well as occupational illnesses are still very much present in current industrial practice. Hence, there is still room for largely improving all management aspects of such negative side of risks, and it can be made much more effective. How this can be theoretically achieved is treated in this book. We thus mainly focus on theories, concepts, techniques, models, frameworks, metaphors and practical examples related to unwanted events leading to undesirable consequences, in other words, in the field of management of negative uncertainties and risks.

> Uncertainties and risks cannot be excluded from our lives. Therefore, we should always be able to deal with them in the best possible way. To this end, a risk should be viewed as a two-sided coin: one side is positive and the other side is negative. Bearing this in mind, we need the right mindset, enough information and the right people to deal with risks and uncertainties.

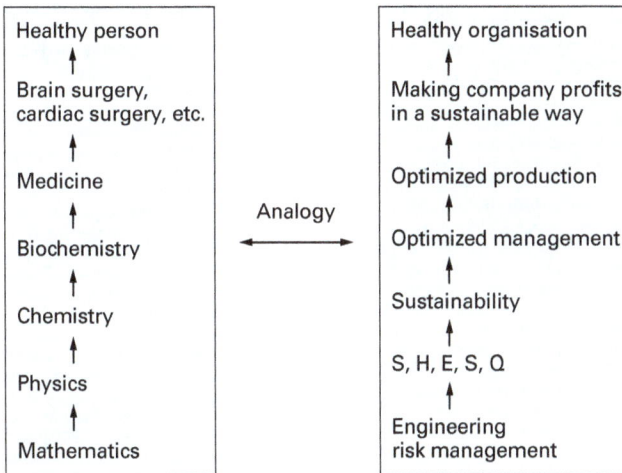

Fig. 2.3: Analogy between exact sciences and industrial practice.

2.2 The complexity of risks and uncertainties

"Risk" means different things to different people at different times. However, as already mentioned, one element characterizing risk is the notion of uncertainty. That is, for the omniscient or omnipotent, the "risk" concept would be incomprehensible. Nonetheless, in the real world, the future of events, handlings, circumstances, etc. cannot be perfectly predicted. Unexpected things happen and cause unexpected events. An "uncertainty sandglass" can thus be drafted, and strengths, weaknesses, opportunities and threats (SWOT, see also Section 4.2) can be identified. Figure 2.4 displays the uncertainty sandglass, with the SWOT elements situated within the concept.

A SWOT analysis is a strategic planning method that specifies the objectives of, e.g., a business venture or a project and aims to identify the internal and external factors that are favorable and unfavorable to achieving those objectives. Using the principles of a SWOT analysis (identifying positive as well as negative uncertainties) allows risk managers to close the gap between line management and corporate or top management regarding taking decisions on existing uncertainties in the organization. By not only focusing on the negative side of risks (weaknesses and threats), but also emphasizing the positive side of risks (strengths and opportunities), an approach is adopted by the risk manager that is more easily understandable by the top management.

As mentioned above, in this book, we only focus on the top triangle of the uncertainty sandglass: negative uncertainties – exposure to uncertainties – not achieving objectives. If we translate this into "(negative) (safety-related) risk management language," the upper triangle should be "hazards–exposure to hazards–unintentional losses." This triangle can be called the "safety risk trias." The "security risk trias" can be seen in an analogous way and consists of "threats (or the intentional misuse of hazards)–vulnerability to threats–intentional losses."

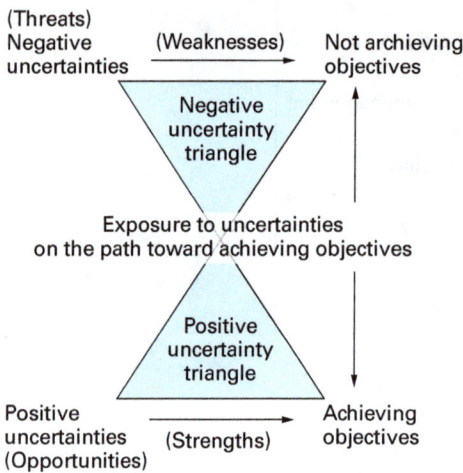

Fig. 2.4: The uncertainty sandglass.

If one of these elements is removed from one of these triangles ("trias" in Latin), there is no safety/security risk. The engineering aspects of RM, discussed in this book, focus on how to omit, diminish, decrease or soften as much as possible one (or a combination) of the three elements of the "safety/security risk trias" (hazards/threats, exposures/vulnerabilities or unintentional/deliberate losses), or a combination thereof.

Roughly, two types of uncertainties can be identified:
- Type I – uncertainties where a lot of historical data is available.
- Type II – uncertainties where little or very little historical data is available.

Remark that the type of uncertainties where no historical data is available can be seen as an extremum of the second type of risks.

From the viewpoint of the negative side of risks, consequences of type I uncertainties mainly relate to individual employees (e.g., most work-related accidents). The outcome of type II uncertainties may affect a company or large parts thereof, e.g., large explosions and internal domino effects – for this sort of accident, the reader is referred to Lees [12], Wells [13], Kletz [14, 15], Atherton and Gil [16] and Reniers [17]. The uncertainties that have unprecedented and unseen impacts upon the organization and society, so-called black swans, can, as mentioned above, be seen as an extremum of type II uncertainties.

Thus, whereas type I "negative" risks lead to most work-related accidents, such as falling, little fires, and slipping, type II "negative" risks can result in catastrophes with major consequences and often with multiple fatalities. Type II accidents do occur on a (semi-)regular basis from a worldwide perspective, and large fires, large releases, explosions, toxic clouds, etc. belong to this class of accident. The extremum of the second type of risks, the black swans, may transpire into "true disasters" in terms of the loss of lives and/or economic devastation. These accidents often become part of the collective memory of humankind. Examples include disasters such as Seveso (Italy, 1976), Bhopal (India, 1984), Chernobyl (USSR, 1986), Piper Alpha (North Sea, 1988), 9/11 terrorist attacks (USA, 2001) and more recently Deepwater Horizon (Gulf of Mexico, 2010), Fukushima (Japan, 2011) and Tianjin (China, 2015).

To prevent type I risks from turning into accidents, RM techniques and practices are widely available. We will discuss some of those techniques in Chapter 4. Statistical and mathematical models based on past accidents can be used to predict possible future type I accidents, indicating the prevention measures that should be taken.

Type II uncertainties and related accidents are much more difficult to predict via commonly used mathematical models because the frequency with which these events happen is too low, and the available information is not sufficient to be investigated via, e.g., regular statistics. The errors of probability estimates are very large, and one should thus be extremely careful while using such probabilities. Hence, managing such risks is based on the scarce data that are available and on extrapolation, assumption and expert opinion, e.g., see [18]. Such risks are also investigated via available RM techniques and practices, but these techniques should be used with much more caution as the uncertainties are much higher for these than for type I risks. A lot of risks (and latent causes) are present that never turn into large-scale accidents because of adequate RM, but very few risks are present that turn into accidents with huge consequences. Hence, highly specific mathematical models (such as QRAs, see Section 4.9) should be employed for determining such risks. It should be noted that the complex calculations lead these risks to appear accurate; nonetheless, they are not accurate at all (because of all the assumptions that have to be made for the calculations), and they also (besides, but even more so, than the type I risks) should be regarded and treated as relative risks (instead of absolute risks). As Balmert [19] indicates, physicist

Richard Feynman's dissenting opinion in the Rogers Commission Report [20] offers the perfect perspective: by NASA's RM protocols, perhaps among the most sophisticated work on risk, the failure rate for the Space Shuttle orbiter was determined to be 1 in 125,000. Feynman used a commonsense approach, asking scientists and engineers the approximate failure rate for an unmanned rocket. He came up with a failure rate between 2% and 4%. With two failures – *Challenger* and *Columbia* – in 137 missions, Feynman's "back-of-the-envelope" calculation was by far the more accurate.

The uncertainties characterizing black swans [21] are extremely high, and their related accidents are simply impossible to predict. No information is available about them, and they are extremely rare. They are the result of pure coincidence and cannot be predicted by past events in any way; they can only be predicted or conceived by imagination. For such accidents, we refer to Section 2.9. Such events can truly only be described by "the unthinkable" – which does not mean that they cannot be thought of, but merely that people are not capable of (or mentally ready for) realizing that such events really may take place.

Type I unwanted events can usually be regarded as "occupational accidents" (e.g., accidents resulting in the inability to work for several days and accidents requiring first aid). Type II (and its extremum which is Black swans) accidents can mostly be categorized as "major accidents" (e.g., multiple fatality accidents and accidents with huge economic losses).

The management implications of the typology of risks and accidents into the two types are that the positive side of the risk coin goes hand in hand with the negative side of the coin, that is, if a manager is interested in making limited profits by starting or increasing an action/activity, he/she should also take into account the possibility of limited negative consequences (type I risks). If this manager is however taking action to make huge profits, he/she should at the same time be aware that these potential profits go together with possible huge losses (type II risks).

Another way of explaining this distinction between possible events is by using a slightly adapted table from the statement of Donald Rumsfeld on the absence of evidence linking the government of Iraq with the supply of weapons of mass destruction in terrorist groups (US DoD) [22]. Figure 2.5 illustrates the different types of events.

Events are classified into different groups, based, on the one hand, on the available knowledge or information on events from the past, and, on the other hand, on the fact that people have an open mind toward the possibility of the event. Looking at the knowledge of events from the past, or the lack of it, in combination with the level of open mindedness of people toward the event, four separate groups can be distinct:

1. Events that we do not know from the past (they have never occurred), but look at with closed minds. These events are called "unknown unknowns" (e.g., 9/11 and Chernobyl disaster).

2. Events that we know from the past (we have certain information or records about them), but look at with closed minds. These events are called "unknown knowns" (e.g., vapor cloud explosion and Covid-19 pandemic).
3. Events that we know from the past and have open minds toward them. These events are called "known knowns" (e.g., transport accidents and falling down the stairs).
4. Events that we do not know from the past, but have open minds toward them. They are defined as "known unknowns" (e.g., electromagnetic radiation).

	Knowledge	Lack of knowledge
Open mind	Known known	Known unknown
Closed mind	Unknown known	Unknown unknown

Fig. 2.5: Table differentiating between known and unknown.

It should be clear that black swans are classified as *unknown unknowns*, e.g., Fukushima (2011) and Deepwater Horizon (2010). Type II events can be regarded as *unknown knowns* (e.g., Tianjin (2015), Buncefield (2005), Toulouse (2001) and the Covid-19 pandemic) and *known unknowns* (e.g., emerging technologies such as nanotechnology or genetically modified food). Type I events are obviously the *known knowns*. The last category, the unknown unknowns are the black swans.

We refer to Chapter 7 for a discussion of some of these accidents. Hence, an open mind and sufficient information are equally important to prevent all types of accidents from occurring.

> Two types of negative risks exist: type I encompasses all risks where a lot of information is available, and type II encompasses all risks where only very scarce information is available. A third possibility, the black swan, can be regarded as an extremum of the second type, and encompasses all risks where no information is available.

2.3 Hazards and risks

From the "safety risk trias," it is obvious that an easy distinction can be made between what is a *hazard* and what is a (safety) *risk*. Remark that an identical reasoning can be followed if talking about security risk. Further in the book, we will only use the expression "risk trias" to indicate both safety and security risk triases.

A *hazard* can be defined as "The potential of a human, machine, equipment, process, material or physical factor to lead to an unwanted event possibly causing harm to people, environment, assets or production." Hence, a hazard is a disposition, condition or situation that may be a source of an unwanted event leading to potential loss. Disposition refers to properties that are intrinsic to a hazard and will be harmful under certain circumstances. Following the definition of hazard, a risk clearly depends on the context of a situation: depending on circumstantial factors, the outcome or the result of the hazardousness may differ greatly.

Although *risk* is a familiar concept in many fields and activities including engineering, law, economics, business, sports, industry and also in everyday life, various definitions exist. To link with the "risk trias," the following definition of *(negative) risk* can be coined:

> The possibility of loss (injury, damage, detriment, etc.) created by exposure to one or more hazards *under a specified scenario*. Risk is an abstract notion which has been developed to understand a variety of potential futures ("scenarios"), and what are possibly the most efficient and effective ways to obtain certain objectives (and to avoid undesired side-effects of these objectives), considering the hazards surrounding a situation. Risks (scenarios) preferably can be calculated using mathematics and are always relative to the position (prioritization) of other risks (scenarios). The significance of risk is a function of the likelihood of an unwanted event, and the severity of its consequences, both under a specified scenario.

The event likelihood may be either a frequency (the number of specified events occurring per unit of time, space, etc.), a probability (the probability of a specified incident following a prior incident) or a qualitative expression (the "likelihood"), depending on the circumstances and the information. The magnitude of the loss determines how the risk is described on a continuous scale from diminutive to catastrophic.

Kirchsteiger [23] points out that in reality risk is not simply a product type of function between likelihood and consequence values, but an extremely complex multiparametric function of *all* circumstantial factors surrounding the event's source of occurrence. However, in order to be able to make a reproducible and thorough risk assessment, and since a risk can be regarded as a hazard that has been quantified, it is important for the usability of the concept, to establish a formula for the quantification of any scenario related to a hazard. In the currently accepted definition, risk is calculated by multiplying the likelihood of an unwanted event by the magnitude of its probable consequences:

$$\text{Negative risk} = (\text{likelihood of unwanted event}) \times (\text{severity of consequences of unwanted event})$$

Thus, if we are able to accurately and quantitatively assess the likelihood of an event/ scenario, as well as the probable severity of the event's/scenario's consequences, we find a quantitative expression of risk of that event (or scenario). A risk can thus be expressed by a number. It should be noted that the *exposure* factor of the risk trias is part of the likelihood assessment. As Casal [18] points out, such an equation of risk is very convenient for many purposes, but it also creates several difficulties, e.g., determining the units in which risk is measured. Risk can be expressed in terms of number of fatalities, the monetary losses per unit of time, the probability of certain injuries to people, the probability of a certain level of damage to the environment, etc. Also, in some cases, especially in type II (and certainly black swan) events, it is obviously very difficult to estimate the likelihood of a given unwanted event and the magnitude of its consequences. To be able to obtain adequate risk assessments, appropriate methods and approaches are used, as explained further in this book.

Examples of the different risk trias elements are given in Tab. 2.1.

Table 2.1 clearly demonstrates that if either the hazard, the exposure or the loss are avoided or decreased, the risk will diminish or even cease to exist. Hence, RM techniques are always aimed at trying to avoid or decrease one or several of the risk trias elements, as mentioned before in this book. The usefulness of risk analysis and RM is also easy to understand. An adult working with a fryer will be less risky than a child working with it. Prevention measures are thus obviously more important in case of the latter situation. Also, a trained and competent worker using the toxic material will lead to a lower risk than an unskilled and incompetent worker. In the same way, an automated system will lead to lower risk levels than a human-operated system, in the case of look-alike product storage.

Tab. 2.1: Illustrative examples of different concepts of the risk trias.

Example	Hazard	Exposure	Loss	Risk
Fryer in use	Heat	Person working with the fryer	Burns	Probability of children having burns of a certain degree
Toxic product	Toxicity	Person working with the toxic product	Intoxication	Probability of worker being intoxicated with a certain severity
Storage of products that look alike	Looking alike	Order by client of one of the look-alike products	Wrong delivery to client	€10,000 claim by client over a certain period of time

Furthermore, it is essential to realize that *a risk value is always relative*, never absolute, and that risk values should always be compared with other risk values, to make RM decisions. Risks should thus be viewed relative to each other – always.

Moreover, by looking at the likelihood values and the consequence values, and taking all the many aspects into consideration, it is possible to put different kinds of risk to proper perspective. This is important as the risk value may depend on the viewpoint from which it is looked at. For example, there often remains a gap between the technical and the nontechnical view of risk [24]. As an example, if the safety record of an individual airplane is, on average, 1 accident in every 10,000,000 flights, engineers may deem air travel to be a reliable form of transport with an acceptable level of risk. However, when all the flights of all the airplanes around the world are accumulated, the probability is that several airplane crashes will occur each year – a situation borne out of observed events. Each crash, however, is clearly unacceptable to the public, in general, and to the people involved in the crashes, in particular.

2.4 Simplified interpretation of (negative) risk

From the above, it is clear that risk is a very theoretical concept and cannot be felt by human senses; it can be only measured or estimated. Risk possesses a value – its criticality – which can be estimated by the association of the constitutive elements of risk in a mathematical model. Always assume that we stay in the area of its probability.

The simplest model defines that the probability of a certain risk depends:

- on the frequency by which the target is exposed to the hazard (sometimes called likelihood of occurrence), supposing that the hazard threatens the target (F = frequency, in the formula below) and
- the evaluation of its consequence corresponding to a measurement of the severity of the mentioned consequences (G = gravity = consequences, in the formula below).

This model for calculating a risk related to a specific scenario is generally expressed by

$$\text{Risk} = \text{frequency} \times \text{severity}; R = F \cdot G$$

Remark that it is also possible to use an aversion factor a in the formula, as an exponent of the severity. Using such an aversion factor puts more or less weight on the consequences and can take the attitude of the decision-maker (see also Section 2.6 in this book) into consideration. An aversion factor lower than 1 reflects a risk-seeking attitude, while an aversion factor higher than 1 represents a risk-averse attitude. In the above case, where $a = 1$, a risk-neutral attitude is assumed.

To be consistent with the concept that measures have been taken against a hazardous situation, the notions of protection and prevention are taken into account in the following formula:

$$R = F \cdot G = \left(\frac{N \cdot T}{\text{Pre}}\right) \cdot \left(\frac{D}{\text{Pro}}\right) \qquad (2.1)$$

The likelihood of occurrence or frequency F depends on
- N – number of set targets
- T – average exposure time of each target at risk
- Pre – prevention implemented to influence N or T

Severity or gravity G is the function of
- D – "crude" hazard of the situation
- Pro – level of protection implemented in the light of this hazard

This formula indicates the possible pathways and solutions to reduce risk. Reducing risk means to act on
- The severity → reduce the hazard and increase protection measures
- The frequency → reduce the exposure time, reduce the number of exposed targets and increase prevention measures

Changing the risk by reducing the occurrence of its components and/or their severity means addressing the following questions:
- Is it possible to reduce the number of exposed targets (N)?
- Is it possible to reduce the time the targets are exposed to the hazard (T)?
- Is it possible to increase the prevention measures (Pre)?
- Is it possible to reduce the hazardousness (D)?
- Is it possible to increase the level of protection (Pro)?

Zero risk is impossible unless the hazard itself is zero or nonexistent, or if there is no exposure to the hazard. The results of the evaluation of risk are often presented in a matrix-like manner. An illustrative example is depicted in Fig. 2.6; the terms of likelihood of occurrence and the severity of the damages are easy to understand. The analyzed situation is quantified in terms of frequency and severity and placed in the corresponding cell (see also Section 4.8).

Thus, quantifying risk does not only convey information about its likelihood and severity. Making risk fully concrete is challenging because it is difficult to fully comprehend. However, with the illustration shown in Fig. 2.7, it is possible to understand the different classes of (simplified) risk. Let us consider the drop of a spherical object – there are several possibilities:
- A – low probability, low damage = low risk
- B – high probability, low damage = medium risk

- C – low probability, high damage = medium risk
- D – high probability, high damage = high risk

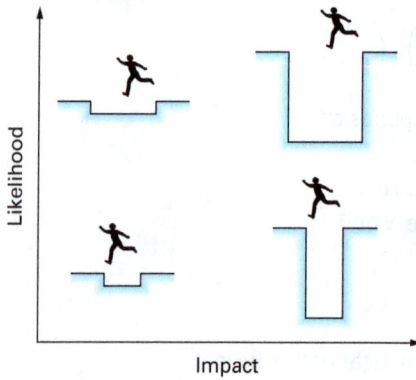

Fig. 2.6: Matrix representation of risk calculation.

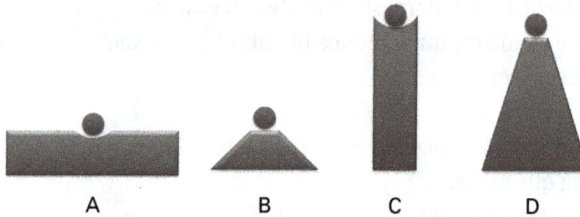

Fig. 2.7: Model of risk classes for a falling spherical object.

In order to be even more concrete, let us illustrate the risk by a concrete example: a fall from a cliff (Fig. 2.8). We could note that the parameters present in eq. (2.1) can be materialized as (only a few examples are presented) follows:
- D = height of the cliff, ground, slippery ground on top of the cliff
- Pre = level of protective fences, warnings signs
- Pro = being secured with a rope, having a loose soil, e.g., water
- T = average time for a walk along the cliff
- N = number of people walking along the cliff

The risk of a person falling will depend on:
- If the surface is slippery
- If there are fences or guardrails
- If the person is secured
- Aggravating factors (weather conditions or physical state of the person)

A cliff might be very risky and
(i)   a little dangerous if
  –   the grass surface is slippery,
  –   there are no guardrails,
  –   the cliff height is low and
  –   there is loose soil at the cliff bottom;
(ii) very dangerous when:
  –   the cliff is without fences,
  –   the cliff is high (even if there are warning signs) and
  –   there is hard ground at the bottom (rocks instead of sand or water).

We could also imagine that some worsening factors (the context) can play a role, e.g., if mist is present and obscuring the edge of the cliff, a strong wind could also worsen that risk. This example emphasizes that even if the reduced formula allows, estimating the risk may be simple, its interpretation and real estimation may be difficult and profoundly depends on the circumstances.

Fig. 2.8: Image of a cliff to illustrate risk determination.

To exist, risks need three factors: hazards, losses and exposure. Decreasing (or even avoiding or taking away) one of these parameters, or (one of) the characteristics of these parameters, in one way or another, leads to lower risks (or even to no risk).

A good and timely exercise for the reader would be to think about the different parameters of the formula in relation to the past Covid-19 pandemic and all its features and possible measures. Another exercise would be to elaborate the parameters with respect to risks at and around the reader's home.

2.5 Hazard and risk mapping

This section is concerned with providing information and increasing the "knowledge and know-how" domain of Fig. 2.2. A hazard map identifies areas that are impacted by or susceptible to a specific hazard. Its purpose is to inform people about the potential extent of damage and relevant accident prevention (and mitigation) measures. Providing clear and understandable information is essential. For example, we could illustrate this by natural disaster prevention. This is necessary to protect human lives, properties and social infrastructure against disaster phenomena. Two types are possible:

– Resident educating: This type of map aims to inform the residents living within the damage forecast area of the risk of danger. The information on areas of danger or places of safety and basic knowledge on disaster prevention are given to residents. Therefore, it is important that such information is represented in an understandable form.

– Administrative information: This type of map is used as the basic materials that administrative agencies utilize to provide disaster prevention services. These hazard maps can be used to establish warning and evacuation systems, as well as evidence for land-use regulations. They may also be used in preventive works.

Hazard mapping provides input to educational programs to illustrate local hazards, to scientists studying hazard phenomena, to land-use planners seeking to base settlement locations to reduce hazard impacts and to combine with other information to illustrate community risks. A map of each type of hazard has unique features that need to be displayed to provide a picture of the distribution of the variations in the size and potential severity of the hazard being mapped. While fieldwork provides data input for many hazard maps, some data will need to be collected from past impacts, from archives maintained of records collected from beneath the built environment, from instruments and/or from the analysis of other data to develop a model of future impacts. The methods for collecting data (e.g., surveys) are still important in providing input to producing hazard maps. Understanding potential hazards typically involves identifying adverse conditions (e.g., steep slopes) and hazard triggers (e.g., rainfall). Together, conditions and triggers help define the potential hazard.

The primary challenge in hazard mapping is quantifying the intensity of the hazard. Focusing on workplace hazards, Marendaz et al. [25] developed an extensive list of specific hazards present in research institutions. A key advantage of the ACHiL platform is that it provides clear and objective criteria for each of the 27 hazard cate-

gories, helping to minimize the impact of individual misperceptions of hazards. The ACHiL platform identifies and categorizes hazards using a four-level scale:
- 0 – Hazard is not present
- 1 – Moderate hazard
- 2 – Medium hazard
- 3 – Severe hazard

Depending on each hazard, quantitative thresholds have been settled from level 1 to 3. This also includes the applied criteria used for level discrimination. Hence, a non-qualified person in health and safety would be less likely to make mistakes when assessing hazards. An illustration of the hazard mapping for a research building is presented in Fig. 2.9. It is obvious that workers at, or visitors of, the lab can easily observe where the most hazardous zones are located and which are the hazards concerned.

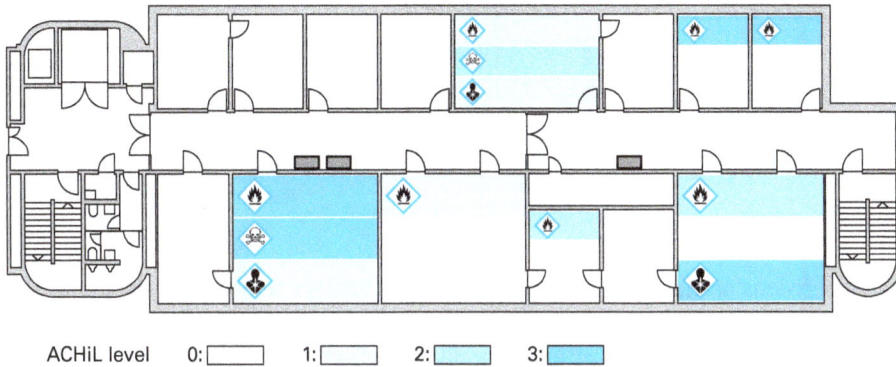

ACHiL level   0:▭   1:▭   2:▭   3:▭

Fig. 2.9: Hazard mapping of a lab research building for flammable solvents and acute and chronic toxics after ACHiL-scale application.

Another scope of hazard mapping or hazard portfolio is to identify where resources for an in-depth risk analysis should be performed. Hence, as resources are not infinite, it would make sense to concentrate on where hazards are most important, or where different hazards could, in a synergetic way, lead to combined hazards being often less obvious to note. In addition, hazard mapping is a powerful decision support tool, which can be used to detect rapidly "hot spots," e.g., laboratories combining several severe dangers.

In a similar manner, risk mapping is a technique used to help present identified risks and determine what actions should be taken toward those risks and to define priorities. Risk mapping is an approach to illustrating and showing the risks associated with an organization, project or other system in a way that enables one to under-

stand it better: what is important, what is not and whether the risk picture is comprehensive. Risk mapping is primarily qualitative and its benefits are as follows:
-   To improve the understanding of the risk profile and the ability to communicate about it.
-   To force people to think rigorously through the nature and impact of the risks that have been identified.
-   To improve risk models by building an intermediate link between the risk register and the model.
-   To improve the risk register by basing it on a more transparent and accurate understanding of the system.

A complete chapter on risk mapping would be out of the scope of this book. Readers interested in having more insights are referred to [26], a book dealing with mapping wildfire hazards and risks.

Since in relation to security issues, "perimeter thinking" or the so-called sanctuary principle is often employed, hazard mapping can be used for developing perimeters and different zones at a location which is being considered by security management. In the discerning zones, different security levels and security countermeasures can then be applied.

2.6 Risk perception and risk attitude

Many decisions on risk-taking, and other things in life, are colored by perception. The perception people have of risks is steered by a variety of factors that determine what risk is considered acceptable or unacceptable. Whether and how people expose themselves personally to certain risks (e.g., skiing and smoking), and how much, is a matter of preference and choice. The subjective factors on which individuals base their decisions on to take a risk or not include the degree to which the risk is known or unknown, voluntary or involuntary, acceptable or avoidable, threatening or attractive, controlled or uncontrolled.

Perception is very important, as if we have the mere perception that a risk is high, we will consciously or unconsciously take actions to reduce the risk. Also, where a person gives more weight to the positive side of a risk, he or she will be more prepared to take risks. Hence, influencing the perception is influencing the risk. The perception is obviously partially influenced by knowledge, which can be measured. Other factors influencing perception are beliefs, assumptions and espoused values, which are much more difficult to measure. Moreover, perception is fully linked with ERM, as the latter leads to an improvement of the perception of reality. By doing just that, ERM induces the taking of more objective decisions and of adequate actions.

Perception is strongly related with attitude. *Risk attitude* can be regarded as the chosen state of mind, mental view or disposition with regard to those uncertainties that could have a positive or a negative effect on achieving objectives. Hillson and Murray-Webster [27] explain that attitudes differ from personal characteristics in that they are situational responses rather than natural preferences or traits, and chosen attitudes may therefore differ depending on a range of different influences. If these influences can be identified and understood, obviously they can be changed and individuals and groups may than proactively manage and modify their attitudes. *Simply put, a person's risk attitude is his or her chosen response to the perception of significant uncertainty.*

Different possible attitudes result in differing behaviors, which leads to consequences. This can be represented by the cyclic KPABC model:

$$\textbf{Knowledge} \rightarrow \textbf{Perception} \rightarrow \textbf{Attitude} \rightarrow \textbf{Behavior} \rightarrow \textbf{Consequences}$$
$$\rightarrow \textbf{Knowledge} \rightarrow \text{etc.}$$

As Hillson and Murray-Webster [27] indicate, although the responses to positive and negative situations suggest at first sight that situation is the foremost determinant of behavior, in fact it is how the situation is *perceived* by each person, because a situation that appears hostile to one may seem benign to another. This leads to the important question of what influences behavior when the situation is *uncertain*. In this case, it is essential to know whether uncertainty is perceived as favorable, neutral, unfavorable or hostile. This reaction to uncertainty is "risk attitude."

Risk attitudes exist on a spectrum. The same uncertain situation will elicit different preferred attitudes from different individuals or groups/organizations, depending on how they perceive the uncertainty. Hence, different people will behave differently in the same situation, as a result of their differing underlying risk attitudes and perceptions. The variety of possible responses to a given level of risk is illustrated in Fig. 2.10. The *x*-axis displays the comfort or discomfort level people relate to uncertainty. The *y*-axis displays the response that people will have on a given level of uncertainty. Different types of risk attitude can be seen in the figure, ranging from risk-paranoid to risk-addicted.

Obviously, the extreme attitudes (risk-paranoid and risk-addicted) are not common at all; therefore, the two well-known polarities are the risk-averse and risk-seeking attitudes. On the one hand, a risk-averse person or group of persons feel uncomfortable with uncertainty, and therefore looks for certainty (safety and security) and resolution in the face of risks. Hazards and threats are perceived more readily and as more severe by the risk-averse, leading to a preference for risk minimization and risk avoidance. Risk-seeking people, on the other hand, have no problem at all with uncertainty and are therefore likely to identify fewer threats and hazards, as they see these as part of normal business. Moreover, hazards and threats are likely to be underestimated both in probability and possible impact, and acceptance will be the preferred response. A risk-tolerant person or group of persons are situated somewhere in the middle of the spectrum and views uncertainty (and risk) as a normal part of life, and is reasonably comfortable with most uncertainty. This attitude may lead to a laissez-faire approach, which in turn may lead to a nonproactive mindset and thus nonproactive measures. As suggested by Hillson and Murray-Webster [27], this may be the most dangerous of all the risk attitudes. No proper RM not only leads to important losses and problems but also missed opportunities and missed hypothetical benefits (see also Chapter 9).

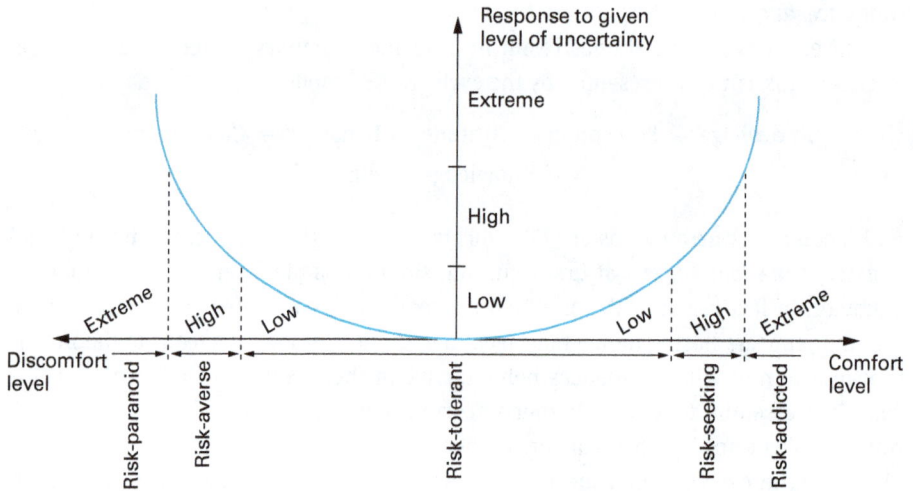

Fig. 2.10: Response to a given level of uncertainty in function of the level of comfort of people.

There are a number of situational factors that can modify the preferred risk attitude:
1. The level of relevant skills, knowledge or expertise: if high → risk-seeking attitude.
2. The perception of likelihood (probability of occurrence): if high → risk-averse attitude.
3. The perception of (negative) impact magnitude: if high → risk-averse attitude.
4. The degree of perceived control or choice in a situation: if high → risk-seeking attitude.
5. The closeness of risk in time: if high → risk-averse attitude.
6. The potential for direct (negative) consequences: if high → risk-averse attitude.

> **!** The way in which people perceive reality is very important: increasing the accuracy of our perception of reality leads to increasing our adequate ability to understand, interpret and efficiently and effectively deal with risk.

2.7 ERM: main steps

Many flowcharts exist in the literature to describe the sequences of RM. If we look through "engineering's goggles," we could draw the main steps involved in the ERM process. The process illustrated in Fig. 2.11 is based on a structured and systematic approach covering all of the following phases: the definition of the problem and its context, risk evaluation, identification, attribution of the risk ownership and examination of the RM options, the choice of management strategy, intervention implementations, process evaluation and interventions as well as risk communication. The phases are represented by circles, and the intersections show their interrelations.

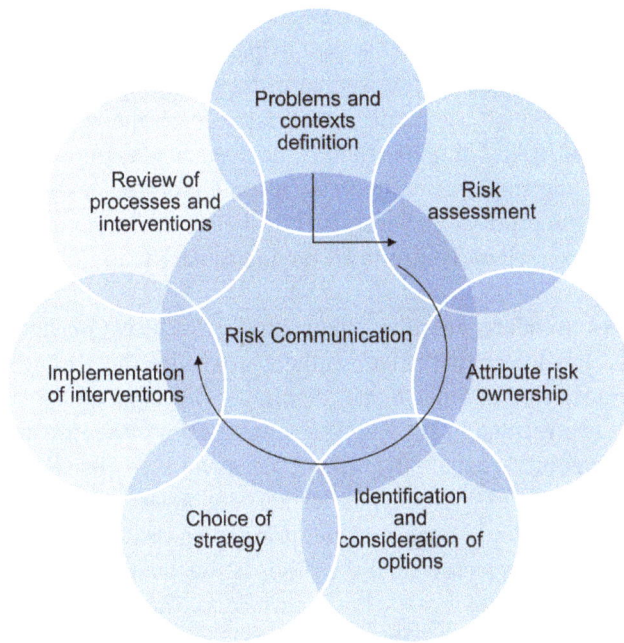

Fig. 2.11: The engineering risk management process.

The process normally starts at the problem definition step and proceeds clockwise. The central position of the risk communication phase indicates its integration into the whole process and the particular attention this aspect should receive during the realization of any of these phases.

The process must be applied by taking into account the necessity of implementing the coordination and concentration mechanism, of adapting its intensity and its extent depending on the situation, and to enable revision of particular phases depending on the needs.

Although phases must generally be accomplished in a successive way, the circular form of this process indicates that it is iterative. This characteristic enables the revision of phases in light of all new significant information that would emerge during or at the end of the process and would enlighten the deliberations and anterior decisions. The made decisions should be, as often as possible, revisable and the adopted solutions should be reversible. Although the iterative character is an important quality of the process, it should not be an excuse to stop the process before implementing the interventions. Selecting an option and implementing it should be realized even if the information is incomplete.

The flexibility must be maintained all along the process in order to adjust the relative importance given to the execution and the revision of the phases, as well as the depth level of analysis to perform or the elements to take into consideration.

The RM process intensity should be adapted as best as possible to the different situations and can vary according to the context, the nature and the importance of the problem, the emergency of the situation, the controversy level, the expected health impact, the socioeconomic stakes and the scientific data availability. As an example, emergency situations requiring a rapid intervention could require a brief application of certain phases of the process. However, even if the situation requires a rapid examination of these phases, all of them, and all the indicated activities in this frame, should be considered.

Because this cycle has been similarly described many times in the literature, here we will go a little bit deeper. It is interesting to look at the RM iterative ring through the questions that must be answered in order to get the process moving forward. A summary of these questions is presented in Fig. 2.12. The starting point is the instruction or mission being the answer of, "What are the tasks of (negative) RM?" Hence, we should identify "What could go wrong?" in the identification step. Answering "What exactly is the risk?" allows for describing, analyzing and prioritizing risks. The question of the risk ownership should then be solved. If the risk is left without a responsible person who will deal with, everything is ready for an unsuccess. Then, in order to control and plan, the question "What are the important risks?" is raised. To implement the adequate measure for risk reduction, we have to answer, "What has to be done to reduce risk?" This allows also for controlling and tracking the implementation. The task is not yet over, as we should not forget to monitor the situation by asking several questions: "What is the risk status?" allows following the time evolution of the considered risk. If something begins to deviate, then, "What has to be changed?" brings us back to the risk identification step. Another important point, often forgotten in ERM, is the answer to "What did we learn?"

In summary, the ERM process is not only an identification and treatment process, but a learning process that never ends and must be continuously performed.

Another characteristic of engineers is to simplify complex systems in order to master them more efficiently. From this perspective, we could imagine a simplification of the ERM process as depicted in Fig. 2.13.

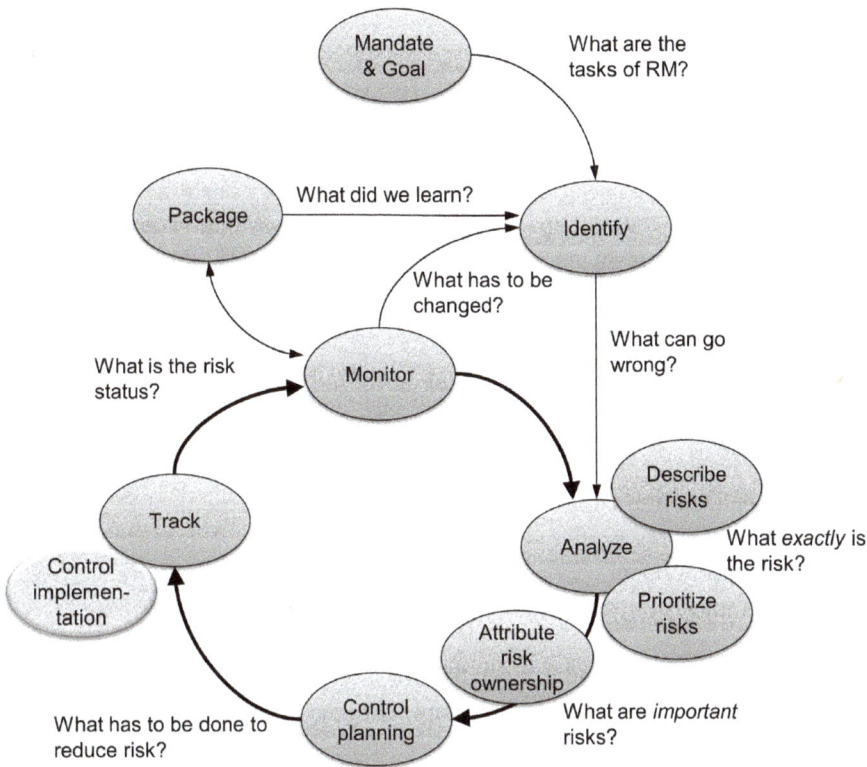

Fig. 2.12: Main questions of the ERM process.

The purpose of ERM is the creation and protection of value. It improves performance, encourages innovation and supports the achievement of objectives. The principles of ISO 3100:2018 [11] provide guidance on the characteristics of effective and efficient ERM, communicating its value and explaining its intention and purpose. The principles are the foundation for managing risk and should be considered when establishing the organization's ERM framework and processes. These principles should enable an organization to manage the effects of uncertainty on its objectives.

Effective ERM requires the following elements:

a) Integrated
   ERM is an integral part of all organizational activities.
b) Structured and comprehensive
   A structured and comprehensive approach to ERM contributes to consistent and comparable results.
c) Customized
   The ERM framework and process are customized and proportionate to the organization's external and internal context related to its objectives.

d) Inclusive

Appropriate and timely involvement of stakeholders enables their knowledge, views and perceptions to be considered. This results in improved awareness and informed ERM.

e) Dynamic

Risks can emerge, change or disappear as an organization's external and internal context changes. ERM anticipates, detects, acknowledges and responds to those changes and events in an appropriate and timely manner.

f) Best available information

The inputs to ERM are based on historical and current information, as well as on future expectations. ERM explicitly takes into account any limitations and uncertainties associated with such information and expectations. Information should be timely, clear and available to relevant stakeholders.

g) Human and cultural factors

Human behavior and culture significantly influence all aspects of ERM at each level and stage.

h) Continual improvement

ERM is continually improved through learning and experience.

In their previous version issued in 2009, ISO 31000 indicates that for ERM to be effective, an organization should at all levels comply with the principles below:

–   ERM creates and protects value. ERM contributes to the demonstrable achievement of objectives and improvement of performance in, e.g., human health and safety, security, legal and regulatory compliance, public acceptance, environmental protection, product quality, project management, efficiency in operations, governance and reputation.
–   ERM is an integral part of all organizational processes. ERM is not a stand-alone activity that is separate from the main activities and processes of the organization. ERM is part of the responsibilities of management and an integral part of all organizational processes, including strategic planning and all project and change management processes.
–   ERM is part of decision-making. ERM helps decision-makers make informed choices, prioritize actions and distinguish among alternative courses of action.
–   ERM explicitly addresses uncertainty. ERM explicitly takes account of uncertainty, the nature of that uncertainty and how it can be addressed.
–   ERM is systematic, structured and timely. A systematic, timely and structured approach to ERM contributes to efficiency and to consistent, comparable and reliable results.
–   ERM is based on the best available information. The inputs to the process of managing risk are based on information sources such as historical data, experience, stakeholder feedback, observation, forecasts and expert judgment. However, decision-

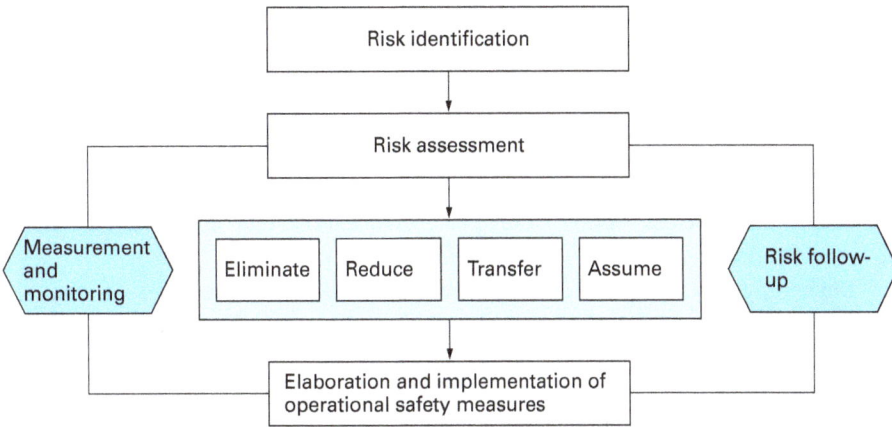

Fig. 2.13: Simplified ERM process.

makers should inform themselves of, and should take into account, any limitations of the data or modeling used or the possibility of divergence among experts.

– ERM is tailored. ERM is aligned with the organization's external and internal context and risk profile.

– ERM takes human and cultural factors into account. ERM recognizes the capabilities, perceptions and intentions of external and internal people that can facilitate or hinder achievement of the organization's objectives.

– ERM is transparent and inclusive. Appropriate and timely involvement of stakeholders and, in particular, decision-makers at all levels of the organization ensures that ERM remains relevant and up to date. Involvement also allows stakeholders to be properly represented and to have their views taken into account in determining risk criteria.

– ERM is dynamic, iterative and responsive to change. ERM continually senses and responds to change. As external and internal events occur, context and knowledge change, monitoring and review of risks take place, new risks emerge, some change and others disappear.

– ERM facilitates continual improvement of the organization. Organizations should develop and implement strategies to improve their ERM maturity alongside all other aspects of their organization.

The success of ERM will depend on the effectiveness of the management framework that provides the foundations and arrangements to embed it throughout the organization at all levels. The framework assists in managing risks effectively through the application of the ERM process at varying levels and within specific contexts of the organization. The framework ensures that information about risk derived from the ERM process is adequately reported and used as a basis for decision-making and accountability at all relevant organizational levels.

Figure 2.14 shows the necessary components of the framework for managing risk and the way in which they interrelate in an iterative manner. We could observe that "communication and consultation" as well as "monitoring and review" are completely integrated in the ERM process. At almost all stages, interrelations happen, the implementation of key factors is present (key performance indicators or key risk indicators) as the analysis of incident is crucial in the monitoring process.

Depending on the level of complexity required, we could use different manners to represent what is ERM. Some users will prefer the simplified model (Fig. 2.13), while others will argue that the innovative representation with answering questions (Fig. 2.12) is preferable; still others will use the complete scheme (Fig. 2.14).

It is not really so important what scheme is used, and the most important aspect is that with time one remains consistent in the use and in the follow-up. It is better to have a simplified system in adequate use rather than a complex scheme that will be only partially used.

> **!** A variety of RM schemes and frameworks are available to be used in industrial practice. A framework should always have a feedback loop built into it, where one is certain that RM efforts never stop. Risk policy, assessment, communication and monitoring should also always be part of the scheme.

2.8 Objectives and importance of ERM

Engineering today's organizations and their technology and systems is sophisticated and complex. Technology, processes, systems, etc. become evermore complex, distributed geographically and spatially, and are linked through evermore complicated networks and subsystems. These operate to fulfill the needs of a variety of stakeholders, and they need to be evermore flexible, balanced in relation to expected performance and risk managed in a responsible and respectful manner. ERM offers an answer to the evermore difficult task of managing the risks of organizational systems, technology, infrastructure, etc.

It is widely recognized that enacted law is by no means the only driver for improved ERM by corporations. Cost imperatives, liability issues, corporate reputation, industry peer pressure, societal concerns, etc. are increasingly important drivers. Community expectations about a better educated and informed workforce, more respect toward all stakeholders and the absence of hazards possibly leading to major risks, enhance the use of ERM to achieve these expectations.

> **!** We define the objectives of ERM as: "the early and continuous identification, assessment, and resolution of non-financial risks such that the most effective and efficient decisions can be taken to manage these risks."

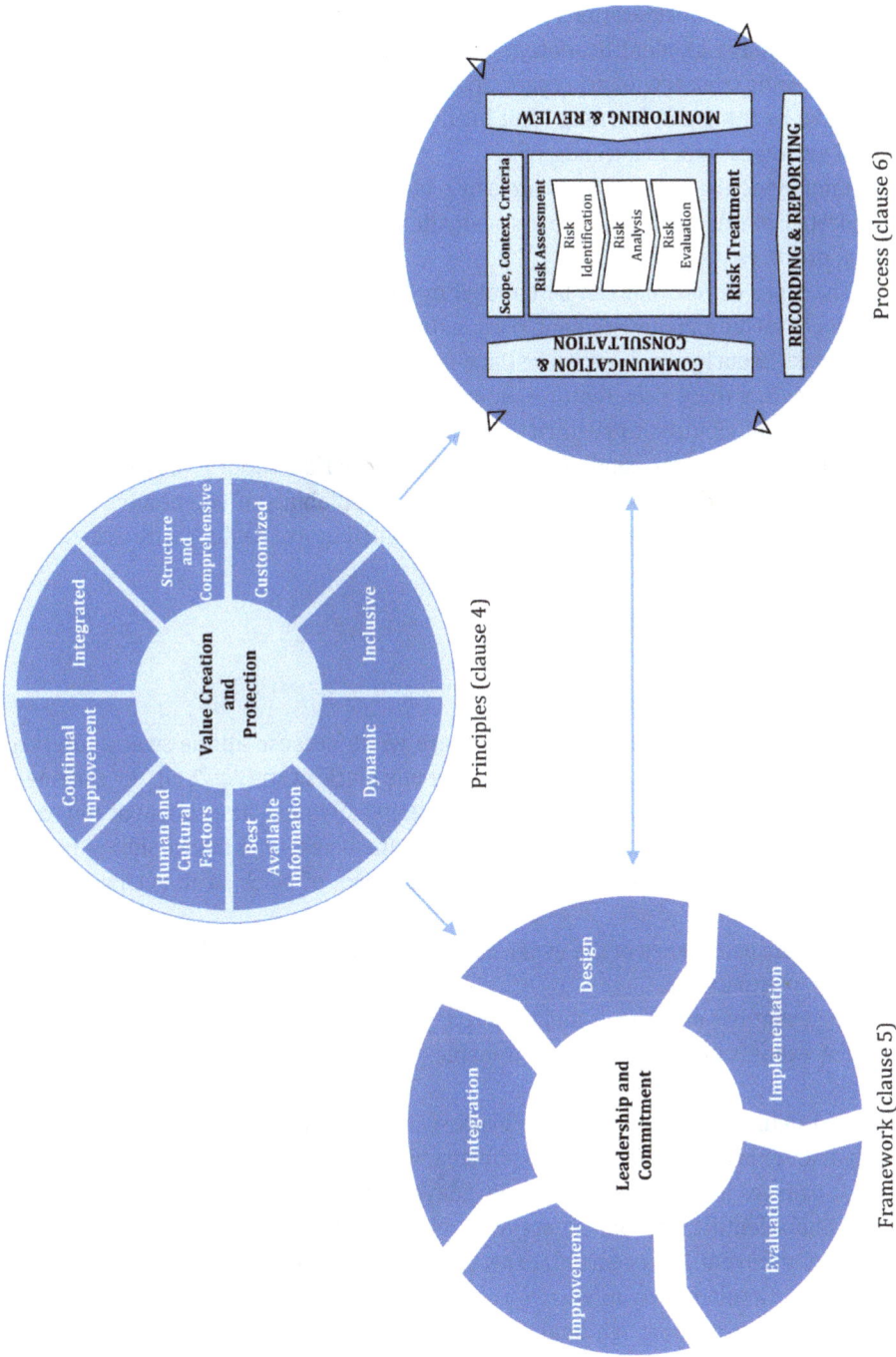

Fig. 2.14: Risk management process according to ISO 31000:2018.

There are many considerations explaining the importance of ERM, including early and continuous risk identification, risk-informed (risk-based) decision-making, gaining a systemic overview of an organization, proactive planning for unwanted events, proactive learning and learning from incidents, intelligent resource allocation, situational awareness and risk trends, etc.

Employing the models, methodologies, ideas, theories, tools, etc., explained and elaborated in this book, leads to a systematic continuous improvement of organizational practices.

The types of risk that can be handled by ERM, and by the methods described in this book, include major accident risks, natural disasters, fire risks, technical risks, social risks, security risks, exposure (long-term health) risks, and labor (occupational) risks. Handling these risks requires an analytical and systematic approach to identify important contributors to such risks, providing input in developing procedures for normal and emergency conditions, providing appropriate input to design processes, providing adequate inputs to the assessment of acceptability of potentially hazardous facilities, updating information on key risk contributors, dealing with societal demands, etc.

2.9 The black swan

Based on the presumption that all swans are white because all the swans observed before were white until someone discovered one black swan (Fig. 2.15), the statement of impossibility "black swan" was used to describe anything "impossible or not existing." Introduced by Nassim Nicholas Taleb [28], the black swan theory shows the possibility to expect the impossible. Three main features of black swan events could be identified:
- After the occurrence of the event, explanations are formulated, making it predictable or expectable.
- The event has extreme or major impact.
- The event is unexpected or not probable.

The first defining characteristic of a black swan event is that human thought tries to explain an enormous amount of phenomena, whereas, in fact, it can explain a lot less than it believes it is able to explain. Nowadays, explanations on almost all imaginable social and scientific phenomena are available through a variety of information sources. Once an event has occurred, experts and analysts waste no time in finding the causes that fit adequately to what was observed. These explanations a posteriori are often the result of thorough analysis and usually generate accounts supposing the event as being evident and explainable. Many models and theories are therefore formulated on the occurrence of past events. However, these models may often be useless in explaining or predicting future ones, especially in case of type II or even more in case of so-called black swan events.

The second characteristic regards the consequences that its occurrence may have. The impact could be classified either positive or negative: positive could be the invention of the steam machine and the discovery of a new vaccine or mobile telecommunication; negative could be the 9/11 terrorist attack or the sinking of the Titanic – a black swan precisely because it was considered "unsinkable." The arrival of Europeans in the new world was a black swan, negative from the viewpoint of the Aztecs, but maybe positive from the contemporary European perspective.

The last characteristic deals with the boundaries of human knowledge and the belief that it encompasses more than it really can. The enormous impact these events have is partly caused by our incapacity to predict their occurrence. According to the most traditional estimation methods, their probability is very low because of their infrequency. Narrowing this phenomenon based on the notion of an extremely low probability, it can be believed that its occurrence is impossible, accounting for the conditions in which previous events developed. This characteristic weakens the traditional techniques available for statistic forecasting. Their apparition does not obey any tendency, nor does it adjust to any known distribution: it simply occurs without apparent reason (until a group of experts and analysts possess a series of explanations to make its occurrence evident).

Fig. 2.15: Picture of a black swan.

A "black swan event" is an event that is nearly impossible to predict. An event that is unpredictable and unexpected in even the most detailed and carefully calculated probability models. It is characterized with the so-called deep uncertainty.

If any conventional statistical method is used, future black swans would be located in the tales of its distribution, that is, those events with extremely high impact and extremely low probability – but given the forecast method, the chance of these events to occur is considered nearly impossible.

It would be erroneous to state that the world is "dominated" by the unknown unknowns as suggested by Taleb. It would be closer to the truth to say the world is "challenged" from time to time by the unknown unknowns.

It is not always true that it is impossible to predict technological and scientific innovations. Prediction requires knowing about technologies that will be discovered in the future. But that knowledge would almost automatically allow us to start developing those technologies right away. For example, Jules Verne and H.G. Wells forecast the motorization of society, the submarines, the advent of airplanes and the landing on the moon but did not have the will, the money, the expertise or the inclination to develop those technologies themselves. The theoretical knowledge required to make a technological prediction does not necessarily and automatically lead in practice to its immediate invention or development.

Black swans have a large impact on ERM; they introduced the concept of "think the unthinkable."

Remark that people, thinking about safety, use quite a few animals as representations of safety concepts. In addition to *Paper Tigers* [32], *Dinosaurs* [33], *Black Swans* [21], *Dragons* [34] and *Elephants* [35], [36] introduced *Gray Rhino*'s in – what by now could be called – the *safety zoo* [31].

Michele Wucker [37] herself issued a *Reader Discussion Guide* to complement her 2016 book about the *Gray Rhino* [36]. Nathan Jaye [38] interviews Michele Wucker on October 23, 2017, asking her about "black swans." In safety science, these stand for the combination of unlikely factors suddenly creating a big problem. Wucker explains that the "gray rhino" is staring you in the face, ready to cause havoc. You'd better find out why you are not responding and use the opportunity to act while you still can. Then Jaye continues asking about the "elephant in the room." Wucker replies with an explanation about how difficult it can be to even see a nearby group of very large elephants in their natural habitat. Jaye and Wucker then engage in a discussion about China and – among other things – about recurring influenza viruses, comparing them with recurring "gray rhino's." Not reacting to a charging rhino is the worst thing to do, even when you are frightened, prone to freezing on the spot or are scared to make a wrong choice. Later, Schleicher [39] writes that the "gray rhino" is *a perfect complement to the Black Swan concept. It is about engaging more thoroughly with reality, recognizing obvious threats and our own biases, and working to overcome them.*

These animal concepts all address safety problems we do not seem to handle very well. We deem these problems as unknown, ignore them, accept their risk, regard them as unlikely and we just don't act on them, even when we really should and actually could. And yes, an *Ostrich* bird springs to mind.

2.10 Risk management and education

ERM cannot stand without a proper education. Continuous and spiral safety and security education reinforce the value and need for safety/security and demonstrate to students the high priority that employees and managers should place on these fields. It will build the knowledge base and the safety/security skills the students will need as they pursue further education and move into the workforce [29]. Risk itself is often not well understood, in terms of objectives and uncertainties. *Un*safety is often misinterpreted as safety. Safety/security performance indicators are unclear and difficult to validate. Safety culture is a debatable subject among scholars and practitioners. Ethical aspects of safety are characterized with opposite views of stakeholders. People therefore really need to be taught and educated about risk, safety, management issues, and what have you.

When starting with the simple formula "risk equals probability times damage," the result is often being expressed in monetary consequences. This simplified vision of risk is widely used, leading to the concept that "risk is probability times effect." However, often it is the case that there is still the problem of finding significant data on the basis of which it will be possible to calculate the likelihood or probability of a particular type of accident occurring and causing a given financial loss. How deep does one go into finding/determining the consequences: direct effects are relatively easy, but indirect and hidden consequences and losses are much more complicated and sometimes more intangible as well. As soon as we try to apply to simple formula, we face the problem of how to quantify probability and effect. Disagreements on the subject of ERM can often be traced back to choices made regarding the definition of the limits of the system to be examined. The discussion is then often about what constitutes a risk-bearing activity, what constitutes damage or loss, how to define the causal link that has to be proven and within what period the effects have to become apparent [30]. As the systems become more and more complex, simpler models tend to fail and emphasizes the central role of safety education in the ERM process.

The value of ERM lies not so much in the ways it serves as a means to investigate social responses to particular forms of possible danger, but more in the ways in which it might sensitize us to use the language in current policy debates surrounding individual responsibilities. This should invite us to have a prospective attitude to reimagine ourselves as subjects with a fluid set of aspirations within the context of risky futures.

The education of engineering risk managers should address the future concerns of modern management of engineering risks. The latter will need to be able to identify and address new risks quickly, be more agile and modular, deliver new technology and techniques rapidly and work increasingly in partnership with finance, operations and the businesses. Engineering risk managers need to become more familiar and aware of the behavioral aspects of ERM. In addition, tools and techniques may need to evolve to reflect the changing nature of the risk framework and landscape. These changes will require them to recruit, develop and retain staff with skills that differ significantly from those that are found in organization today. These skills should at least be the concern of modern education at all levels: undergraduate, graduate and continuing education.

2.11 Tips for managing risks

Risk is inherent in everything we do. Whether it is driving your bike to work, choosing your next holidays, crossing the road, making dietary choices or investing in financial products, life is full of risk. You cannot remove the negative side of it, but you can contain or mitigate it. The solution is the management of risk. It is evolutionary in nature – the level of risk can change and so too can our perceptions of it. It involves understanding and analyzing risk to ensure organizations and/or individuals meet their objectives. The following tips can help on how to improve ERM:

- Define the goal and mandate: Set the objectives in order that everyone knows exactly what part of the business and what activities and tasks they are responsible for.
- Identify risks at an early stage: It is never too early to start thinking about risks. The sooner it is done, the easier it will be to manage them. Use early warning indicators when possible to identify raising risks. ERM should be integrated into all work processes and corporate culture.
- Look at the positive sides of risks: Not all risks have only a negative side (such as nature-related disaster risks), risks in organizations always have a positive side (cf. the uncertainty sandglass), presenting opportunities and enabling us to take advantage of a given event or situation.
- Describe the risk appropriately: A good practice in the risk assessment process is to create a "chain" of risks, distinguishing between cause and effect.
- Evaluate, rank and prioritize risk, for both the type I risks and the type II risks. Make two separate prioritization lists. Assess and prioritize all known type I and type II risks. The simplified model to calculate the severity of a risk by looking at both the likelihood of occurrence and impact (severity) can be used as a starting assumption.
- Taking responsibility and ownership: If something wrong is observed, such as potential safety issues, fraud, malevolence or security breach, responsibility should be taken rather than waiting for someone else to fix it. ERM is more effective when everyone is empowered to take action.
- Learn from the past: Use historical data and anecdotes ("story telling") to learn from past mistakes to ensure they will never happen again. Take measures especially at the management system level.
- Use appropriate strategies to manage risk: Eliminate, reduce, transfer or assume (see also Chapter 5).

- Document all risks in one file (the "risk register"): This will improve information sharing and accountability.
- Keep monitoring and reviewing: This is like the Plan-Do-Check-Act (PDCA) model, a continuous improvement methodology. The level of risk we all face is constantly evolving, with new risks emerging and others becoming less important. By being proactive and regularly monitoring them, we will be ready to act when the time comes.

2.12 Conclusions

To have a thorough understanding of what constitutes *ERM* and how the theories, concepts, methods, etc. can be employed in industrial practice, some terms and concepts (uncertainty, hazard, threat, risk, risk type, exposure, losses, risk perception, risk attitude, etc.) have to be unambiguously defined. In this chapter, we further focus on the negative side of engineering risks, and in what possible operational ways such safety/security/environment/ethics/quality risks can be decreased. The strategic part of ERM is explained by presenting different possible ERM frameworks.

References

[1] ISO 9001:2008. (2008). Quality Management Systems – Requirements. International Organization for Standardization Geneva, Switzerland.
[2] EFQM. (2009). EFQM Excellence Model 2010, Excellent Organizations Achieve and Sustain Superior Levels of Performance that Meet or Exceed the Expectations of All Their Stakeholders. Brussels, Belgium: EFQM Publication.
[3] ISO 9004:2009. (2009). Guidelines for Performance Improvement. International Organization for Standardization Geneva, Switzerland.
[4] ISO 14001:2004. (2004). Environmental Management Systems – Requirements with Guidance for Use. International Organization for Standardization Geneva, Switzerland.
[5] EMAS (Eco-Management and Audit Scheme). (2009). Regulation (EC) No 1221/2009 of the European Parliament and of the Council of 25 November 2009. Brussels, Belgium.
[6] OHSAS 18001:2007. (2007). Occupational Health and Safety Management Systems – Requirements. London, UK: British Standardization Institute.
[7] SA 8000:2008. (2008). Social Accountability Standard. Social Accountability International. New York.
[8] Van Heuverswyn, K. (2009). Leven in de Risicomaatschappij Deel 1, Negen Basisvereisten voor Doeltreffend Risicomanagement. Antwerpen: Maklu/Garant.
[9] Government of Canada Treasury Board. (2001). Integrated Risk Management Framework. Ottawa, Canada: Treasury Board of Canada.
[10] AS/NZS 4360:2004. (2004). Australian-New Zealand Risk Management Standard. Sydney, Australia: Standards Australia International Ltd.
[11] ISO 31000:2018. (2018). Risk Management Standard – Principles and Guidelines. International Organization for Standardization Geneva, Switzerland.
[12] Lees, F.P. (1996). Loss Prevention in the Process Industries. 2nd. vol. 3. Oxford, UK: Butterworth-Heinemann.

[13]   Wells, G. (1997). Major Hazards and Their Management. Rugby, UK: Institution of Chemical Engineers.
[14]   Kletz, T. (1999). What Went Wrong? Case Histories of Process Plant Disasters. 4th edn. Houston, TX: Gulf Publishing Company.
[15]   Kletz, T. (2003). Still Going Wrong. Case Histories of Process Plant Disasters and How They Could Have Been Avoided. Burlington, VT: Butterworth-Heinemann.
[16]   Atherton, J., Gil, F. (2008). Incidents that Define Process Safety. New York: John Wiley & Sons.
[17]   Reniers, G.L.L. (2010). Multi-plant Safety and Security Management in the Chemical and Process Industries. Weinheim, Germany: Wiley-VCH.
[18]   Casal, J. (2008). Evaluation of the Effects and Consequences of Major Accidents in Industrial Plants. Amsterdam, The Netherlands: Elsevier.
[19]   Balmert, P.D. (2010). Alive and Well at the End of the Day. The Supervisor's Guide to Managing Safety in Operations. Hoboken, NJ: John Wiley & Sons.
[20]   Rogers Commission Report. (1986). Report of the Presidential Commission on the Space Shuttle Challenger Accident. Washington: US Government Printing Office.
[21]   Taleb, N.N. (2007). The Black Swan. The Impact of the Highly Improbable. New York: Random House.
[22]   US DoD. (2002). Department of Defense, Office of the Assistant Secretary of Defense (Public Affairs). News Transcript, February 12.
[23]   Kirchsteiger, C. (1998). Absolute and relative ranking approaches for comparing and communicating industrial accidents. J. Hazard. Mater. 59: 31–54.
[24]   Fuller, C.W., Vassie, L.H. (2004). Health and Safety Management. Principles and Best Practice. Essex, UK: Prentice Hall.
[25]   Marendaz, J.L., Suard, J.C., Meyer, T. (2013). A systematic tool for Assessment and Classification of Hazards in Laboratories (ACHiL). Safety Sci. 53: 168–176.
[26]   Sampson, R.N., Atkinson, R.D., Lewis, J.W. (2000). Mapping Wildfire Hazards and Risks. New York: CRC Press.
[27]   Hillson, D., Murray-Webster, R. (2005). Understanding and Managing Risk Attitude. Aldershot, UK: Gower Publishing Ltd.
[28]   Taleb, N.N. (2010). The Black Swan. The Impact of the Highly Improbable. 2nd ed. USA: Random House Inc.
[29]   Meyer, T., Reniers, G., Cozzani, V. (2019). Risk Management and Education. Berlin, Germany: De Gruyter.
[30]   Helsloot, I., Jong, W. (2006). Risk management in higher education and research in the Netherlands. J. Contingencies Crisis Manage 14(3): 142–159.
[31]   Lindhout and Reniers. (2020). "Reflecting on the safety zoo: Developing an integrated pandemics barrier model using early lessons from the Covid-19 pandemic", Saf. Sci. 130.
[32]   Dekker, S.W. (2014). The bureaucratization of safety. Saf. Sci. 70: 348-357.
[33]   Cohen, H.B. (2005). The Dinosaur in the Living Room: Achieving Positive Change by Tackling the Obvious. Author House.
[34]   Elahi, S. (2011). Here be dragons… exploring the 'unknown unknowns'. Futures, 43(2): 196-201.
[35]   Srivatsa, K.M. (2018). Elephant in the doctors room. Arch. Infect. Dis. Therapy. 2(1): 1-8.
[36]   Wucker, M. (2016). The Gray Rhino: How to Recognize and Act on the Obvious Dangers We Ignore. Macmillan. St. Martins Press, April 2016, ISBN 9781250053824.
[37]   Wucker, M. (2017). Reader Discussion Guide, Why Read THE GRAY RHINO? https://www.thegrayrhino.com/wp-content/uploads/Gray-Rhino-Readers-Guide.pdf
[38]   Jaye, N. (2017). Do "Gray Rhinos" Pose a Greater Threat Than Black Swans? 23 October 2017. https://blogs.cfainstitute.org/blog/2017/10/23/do-gray-rhinos-pose-a-greater-threat-than-black-swans/
[39]   Schleicher, D. (2020). Book review "The Gray Rhino: How to Recognize and Act on the Obvious Dangers We Ignore by Michele Wucker". ISBN: 9781 250053824. https://www.porchlightbooks.com/blog/editors-choice/the-gray-rhino-how-to-recognize-and-act-on-the-obvious-dangers-we-ignore.

3 Risk management principles

3.1 Introduction to risk management

The modern conception of risk is rooted in the Hindu-Arabic numbering system that reached the West some 800 years ago [1]. Serious study began during the Renaissance in the seventeenth century. At that time, large parts of the world began to be discovered by the Europeans, and they began to break loose from the constraints of the past. It was a time of religious turmoil and a steep increase in the importance and rigorousness of mathematics and science in Europe. As the years passed, mathematicians transformed probability theory from a gambler's toy into an instrument for organizing, interpreting and applying information. By calculating the probability of death with and without smallpox vaccination, Laplace developed the basis of modern quantitative analysis in the late eighteenth century.

The word "risk" seems to have been derived from the early Italian *risicare*, which means "to dare." In this sense, risk is a choice rather than a fate. The actions we dare to take, which depend on how free we are to make choices, are what "risk" should be about. The modern term "risk management" seems to have been first used in the early 1950s [2].

Whereas determining risk values as accurately as possible and obtaining a correct perception of all risks present (see Chapter 2) are both very important aspects of decreasing or avoiding the consequences of unwanted events, the aspect of making decisions about how to do this and what measures to take is at least as important. In contrast to risk assessment (including risk identification and analysis), which is a rather technical matter, risk management is largely based on company policy and can be seen as a response to perceptions. Therefore, risk management significantly differs across organizations, mainly as a result of the different values and attitudes toward specific risks in different organizational culture contexts.

As mentioned in Chapter 2, risk assessment connotes a systematic approach to organizing and analyzing scientific knowledge and information for potentially hazardous activities, machines, processes, materials, etc. that might pose risks under specified circumstances. The overall objective of risk assessment is to estimate the level of risk associated with adverse effects of one or more unwanted events from one or more hazardous sources. By doing so, it supports the ability to proactively and reactively deal with minor accident scenarios as well as major accident scenarios through "risk management." Hence, risk-based decision-making, and more broadly risk management, consists of risk assessment and risk treatment. The former is the process by which the results of a risk analysis (i.e., risk estimates) are used to make decisions either through relative ranking of risk reduction strategies or through comparison with risk objectives; while the latter consists of the planning, organizing, leading and controlling of an organization's assets and activities in ways that minimize the

https://doi.org/10.1515/9783111493633-003

adverse operational and financial effects of losses upon the organization. Risk assessment and risk treatment are only some elements belonging to the larger domain of risk management. Other elements that are part of risk management are nonexhaustive, safety training and education, training-on-the-job, management by walking around, emergency response, business continuity planning, economic analyses with respect to risk, risk communication, risk perception, psychosocial aspects of risk, emergency planning, risk governance and ethical aspects of risk. We may define "risk management" as *the systematic application of management policies, procedures and practices to the tasks of identifying, analyzing, evaluating, treating and monitoring risks* [3]. Figure 3.1 illustrates the engineering risk management set a portfolio of a nonexhaustive list of research domains all being important for the ERM manager.

A simplified overview of the risk management process to cope with all the tasks related to the risk management set, according to ISO 31000:2009 [4], is illustrated in Fig. 3.2.

Engineering risk management

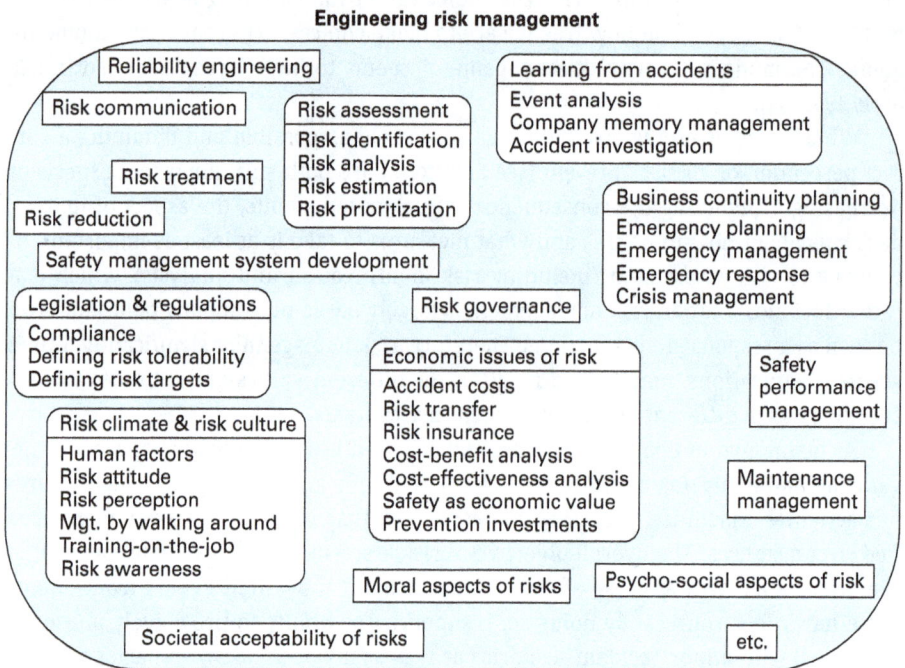

Fig. 3.1: The risk management set.

The sequence of the various process steps in assessing the risks originating from a specified system, that is, establishing the context, identification, analysis, assessment, handling, monitoring, management and decision-making, is very similar at a generic level across different industries and countries. This observation also holds for different kinds of risk – safety-related risks, health-related risks, but also environmental risks, security risks, quality risks and ethical risks. Section 3.2 explains more in-depth how risk management should ideally be elaborated in a generic and integrated way.

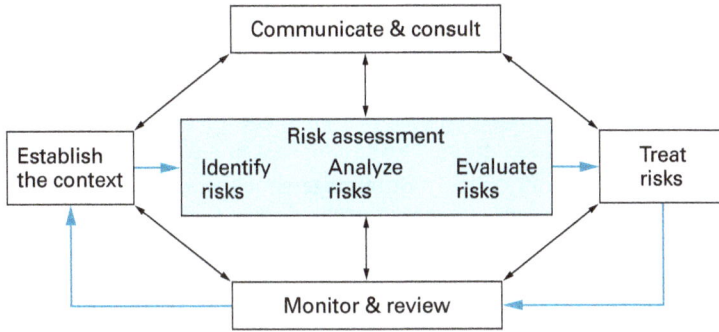

Fig. 3.2: Risk management process – simplified overview.

3.2 Integrated risk management

When comparing a specific management system (designed for one domain, e.g., safety or environment or quality) with an integrated management system (designed for all risk domains), five common aspects can be found:

1. Both systems provide guidelines on how to develop management systems without explicitly prescribing the how-exercise in detail.
2. They consider a risk management system to be an integral part of the overall company management system. This approach guarantees the focus and ability to realize the company's general and strategic objectives.
3. They all have two common aims: (i) to realize the organization's objectives, taking compliance into full consideration; (ii) continuous improvement of an organization's achievements and performance.
4. The process approach is employed.
5. The models can be applied to every type of industrial organization.

The comparison also leads to the identification of several differences:

- Integrated risk management systems recognize the positive as well as the negative possible outcomes of risks. Hence, both damage and loss on the one hand, and opportunities and innovation on the other hand, are simultaneously considered.

- All kinds of risks are considered: operational, financial, strategic, juridical, etc., and hence, a balanced equilibrium is strived for.
- The objectives of integrated management systems surpass compliance and continuous improvement.

The question that needs to be answered is how the integration of management systems can be realized in industrial practice, taking the similarities and differences of specific risk management systems into account. A suggested option is to integrate all management systems according to the scheme illustrated in Fig. 3.3.

It is essential that an organization initially formulates a vision with respect to the system that is used to build the integrated risk management system. Such a vision demonstrates the intentions of the organization with respect to its risk management system in the long term.

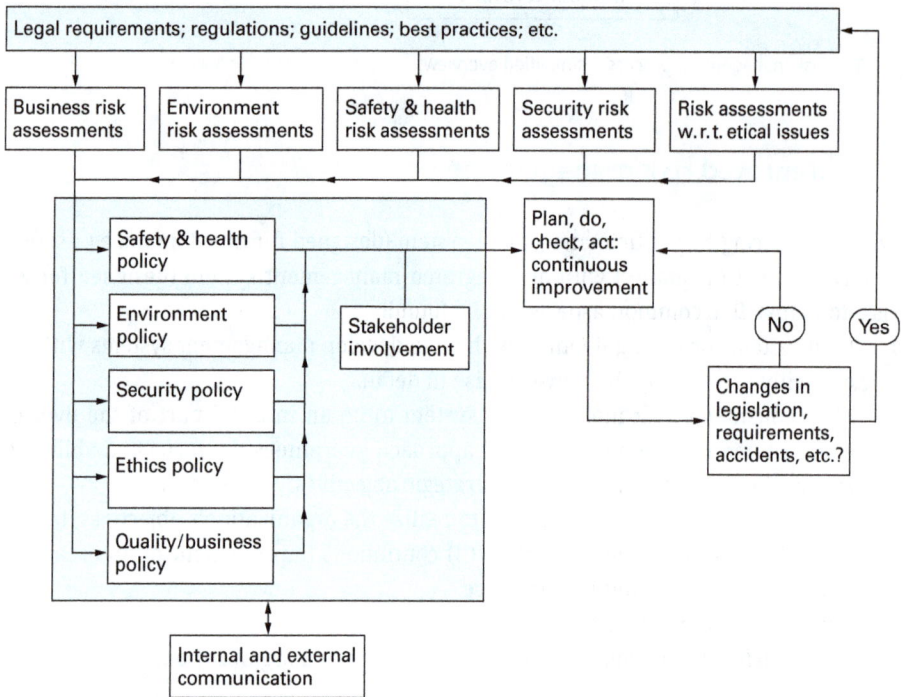

Fig. 3.3: Scheme of integrated risk management.

> ! Risk management requires an integrated view. Risks very often do not only have an impact on one domain, for example, health and safety, but they also affect other domains such as environment or security or quality. As such, the generic risk management process should be elaborated at – and implemented on – a helicopter level, for all domains at once.

3.3 Risk management models

Risk management models are described and discussed in this section, providing insights into organizational risk management. The models visualize the various factors and the different domains that the structure and approaches of adequate risk management should take into account within any organization.

Let us look at some very brief and basic insights into the history of safety science and the development of models, theories, concepts and metaphors with respect to safety. It is obvious that safety science is a modern science with a limited history when compared with, for instance, physics, chemistry and mathematics. The study of safety, how to be "not-unsafe," and how to avoid occupational accidents indeed only emerged in a systematic way at the beginning of the twentieth century.

Up until somewhere around 1960, research was mainly conducted by people from industry and insurance companies and was focused on incidents and accidents happening to individual employees and how to ensure that a safe workplace is created. This is the era of the "safety first movement." Models and theories are mainly ratio relationships based on thousands of accident reports (such as the accident pyramid, see Section 3.3.1) developed by safety pioneers such as Herbert Heinrich, sometimes also called the "father of industrial safety." The central thinking and modeling were around safety behavior, and a very popular theory in this period was the so-called accident proneness theory. Basically, in the first half of the twentieth century, it was believed that all the focus should be on changing the behavior of people, and people could be seen as merely a pair of hands to do the work, but not to think while working. The focus of the studies was type I risks and accidents.

From the 1960s on, a new safety era emerged, driven by the mindset of the nuclear and chemical industrial age and all its accompanying risks, which we can call the "loss prevention and management" period. In this period, research institutes and academia became involved, as well as governmental agencies. Research was not conducted anymore solely by foremost practitioners, but also by academic and governmental researchers, with more focus on understanding the phenomena and on validation of theories and models. The focus of the safety research carried out in this period was on major accidents and catastrophes, and developing models, metaphors and theories to understand the underlying phenomena of disasters and to deal with type II risks and accidents. Protecting the surrounding communities of industrial areas as well as the workers of the neighboring companies became the goal of safety research, besides improving an organization's employees' safety and health. This second period lasted roughly until somewhere around 2020 (see Fig. 3.4).

From the 2020s (and starting earlier), we see yet a new safety era emerging, instigated by the revolution of communication technology (internet and social media), and all the possibilities in that regard. Society has put more emphasis on ethical issues and transparency, and at the same time, individual citizens have become much more knowledgeable about risk, policies and health. Citizen networks appear and mingle in the debate about health and safety policies in organizations, using social media, and the opinions, films and photos that can be shared easily, as leverage to influence decision-making and to further advance (or sometimes not) safety research. The focus of safety studies is being enlarged in this latest safety era toward society as a whole, besides the still ongoing research to better the health and safety of employees and citizens working within industrial parks and living in the surrounding communities.

Let us, in the next sections, discuss some of the most known models, metaphors and theories from the past century of safety research.

3.3.1 Model of the accident pyramid

Heinrich [5], Bird [6] and Pearson [7], among other researchers, determined the existence of a ratio relationship between the numbers of incidents with no visible injury or damage, those with property damage, those with minor injuries and those with major injuries. This accident ratio relationship is known as "the accident pyramid" or "the safety triangle" (see an example in Fig. 3.5). Remark that this accident ratio relationship may also be applied to security/quality/environmental incidents. Generally speaking, from a management perspective, accident pyramids indicate that accidents are "announced." Hence the importance of awareness and incident analyses.

Different ratios were found in different studies (varying from 1:300 to 1:600) depending on the industrial sector, the area of research, cultural aspects, etc. However, the existence of the accident pyramid should be seen from a qualitative point of view. It may be possible to prevent more serious type I accidents by taking preventive measures aimed at near-misses, minor accidents, etc. These "classic" accident pyramids clearly provide an insight into type I accidents where a lot of data is at hand.

In brief, the assumptions of the "old" safety paradigm emanating from the accident pyramid (see Fig. 3.5) hold that
(i)   As injuries increase in severity, their number decreases in frequency.
(ii) All injuries of low severity have the same potential for serious injury.
(iii) Injuries of differing severity have the same underlying causes.
(iv) One injury reduction strategy will reach all kinds of injuries equally (i.e., reducing minor injuries by 20% and also reducing major injuries by 20%).

Using injury statistics, Krause [8] indicated that while minor injuries may decline in companies, serious injuries can remain the same, hence casting credible doubts on the validity of the accident pyramid or safety triangle concept. In fact, research indicated that only about 20% of the type I incidents have the potential to lead to a serious type I accident. This finding implies that if there is only focus on "the other 80%" of the type I incidents (80% of the incidents do not have the possibility to lead to serious injury), the causative factors that create potential serious accidents will continue to exist and so will serious accidents themselves.

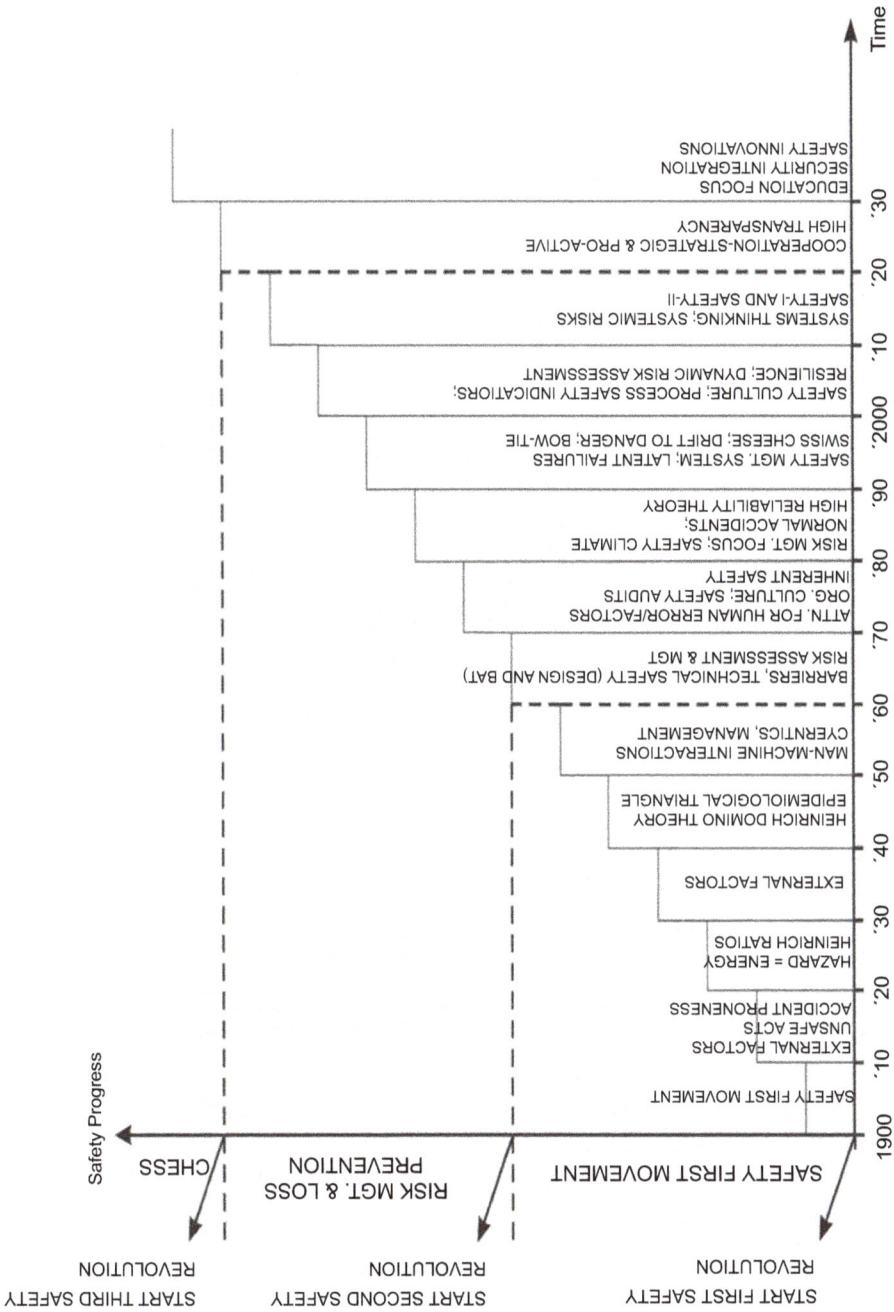

Fig. 3.4: History of safety science.

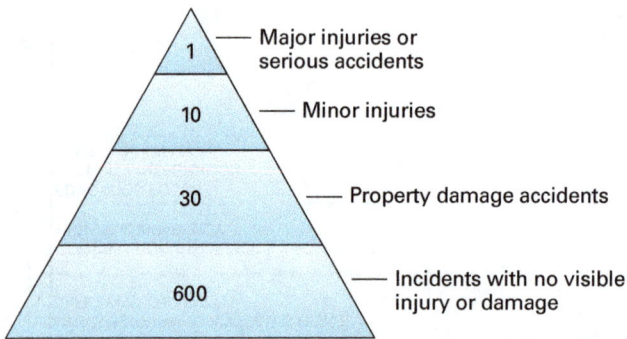

Fig. 3.5: The bird accident pyramid ("Egyptian").

Therefore, Krause and his colleagues proposed a "new" safety paradigm with the following assumptions:
(i)   All minor injuries are not the same in their potential for serious injury or fatality. A subset of low-severity injuries comes from exposures that act as a precursor to serious accidents.
(ii) Injuries of differing severity may have differing underlying causes.
(iii) Reducing serious injuries requires a different strategy from reducing less serious injuries.
(iv) The strategy for reducing serious injuries should use precursor data derived from accidents, injuries, near-misses and exposure.

Based on research by Krause et al. on the different types of uncertainties/risks available, and based on several disasters (such as the BP Texas City Refinery disaster of 2005), the classic single pyramid shape thus needs to be improved. Instead of a single pyramid model (as shown in Fig. 3.5), a double pyramid model with two pyramids next to each other and a small overlap area can better be used [9]. One pyramid stands for the type I accidents and the other (higher pyramid) for the type II accidents. The double pyramid model shows that there is a difference between type I risks and type II risks – in other words, "regular accidents" (and the incidents going hand-in-hand with them) should not be confused with "major accidents." Not all type I near-misses have the potential to lead to disaster, but only a minority of unwanted type I events (20% of them) may actually eventually end up in a catastrophe. The other 80% of near-misses for type II accidents are type II specific and should be searched for with type II precursors and indicators.

Obviously, to prevent disasters and catastrophes, risk management should be aimed at both types of risks and certainly not only at the large majority of "regular" or obvious risks.

Thinking along the lines of the classic Egyptian accident pyramid – resulting from the studies by Heinrich, Bird, Pearson and others – has had an important influence on dealing with safety in organizations, and it still does. However, we should realize that this way of thinking is the cause as well as a symptom of *blindness toward disaster*. Therefore, the bipyramid model is essential, and it provides a much better picture of how to deal with safety. Hopkins [9] indicates that the airline industry was the pioneer regarding this kind of safety thinking. In this industry, taking all necessary precautionary measures to ensure flight safety is regarded as fundamentally different from taking preventive measures to guarantee employee safety and health. Two databases are maintained by airline companies: one database is used to keep data of near-miss incidents affecting flight safety, and another database is maintained to store information regarding workforce health and safety. Hence, in this particular industry, it is understood that workforce injury statistics tell nothing about the risk of an aircraft crash. This line of thinking should be implemented in every industrial sector: railway industry, marine, shipping and fishing, chemical and petrochemical industry, nuclear, retail, education, administration, building and construction, food, agriculture, pharmaceutical and what have you.

In summary, the correct pyramid shape should be double, making a distinction between the different types of risk. Type II risks not only require constant individual mindfulness based on statistical modeling but also collective mindfulness, a focus on major accident, safety, an open mind toward extremely unlikely events, etc., based on qualitative and semiquantitative risk approaches and assessments.

3.3.2 The P2T model

If accidents result from holes in the safety system, then it is important to "close the holes" in time. Based on the OGP model for human factors [10], Reniers and Dullaert [11] discern three dimensions in which measures can be taken to avoid and prevent unwanted events and mitigate their consequences. The three dimensions are *people, procedures and technology* (see Fig. 3.6), and the model is therefore called the "P2T" model.

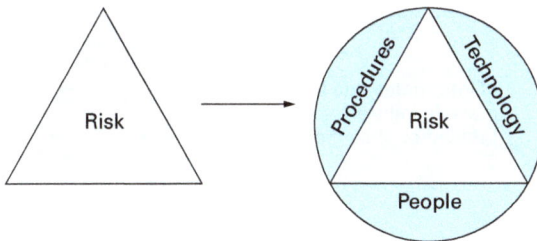

Fig. 3.6: People, procedures and technology to manage and control risks.

Applied to risk management within an organization, the first dimension, *people*, indicates how people (individually and in a group) deal with risks and think about risks in the organization (including domains such as training, competence, behavior and attitude). The second dimension, *procedures*, concerns all management measures taken in the organization to tackle risks in all possible situations and under all conceivable circumstances (including topics such as work instructions, procedures and guidelines). The third component, *technology*, comprises all technological measures and solutions taken and implemented with respect to risk management (including risk software and safety instrumented functions (SIFs)). In this way, the sharp ends of the risk have become part of a soft circle.

3.3.3 The Swiss cheese model and the domino theory

The "Swiss cheese" model was developed by the British psychologist Reason [12] to explain the existence of accidents by the presence of "holes" in the risk management system (see Fig. 3.7). A solid insight into the working of the organization allows for the possibility of detecting such "holes," while risk assessment includes the identification of suitable measures to "close the holes." The cheese itself can be considered the positive side of risks, and thus the more cheese, the more gains the risks may provide.

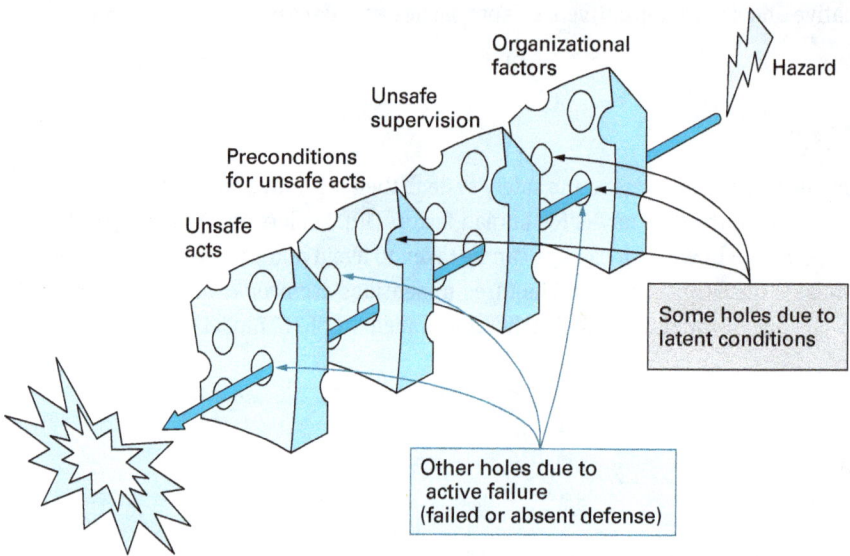

Fig. 3.7: The Swiss cheese model.

It is important to note that the Swiss cheese is dynamic: Holes may increase (e.g., caused by mistakes, errors, violations, lack of maintenance and security issues), but they may also decrease (because of solid risk management and adequate preventive and protective measures).

This model is very powerful in its use of "barrier" thinking (or "layer of protection" or "rings of protection" thinking). The holes within the barriers should be made as small as possible through adequate risk management, and this should be done for type I as well as type II risks. Figure 3.8 illustrates the theoretical safety continuous amelioration picture in an industrial area, combining the P2T (people, procedures and technology) model from Section 3.3.2 with the Swiss cheese model.

The upper half of Fig. 3.8 shows the stepwise progress that risk management science has made over time and is still making. Taking two companies, A and B, as an example, every distinct progress step (characterized by new insights and representing the dimensions from the P2T model) is represented by a rhombus. This rhombus can actually be considered a kind of safety layer. The hypothetical development of five potential accidents for the two plants is shown. Every potential accident is prevented by one of the dimensions, elaborated on an individual plant level or on a multiplant scale, except for accident number 3. Accident number 1, e.g., was stopped by the accident prevention layer marked by the existing procedures on a plant level. The lower half of the figure illustrates the needed ceaseless stepwise safety improvement resulting from new insights (captured over time) dealing with risks. Each safety layer (itself composed of a number of safety barriers) is considered as increasing safety effectiveness.

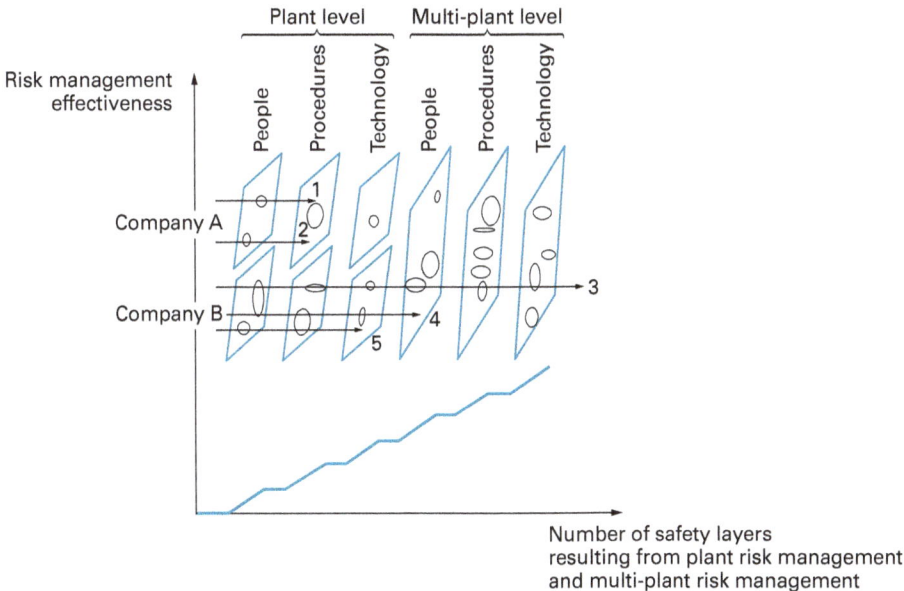

Fig. 3.8: Preventing unwanted events by combining the Swiss cheese model with the P2T model.

Linked with the Swiss cheese model is the "domino model" by Heinrich [13–15]. Heinrich indicates that the occurrence of an accident invariably results from a completed sequence of factors, the last one of these being the accident itself. The accident further results in loss (injury, damage, etc.). The accident is invariably caused by the unsafe act of a person and/or a mechanical or physical hazard. Figure 3.9 illustrates the Heinrich domino theory model and shows that the sequential accident model has a clear assumption about causality, specifically that there are identifiable cause–effect links that propagate the effects of an unwanted event. Based on the model/metaphor, Heinrich indicated that, completely in the sphere of his time (first half of the twentieth century), the behavior of workers (the "unsafe act" domino block) should be changed to avoid injuries and losses (cf. accident proneness theory).

An *accident can thus be visualized by a set of domino blocks* lined up in such a way that if one falls, it will knock down those that follow. This basic principle is used in accident investigations. It should be noted that the domino model by Heinrich should not be confused with other concepts with terminology such as "domino accidents" and "domino effects," indicating "chains of accidents" or escalating events in the chemical industry, whereby one accident leads to another accident and so on.

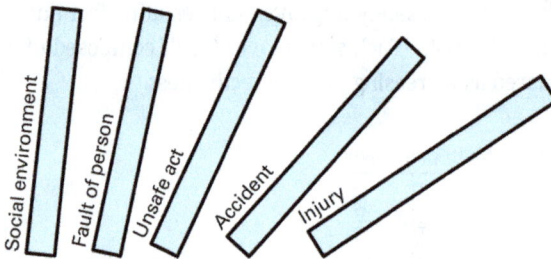

Fig. 3.9: The domino theory model (source: [13]).

Although the domino model by Heinrich only applies to type I risks and is somewhat antiquated (because it has been proven that the accident-proneness theory is wrong, and present insights in risk management indicate that systems thinking (see Section 3.10.1) should be used, and not only a cause-and-effect model), it still has its merits for presenting a complex accident in a simple way.

3.4 The anatomy of an accident: SIFs and SILs

The best way to achieve safety within any organization is to have inherently safe processes and to operate in correctly designed, controlled and maintained environments (from the perspective of people, procedures and technology). However, it seems inevitable that once designs have been engineered to optimize organizational safety, a spectrum of risks remains in many operations. To deal with these risks, a comprehensive safety management system is developed. This safety management system addresses hazard assessment, specification of risk control measures, evaluation of the consequences of failures of these controls, documentation of engineering controls, scheduled maintenance to assure the ongoing integrity of the protective equipment, etc. SIF[1] can, e.g., be considered a prevention measure. An SIF is a combination of sensors, logic solvers and final elements with a specified safety integrity level (SIL) that detects an out-of-limit condition and brings a process to a functionally safe state. However, these methodologies do not investigate the underlying physical and chemical hazards that must be contained and controlled for the process to operate safely, and thus they do not integrate inherent safety into the process of achieving safe plant operation. Most action items refer to existing safety procedures or to technical safeguards or require the addition of new levels of protection around the same underlying hazards. In other words, they represent "add-on safety" (see also Chapter 5). However, taking preventive measures in the conceptual design phase is extremely important to optimize and help achieve organizational safety.

The most desirable requirement of equipment is that it is *inherently safe*. Achieving such inherent safety starts in the design phase of the equipment. An inherently safe design approach includes the selection of the equipment itself, site selection and decisions on dangerous materials inventories and company layout. Complete inherent safety is rarely achievable within economic constraints. Therefore, potential hazards remaining after applying such an approach should be addressed by further specifying independent protection layers (IPLs) to reduce the operating risks to an acceptable level.

In current industry practice, chemical facilities processing dangerous substances are designed with multiple layers of protection, each designed to prevent or mitigate an undesirable event. Multiple IPLs addressing the same event are often necessary to achieve sufficiently high levels of certainty that protection will be available when needed. Powell [16] defines an IPL as having the following characteristics:

– Specific – designed to prevent or mitigate specific, potentially hazardous events.
– Independent – independent of the other protective layers associated with the identified hazard.
– Dependable – can be counted on to operate in a prescribed manner with acceptable reliability. Both random and systematic failure modes are addressed in the assessment of dependability.
– Auditable – designed to facilitate regular validation (including testing) and maintenance of the protective functions.
– Reducing – the likelihood of the identified hazardous event must be reduced by a factor of at least 100.

An IPL can thus be defined as a device, system or action that is capable of preventing a scenario from proceeding to its undesired consequence, independent of the initiating event or the action of any other layer of protection associated with the scenario.

---

1 This is a newer, more precise term for a safety interlock system (SIS).

Figure 3.10 illustrates safety layers of protection that are specifically used in the chemical industry. Choosing the process that one wants to use to produce a product or to provide a service (among several options) can be considered the first layer of protection. Detailed process design provides the second layer of protection. Next come the automatic regulation of the process heat and material flows and the provision of sufficient data for operator supervision, in the chemical industry together called the "basic process control systems." A further layer of protection is provided by a high-priority alarm system and instrumentation that facilitate operator-initiated corrective actions. An SIF, sometimes also called the "emergency shutdown system," may be provided as the fifth protective layer. Until this protection layer, we call all the measures "prevention measures" since they are, in fact, lowering the likelihood that something will go wrong. The further layers of protection, IPL 6–9, consist of so-called protection measures since they lower the (potential) consequences and the outcome. The SIFs are protective systems that are only needed on those rare occasions when normal process controls are inadequate to keep the process within acceptable bounds. Any SIF will qualify as one IPL. Physical protection may be incorporated as the next layer of protection by using venting devices to prevent equipment failure from overpressure. Should these IPLs fail to function, walls or dikes may be present to contain liquid spills. Plant and community emergency response plans further address the hazardous event.

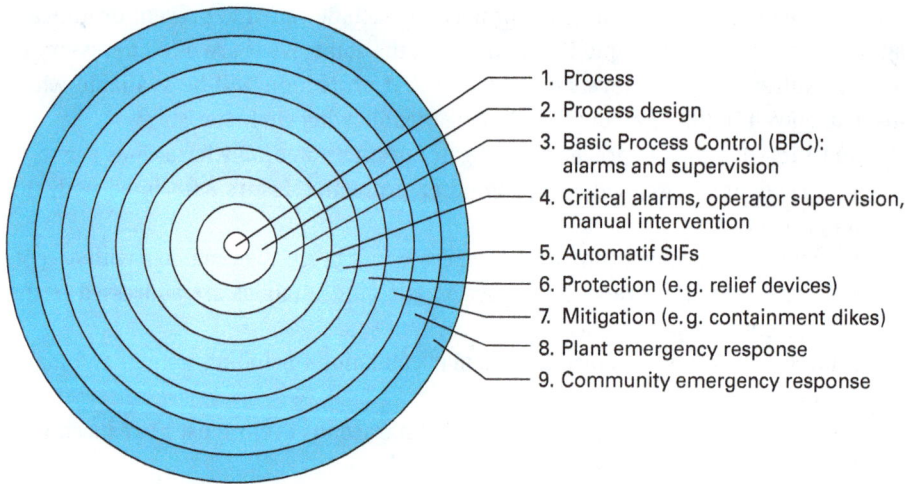

1. Process
2. Process design
3. Basic Process Control (BPC): alarms and supervision
4. Critical alarms, operator supervision, manual intervention
5. Automatif SIFs
6. Protection (e.g. relief devices)
7. Mitigation (e.g. containment dikes)
8. Plant emergency response
9. Community emergency response

Fig. 3.10: Typical layers of protection found in modern chemical plants.

By considering the sequence of events that might lead to a potential accident (see also the "domino" model by Heinrich in the previous subsection), another representation can be developed that highlights the efficiency of the protection layers, as shown in Fig. 3.11 (adapted from [17]).

Figure 3.11 illustrates the benefits of a hazard identification approach to examine inherent safety by considering the underlying hazards of a hazardous substance, a process or an operation. Inherently safer features in a design can reduce the required SIL (see Tab. 3.1) of the SIF or can even eliminate the need for a SIF, thus reducing the cost of installation and maintenance. Indeed, the Center for Chemical Process Safety [18] suggests that added-on barriers applied in noninherently safer processing conditions have some major disadvantages, such as the barriers being expensive to design, build and maintain, the hazard still being present in the process and the accumulated failures of IPLs still having the potential to result in an incident. An incident can be defined as "an undesired specific event or sequence of events that could have resulted in loss," whereas an accident can be defined as "an undesired specific event or sequence of events that has resulted in loss."

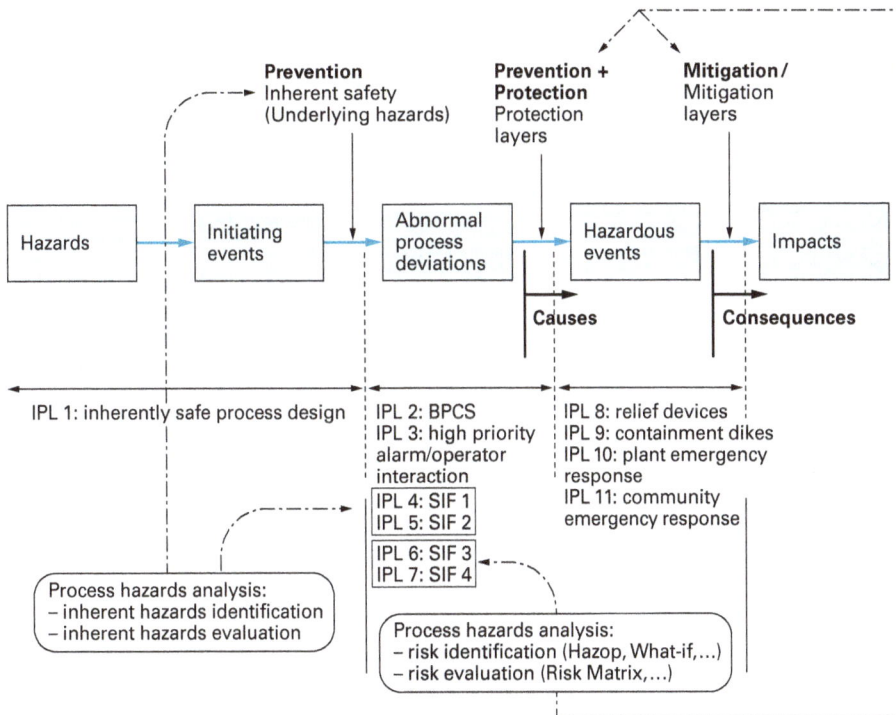

Fig.3.11: Anatomy of an accident (for a process under consideration) (with four SIFs).

The primary purpose of risk evaluation methods currently employed is to determine whether there are sufficient layers of protection against an accident scenario. A scenario may require one or more protection layers depending on the process complexity and the potential severity of a consequence. Note that for a given scenario, only one layer must work successfully for the consequence to be prevented (see also the Swiss cheese model and the Heinrich domino model). However, as no layer is perfectly effective, sufficient protection layers must be provided to render the risk of the accident tolerable. Therefore, it is very important that a consistent basis is provided for judging whether there are sufficient IPLs to control the safety risk of an accident for a given scenario. Especially in the design phase, an approach for drafting inherent safety into the design by implementing satisfactory IPLs is needed for effective safety. In many cases, the SIFs are the final independent layers of protection for preventing hazardous events. Moreover, all SIFs are required to be designed such that they achieve a specified SIL.

Remark that in the case of any barrier, whether it is human or technological, three features are necessary and should be present and working: *detection* of a potential problem, *decision-making* about the problem at hand, and *acting* upon the situation to solve the problem. In each of the three features, or phases, something might be/go wrong: the barrier can fail to detect a problem, it can make the wrong decision and/or it can take the wrong (or no) action. In the case of a technological barrier (e.g., an SIF), we use the terminology "sensor" for detection, "logic solver" for decision-making and "actuator" for taking the actions. In the case of human barriers, the human can fail to see the problem (detection failure), can make a wrong decision, for instance, about the cause(s) or the seriousness of the problem (decision-making failure) or might take the wrong action (act failure).

The SIL is the quantification of the probability of failure on demand (PFD) of an SIF into four discrete categories. The PFD can be defined as "the probability that a system will fail to perform a specified function on demand." Table 3.1 gives an overview of such levels.

Tab. 3.1: Safety integrity.

Safety integrity level (SIL)	Safety availability	Probability of failure on demand (PFD)	Equivalent risk reduction factor (1/PFD)
SIL 4	>99.99%	$\geq 10^{-5}$ to $<10^{-4}$	10,000–100,000
SIL 3	99.9–99.99%	$\geq 10^{-4}$ to $<10^{-3}$	1,000–10,000
SIL 2	99–99.9%	$\geq 10^{-3}$ to $<10^{-2}$	100–1,000
SIL 1	90–99%	$\geq 10^{-2}$ to $<10^{-1}$	10–100

Source: Based on International Standard, 2003 [19].

Thus, four corresponding degrees of reduction in hazardous event likelihood are produced by the SILs. SIL 1 provides about two orders of magnitude of event likelihood reduction, SIL 2 about three orders of magnitude, SIL 3 about four orders of magnitude and SIL 4 more than four orders of magnitude. Obviously, the availability targets for SIL 3 and SIL 4 are extremely stringent and the design practices to achieve and maintain these high levels are extensive and costly. Note that the SIL levels, as defined in Tab. 3.1, only concern so-called low-demand SIFs, which are designed to function in low-demand mode. SIFs operating in high-demand mode require the use of PFH or probability of dangerous failure per hour instead of a PFD as a design parameter.

Gardner [20] points out that methods used to select SILs are based on an evaluation of three characteristics of the process and the hazardous event associated with the SIF: the severity of the hazardous event consequences (minor, serious and extensive), the likelihood that an upset situation will occur that could lead to these consequences (low, moderate, high) and the number of IPLs. Before an SIL can be selected, the inherent risk of the process must be evaluated. Next, credit for all non-SIF mitigation measures (e.g., relief valves and dikes) must be accounted for to determine the baseline risk of the process, which is the starting point of the SIL selection. All of the SIF design, operation and maintenance choices have to be verified against the target SIL. The safety design engineer has to realize that further mitigation with an SIF solely reduces the likelihood of an incident. For example, if the baseline likelihood is 10^{-2} per year, an SIL 2 would reduce the likelihood up to 10^{-5} per year. The risk reduction process is illustrated in Fig. 3.12, in which risk criteria are represented in the form of *FN* limit lines (see also Section 3.5.2).

Determining the necessity of IPLs and SIFs, and the required level of their safety integrity, is performed using risk identification and evaluation methods. No one approach for the selection of an SIL is appropriate in every situation.

Summarizing, in highly technological environments such as, for instance, the nuclear industry or the chemical industry, the SIL is chosen to reduce the incident frequency/probability to a "tolerable" level for the company. It is the design basis for all engineering decisions related to the SIF. When the design is complete, it must be validated against the SIL. Therefore, the SIL closes the design cycle: starting with hazard identification, then requirements quantification and ending with design validation.

Achieving inherent safety is not always possible. When processing, storing or transporting hazardous materials, residual risks often remain to a greater or lesser extent. To identify and assess residual risks in an optimal and effective way, the analyses aimed at remaining risks should be thoroughly carried out.

The fundamental basis of security management can be expressed in a similar way to the layers of protection used in chemical process plants to illustrate safety barriers (see Fig. 3.10). In the similar concept of concentric rings of protection [21], the spatial relationship between the location of the target asset and the location of the physical countermeasures is used as a guiding principle. Figure 3.13 (adapted from [17]) exemplifies the rings of protection and their component countermeasures, illustrating the responsibilities and the distinction between indoor and outdoor security guards.

In security terms, critical infrastructure is broadly defined as people (employees, visitors, contractors, nearby members of the community, etc.), information (formulae, prices, processes, substances, passwords, etc.) and property (buildings, vehicles, production equipment, storage tanks and process vessels, control systems, raw materials, finished products, hazardous materials, natural gas lines, rail lines, personal possessions, etc.) which are believed crucial to prevent major business disruption and resulting substantial economic and/or human and societal damage.

By considering the sequence of events that might lead to a potentially successful attack, another representation can be given, illustrating the effectiveness of the rings of protection (see Fig. 3.14).

First, companies can clearly protect themselves much better against external attacks than against attacks from within the company itself because in the latter case, there exists only indoor security to avert the threat. Second, as the effective prevention, protection and mitigation of attacks depend on meticulously carrying out security risk assessments, the latter is of crucial importance to deter, detect and delay possible threats within a single company as well as within a cluster of companies.

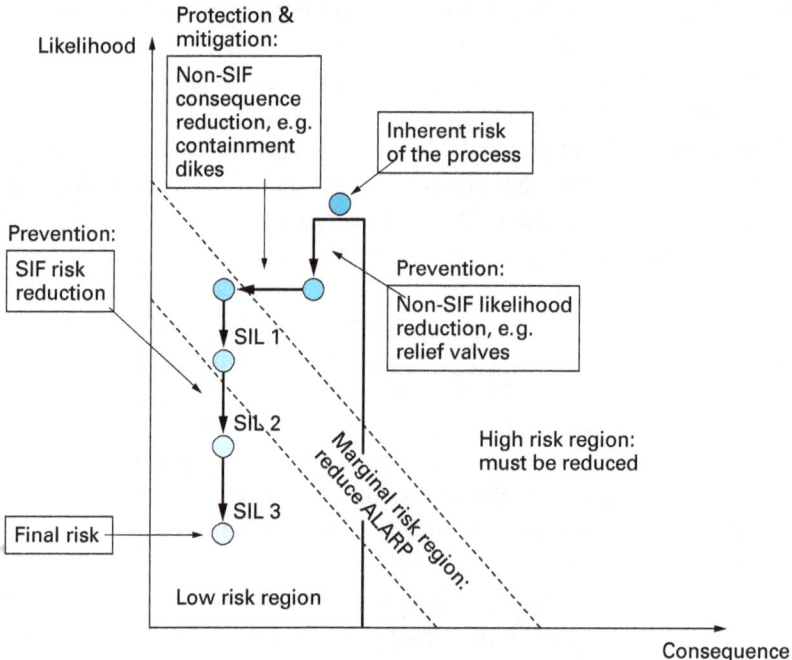

Fig. 3.12: The effect of risk reduction measures (adapted from [17]).

Once the entire sequence of events has taken place, either safety-related or security-related, there is a loss. Regardless of the particular business activity in which the loss may have occurred, losses can be considered – minor, serious, major, or catastrophic. Depending upon the particular business activity, the determining factors for rating the severity of an accident loss can be the degree of physical harm and/or property damage and any resulting human or economic aspects. When evaluating either these human or economic effects, investigators should be particularly aware that those factors readily apparent are usually indicative of much more serious and far-reaching aspects that are not so obvious. Much like the tip of an iceberg, the extent and size of the problems associated with accidents and losses are not easily seen or determined on the surface, but they are, nonetheless, there [22]. We also refer to Chapter 9 on economic issues related to safety.

> Models used to deal with risks are very diverse. The models have been built after decades of experience and research, within a variety of academic disciplines and encompassing diverse industrial sectors. Incidents and accidents usually were a driver and an inspiration for the builders of the models.

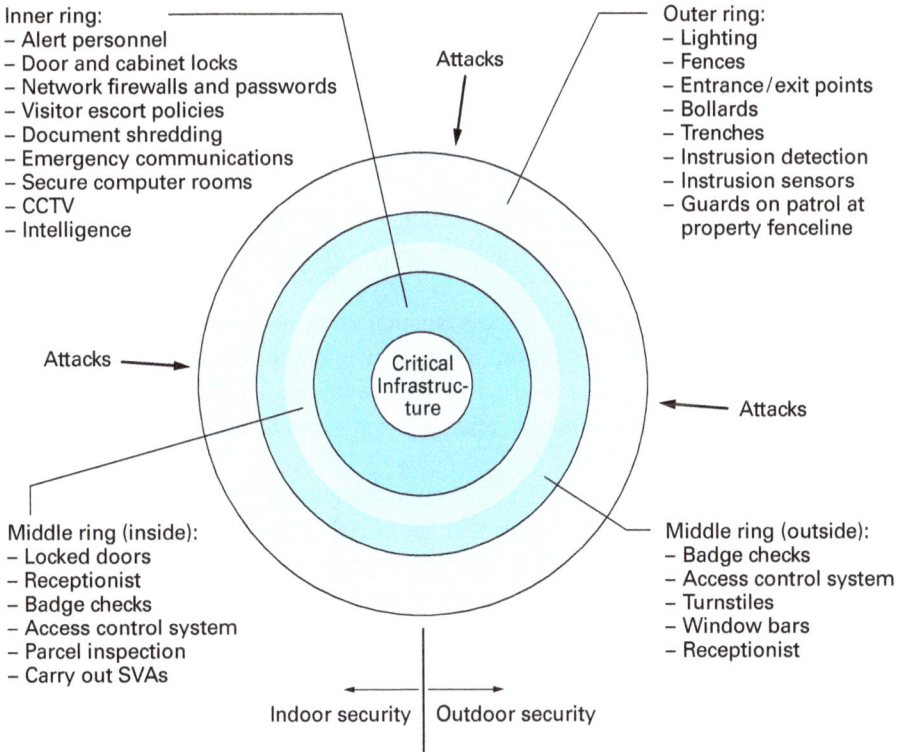

Inner ring:
- Alert personnel
- Door and cabinet locks
- Network firewalls and passwords
- Visitor escort policies
- Document shredding
- Emergency communications
- Secure computer rooms
- CCTV
- Intelligence

Attacks

Outer ring:
- Lighting
- Fences
- Entrance/exit points
- Bollards
- Trenches
- Instrusion detection
- Instrusion sensors
- Guards on patrol at property fenceline

Attacks

Critical Infrastruc-ture

Attacks

Middle ring (inside):
- Locked doors
- Receptionist
- Badge checks
- Access control system
- Parcel inspection
- Carry out SVAs

Middle ring (outside):
- Badge checks
- Access control system
- Turnstiles
- Window bars
- Receptionist

Indoor security | Outdoor security

Fig. 3.13: Security rings of protection.

3.5 Individual risk, societal risk, physical description of risk

3.5.1 Location-based (individual) risk

A so-called location-based risk or individual risk can be defined in general as "the frequency with which a person may expect to sustain a specified level of harm as a result of an adverse event involving a specific hazard" [3]. Hence, individual risk can be used to express the general level of risk to an individual in the general population or to an individual in a specified section of the community. There is widespread agreement in many countries that an additional risk from industrial activities of 10^{-6} per year to a person exposed to this risk is at a very low level compared with risks that are accepted every day. The reasoning is thus. On the one hand, the individual risk of getting killed from driving a car is estimated as 10^{-4} per year (order of magnitude). If the level of (individual) risk is higher than driving a car, it is perceived as unacceptable. On the other hand, the individual risk of being struck by lightning is estimated as 10^{-7} per year (order of magnitude), and this risk level is perceived to be so low that it can be accepted.

Legislators often use levels of individual risk with a "specified level of harm" equal to "fatality" (such risks are called "individual fatality risks") as a regulatory approach to setting risk criteria. In the process industries, the individual (fatality) risk is defined as the risk that an unprotected individual would face from a facility if he or she remained fixed at one spot, 24 h a day, 365 days per year. Therefore, the risk is also called location-based risk. A location-based risk describes the geographic distribution of individual risk for an organization. It is shown using so-called iso-risk curves and is not dependent on whether people or residences are present (see Fig. 3.15).

Location-based risk is used to assess whether individuals are exposed to more than an acceptable risk in the locations where they may spend time (e.g., where they live or work). It does not directly provide information on potential loss of life, nor does it distinguish between exposure affecting employees or the general (surrounding) population.

Individual fatality risks are calculated by multiplying the consequences and the frequency. For example, if the severity of an industrial accident is such that there is a $p\%$ probability of killing a person at a specified location (the probability merely takes into account the level of lethality due to the accident, e.g., due to heat radiation or a pressure wave, but does not consider population figures) and the accident has a frequency of f per year, then the individual fatality risk at this particular location is $p \times f$ per year. Where there is a range of incidents that expose the person at that point to risk, the total individual fatality risk is determined by adding the risks of the separate incidents. Iso-risk contours – contours on which each point has an identical risk level – can then be plotted around an industrial activity and can be used to present the risk levels surrounding the activity. Sometimes different levels of individual fatal-

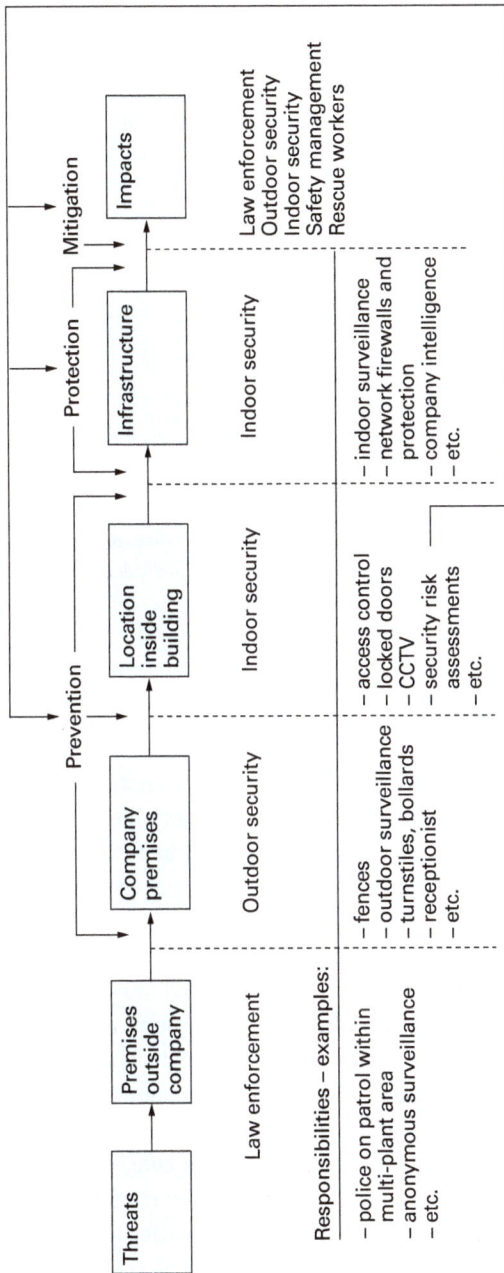

Fig. 3.14: Anatomy of an attack.

ity risk are defined to be allowed by the authorities for different types of locations. For example, a distinction is made between industrial areas, commercial areas, parks and sports fields, schools, rest and nursing homes, hospitals, etc. Individual risks are often calculated using "quantitative risk assessment" (QRA; see Section 4.9).

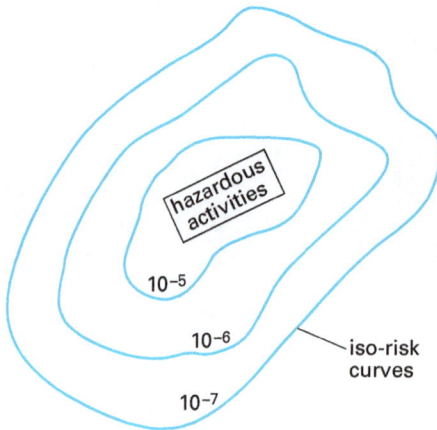

Fig. 3.15: Example of iso-risk curves showing the distribution of location-based (individual) risk surrounding an enterprise.

3.5.2 Societal risk or so-called group risk

Calculating the individual fatality risk around a specific industrial activity does not make a distinction between the activity taking place, e.g., somewhere in the desert or within the center of a major city. The calculated individual risk will be the same, regardless of the number of people exposed to the activity. However, it is evident that in reality, the level of risk will not be identical for both situations: the level of risk will be higher if the industrial activity takes place in the city rather than in the desert. This results from people living in the neighborhood of the activity and thus being exposed to the danger. To take the exposed population figures into account, a "societal risk" is calculated. This is "the probability that a group of a certain size will be harmed – usually killed – simultaneously by the same event or accident" [23]. It is presented in the form of an FN curve (also called Group Risk curve). Each point on the line or curve represents the probability that the extent of the consequence is equal to or larger than the point value. These curves are found by sorting the accidents in descending order of severity and then determining the cumulative frequency. As both the consequence and the cumulative frequency may span several orders of magnitude, the FN curve is usually plotted on double logarithmic scales (see also [24]). A log–log graph is obtained, as depicted in Fig. 3.16.

The societal risk is designed to display how risks vary with changing levels of severity. For example, a hazard may have an acceptable level of risk for just one fatality but may be at an unacceptable level for 100 fatalities. In some jurisdictions, there are rigidly defined boundaries on the societal risk graph between the zones of high, intermediate and low risk (see Fig. 3.17). It is common, where an industrial activity is calculated to generate risks in the intermediate zone between high and low, to require the risks to be reduced to a level that is "as low as reasonably practicable" (ALARP), provided that the benefits of the activity that produces the risks are seen to outweigh the generated risks.

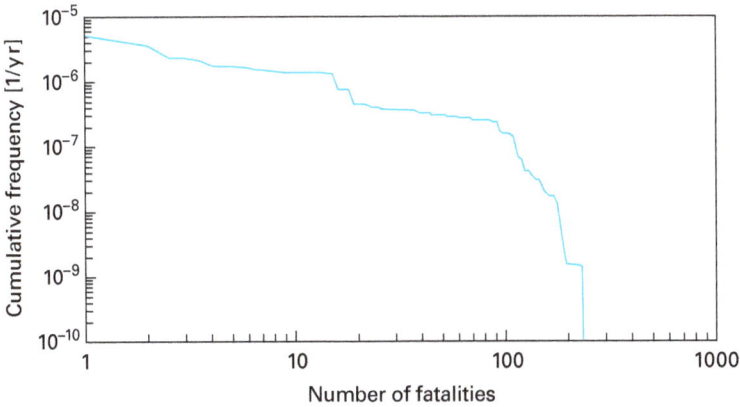

Fig. 3.16: FN curve – an illustrative example.

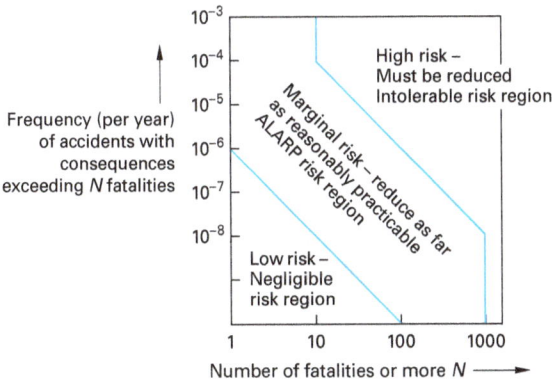

Fig. 3.17: Example of an approach to defining societal risk criteria.

As the severity of the event increases, people become more risk-averse. Particularly, once the death threshold is passed, it appears that the community has a much greater aversion to multiple fatality accidents. In many countries, this seems to amount to a 100-fold decrease in the likelihood of the event for a 10-fold increase in the severity of the consequences measured in fatalities. This is shown in the FN curves defined by the authorities from different countries to indicate the region where societal risks must be reduced and where they can be tolerated, depicted in Fig. 3.18.

One measure of the societal risk from an installation could be obtained by calculating the fatal accident rate (FAR) of the number of fatalities per year from accidents involving dangerous substances. If the *FN curve* of the installation is known, then the value of FAR can be calculated as follows:

$$\text{FAR} = \sum_{N=1}^{N_{\max}} f(N) \cdot N$$

However, using the FAR as a criterion for societal risk attracts criticism, it does not include an allowance for aversion to multifatality accidents. It gives equal weight to the frequencies and consequences of accidents. By not distinguishing between one accident causing 50 fatalities (type II event) and 50 accidents each causing one fatality (50 type I events) over the same period of time, the FAR fails to reflect the importance society attaches to major accidents [25]. Moreover, several industrial activities from nearby companies in the same industrial area may each generate a low level of societal risk, whereas their combined societal risk might fall within the high-risk zone of the chart in Fig. 3.17 (or Fig. 3.18), if these industrial activities were all to be grouped for the purposes of the calculation.

For calculating consequences, often so-called probit functions are employed, providing the probability of specific consequences (e.g., death of an average person or failure of a certain material structure) within a specified exposure time and based on a concentration (in the case of toxic substances) or intensity (in the case of heat radiation), for instance. In the case of toxicity, the formula has the following form:

$$\text{Pr} = a + b \cdot \ln(C^n \cdot t)$$

Constants a, b and n in this equation are substance-specific, C represents the concentration of the toxic substance and t the exposure time. A probit value of 5 is lethal for 50% of the population exposed. The concentration at this 50% fraction is known as LC50 and the dose, $C \times t$, as LD50. Probit 2.67 corresponds to a lethal dose for 1% of the exposed population and probit 7.33 for 99%. Probit tables are usually readily available. Outcome uncertainty is high due to the tables being drawn for the "average person," and hence, not directly applicable to more vulnerable groups. Through international cooperation, probit coefficients are still being updated from time to time. For more information, see, for example, [27].

In enterprises encountering numerous hazards with severe potential consequences, computer programs are used to calculate the risk levels on a topological grid, after which they are used to plot contours of risk on the grid. These contours are used to display the frequency of exceeding excessive levels of hazardous exposure. For example, Reniers et al. [26] discuss software tools that are available that will prepare contours,

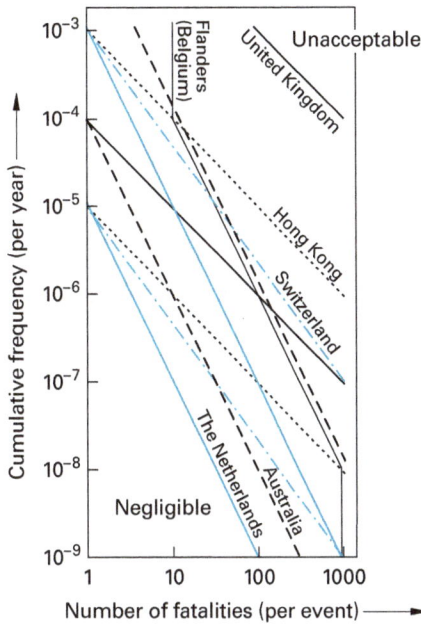

Fig. 3.18: Societal risk criteria in different countries.

e.g., for the frequency of exposure to nominated levels of heat radiation, explosion over-pressure and toxic gas concentration. QRA software is employed to plot *FN* curves.

We should keep in mind that all these software tools provide insight into the pos-sible scale of a disaster and the possibility of its occurrence, but they do not offer ade-quate information for the *optimal prevention* of catastrophic accidents. Moreover, if the number of events observed in the past is not sufficient to estimate significant fre-quency values, as in the case of type II events, a simple histogram plotting the abso-lute number of past events versus a certain type of consequence is often used instead of a risk curve. However, type II event predictions are extremely difficult to make simply because of the lack of sufficient data. Although it is thus not possible to take highly specific precautionary measures based on statistical predictive information in such cases, engineering risk management, leading to a better understanding of rela-tive risk levels and to an insight into possible accident and disaster scenarios, is actu-ally essential to prevent such disastrous accidents and therefore should in one way or another be fully incorporated in industrial activities worldwide.

In summary, there are different possible ways to calculate risk. Two well-known approaches, widely used, are the calculation of the individual/location-specific risk and the societal risk. A location-specific risk provides an idea of the hazardousness of an industrial activity. A societal risk takes the exposure of the population in the vicinity of the hazardous activity into account in the calculation.

3.5.3 Physical description of risk

In the previous chapter, we defined a "negative risk" and indicated that all such risks are characterized by three factors: hazards–exposure–losses, together forming the "risk trias." Risk is a theoretical concept and can be described in yet another way. To have a profound understanding of risk, we also discuss this second, more physical, approach.

In order to physically describe what a risk is, we must define some of its key components. The notion of the "target" needs to be introduced. A target is someone or something that can suffer losses or be subject to depreciation of value. By definition, the target can be represented by

- a human,
- the environment,
- a natural monument,
- a process in a company,
- a company,
- the brand image, etc.

A threat is the potential of a hazard to cause damage. A threat can be intentional – then it is always a human threat related to the field of security – or it can be accidental or by coincidence. A threat is the direct consequence that arises from its link with the hazard; if the threat is non-intentional, it is subordinated to the law of probability. Simply put, a safety-related threat results from a hazard that gets out of control (one could see it as energy getting out of its cage) by coincidence, while a security-related threat results from a hazard that is deliberately misused by a person who has the intention to cause losses.

Risk exists as soon as a hazard affects one or many possible targets. An identified hazard that does not affect any target does not represent a risk. For example, life on Mars may be very hazardous, but as long as nobody lives on Mars, there are no losses, and hence no risk. Risk is found at the interface, or at the cross section, of a hazard and a target, as illustrated in Fig. 3.19.

Basically, a risk is physically characterized by four elements:

1. A hazard
2. One or many targets threatened by the hazard
3. The evaluation of the threat
4. The measures taken to reduce the threat.

These elements, depicted in Fig. 3.20, show that a protection and/or prevention barrier is required in order to prevent a threat from reaching the target.

The main difference between an incident and an accident or a disaster is generally defined by the (increasing) importance of the caused or sustained damage.

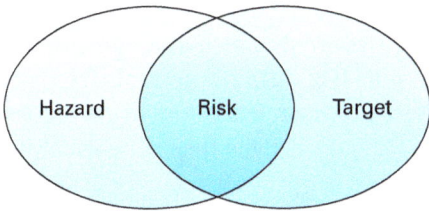

Fig. 3.19: Physical risk model.

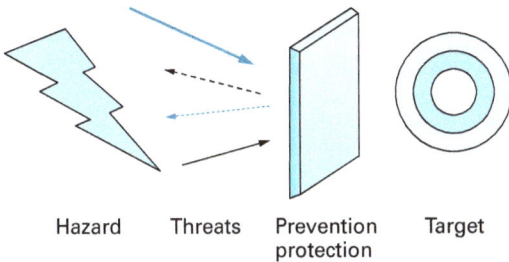

Hazard   Threats   Prevention   Target
                     protection

Fig. 3.20: Constitutive elements of risk.

We defined "incident" and "accident" in Section 3.4. To have a more physical description of these terms, the norm OHSAS 18001 [28] can be used. It defines the notion of "accident" as follows:
- Incident: an event that leads to or could have led to an accident.
- Near-accident or near-miss: an incident that does not damage health or lead to any deterioration or losses (but did have the potential to do so).
- Accident: an unexpected event that leads to health deterioration, injuries, damages or other losses.

A disaster is a major accident. It is an event that is brutal, sudden and of enormous dimension. It has severe consequences that are accompanied by the destruction of goods and/or death.

Sometimes, an "incident" is used as a term to indicate any kind of event that leads, or could have led, to negative consequences. Before an incident takes place, there should be a "near-miss," and even earlier in the sequence order is "poor or bad behavior," and prior to that, the "wrong attitude" and a "faulty perception" (see also the KPABC model discussed in Section 2.6).

There are two possibilities to physically describe the risk: static and dynamic modeling.

3.5.3.1 Static model of an accident

If risk is a potential, the accident is a reality. It is realized as soon as a threat comes into contact with a target, allowing damage creation (Fig. 3.21). The prevention or protection barrier has only partially done its job as a protector. The failure of the barrier is represented by the holes in the wall. For a static model, the time scale is not included.

3.5.3.2 Dynamic model of an accident

While risk is not an event, the accident is one. Often, it is the sequence of other events, the succession of an incident, which leads to damage. These successive incidents induce situations that are more and more hazardous, in which the likelihood of occurrence increases to a critical level and finally overcomes, as illustrated in Fig. 3.22.

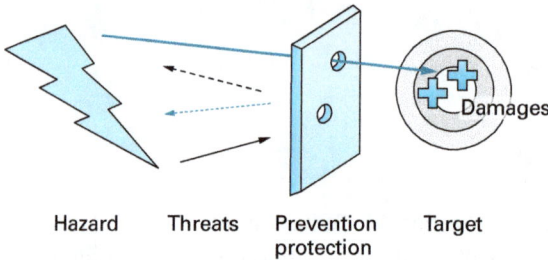

Hazard   Threats   Prevention   Target
protection

Fig. 3.21: Static model of an accident.

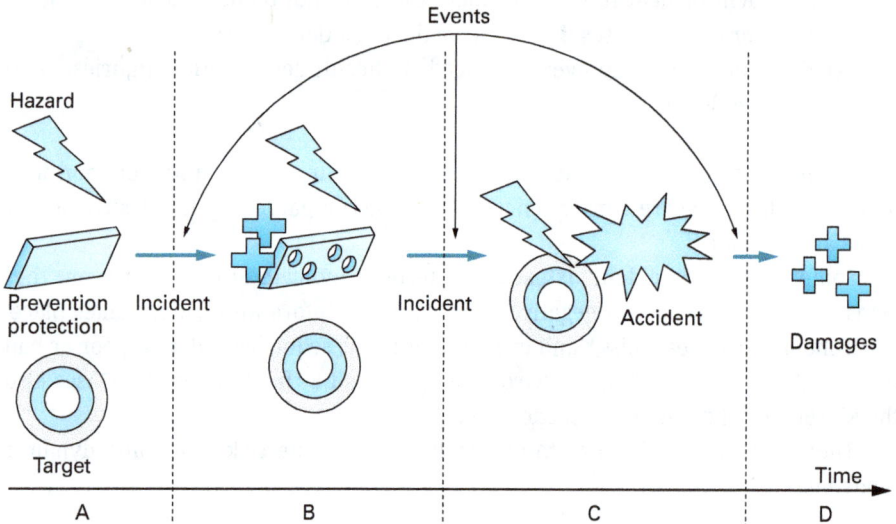

Fig. 3.22: Dynamic model of an accident (freely inspired from [29]).

We observe four zones:

A. At time zero, the protective/preventive barrier is fulfilling its purpose and prevents any threats from reaching the target.

B. Small incidents happen over time, decreasing the protective/preventive level or efficiency of the barrier. This is the first observable sign of precursors for a future accident.

C. As time continues, the degradation of the barrier is now sufficient for the hazard (its threats) to reach the target. We now speak of an accident!

D. Finally, the consequences of the accident are losses and damages.

The accident can also be the result of a situation that has changed continuously without the safety barrier measures being adapted to the changed situation. This is the case in many companies: the established measures have not been increased even if the company grew during a certain period. The protective and preventive measures are not adequate for the new size, and therefore the risk becomes more and more important (Fig. 3.23).

Again, four zones are observed:

A. At time zero, the protective/preventive barrier is fulfilling its purpose and prevents any threats from reaching the target.

B. The target begins to grow with time (it is rare that the barrier shrinks with time), but the barrier is still sufficient.

C. As time continues, the actual target is now larger than the size of the target against which the barrier was designed to protect. We now speak about an accident!

D. Finally, the consequences of the accident are losses and damages.

Dynamic modeling taught us that:

– An accident often results from a sequence of consecutive events or incidents, whether or not they cause immediate damage. Do not downplay the importance of any incident!

– An accident may occur due to a situation that has gradually changed without a reassessment of the original safety measures. Ensure that measures are adapted as situations evolve.

For risk management to effectively serve its protective role, it must be a dynamic process – both proactive and reactive – rooted in thorough and continuous monitoring of risk situations.

Finally, risk is not only a matter of technology but mainly involves dealing with humans. Therefore, in order to achieve a safe situation, one must combine the safe place and the safe person, as illustrated in Fig. 3.24.

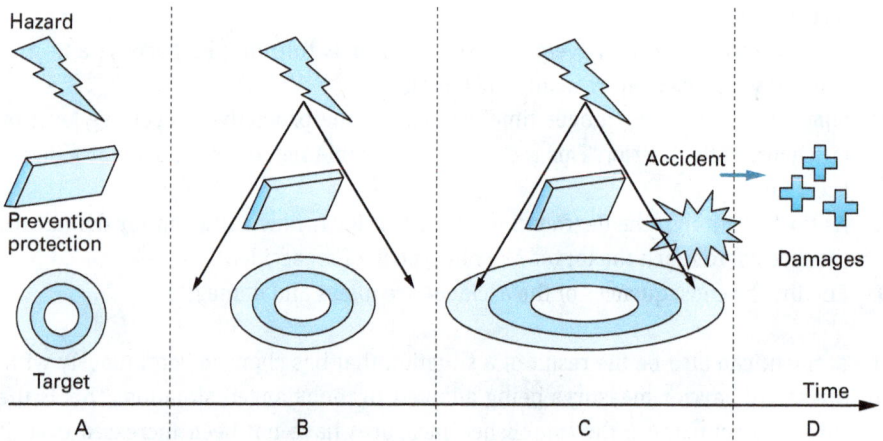

Fig. 3.23: Dynamic model of an accident (freely inspired from [29]).

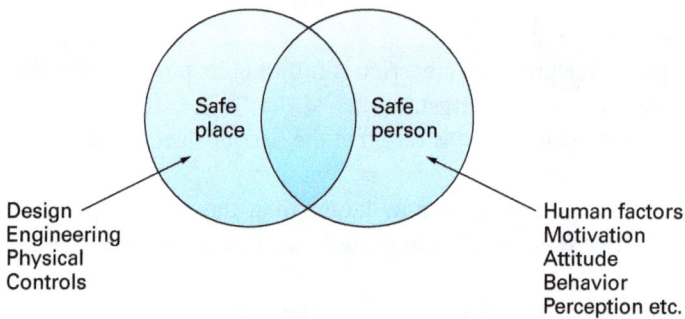

Fig. 3.24: Safe situation according to the classical risk modeling.

> A risk was theoretically characterized in the previous chapter as hazard, exposure and loss. Risk management aims to decrease one of these elements or a combination thereof. When physically describing a risk, it can be characterized as hazard, threat, prevention/protection and target. In industrial practice, risk management aims to identify all of these, and by manipulating one of these elements, or a combination thereof, risks are managed.

3.6 Safety culture and safety climate

3.6.1 Organizational culture and climate

An organizational culture is: "a pattern of shared basic assumptions learned by a group as it solves its problems of external adaptation and internal integration, which has worked well enough to be considered valid and, therefore, to be taught to new members as the correct way to perceive, think and feel in relation to those problems" [30]. An organizational culture has an impact on the behavior of the employees, the operations and the results of the organization. In order for this impact to be positive, it is important that there is a good fit between the strategy and the culture of the organization [31]. However, a company's culture will not be the only factor influencing the achievements and the excellence of a company. Two other influential factors are the organization's structure and its processes [32]. For an extensive overview of definitions for "organizational safety culture," see [33].

Organizational culture can be analyzed at several different degrees or levels at which the cultural phenomenon is visible to the observer [30]. The three levels are:
1. Artifacts (visible structures and processes as well as observable behaviors).
2. Espoused beliefs and values (ideals, values, aspirations, ideologies and rationalizations).
3. Basic underlying assumptions (unconscious, taken-for-granted beliefs and values).

It is essential to realize that a correct interpretation of the (most visible) artifacts depends on the knowledge of the pattern of basic assumptions. Although the essence of a group's culture is its pattern of shared, basic, taken-for-granted assumptions, the culture will manifest itself at the level of observable artifacts and shared espoused values, norms and rules of behavior.

In addition to a company's culture, another important concept is a company's climate (e.g., see [34–38]). Although both concepts are closely linked, it is imperative to make a clear distinction. Generally speaking, a company's climate can be thought of as "the product of some of the underlying assumptions and hence, it is the way in which a company's culture is visible to the outside world." Therefore, a company's climate can be seen as the outer layers of a company's culture and actually the manifestation of the culture. It is reflected in the perception of people working at the company at a certain moment. As a result, a company's culture emphasizes continuity, while its climate is comparable to a snapshot of its culture. An important difference between these two concepts is the way in which they are measured. A company's climate corresponds to the outer and more visible layers of its culture, reflected in the perceptions of people as mentioned, and can therefore be measured with standardized questionnaires. A company's culture is more fundamental and is, for that reason, much more difficult to measure. A combination of measuring methods needs to be used, e.g., observations, questionnaires and in-depth interviews in one go [50]. In any case, both an organization's culture and its climate should be integrated into a single model to truly advance an organizational domain through the process of goal setting, measuring and continuously improving. To achieve excellence in a particular domain (such as safety and security), an organization needs to set goals and measure if it has reached those goals, i.e., it has to manage its performances in the domain. Organizational consensus must be achieved on what to measure, how to measure it and what to do when corrections are needed. In order to determine what to measure, first, the most important dimensions and subdimensions of a company's culture and climate need to be identified. A dimension or subdimension is a part of the company that is of critical importance to realize the corporate mission and strategy. Next, these dimensions and subdimensions should be translated into measurable performance indicators to be able to objectively measure them [40]. If the company's track record is found to be inadequate in whatever domain, the organizational culture and/or climate needs to be changed or adapted by taking corrective measures and/or by setting new corporate objectives in the domain.

3.6.2 Safety culture models

The safety culture and its policy are defined as "the core values and behaviors resulting from a collective commitment by leaders and individuals to emphasize safety over competing goals to ensure the protection of people and the environment." It should emphasize the importance of fostering and maintaining an open, collaborative work environment that encourages all employees and contractors to promptly speak up and share concerns and differing views without fear of negative consequences. The safety culture of an organization is the product of individual and group values, attitudes, perceptions, competencies and patterns of behavior that determine the commitment to, and the style and proficiency of, an organization's health and safety management.

There is no single definition of "a safety culture." The term first arose after the investigation of the Chernobyl nuclear disaster in 1986, which led to safety culture being defined as an organizational atmosphere where safety and health are understood to be, and are accepted as, the number one priority. In high-risk industries like aviation, nuclear power, chemical manufacturing and fuel transportation, this makes sense. However, the problem is that safety and health do not exist in a vacuum isolated from other aspects of organizations, such as people and financial management, as they both influence and are influenced by them, so safety culture is really a part of the overall corporate culture. On this basis, a more realistic definition may be "a safety culture is an organizational atmosphere where safety and health is understood to be, and is accepted as, a high priority." Some indicators for safety culture could be expressed as

- Commitment at all levels.
- Safety and health are treated as an investment, not a cost.
- Safety and health are part of continuous improvement.
- Training and information are provided for everyone.
- A system for workplace analysis and hazard prevention and control is in place.
- The environment in which people work is blame-free.
- The organization celebrates successes.

To develop a safety culture, change needs to be driven from the highest levels. The extent to which one can influence the organization largely depends on their place within the hierarchy. The recognition of the importance of a safety culture in preventing accidents has led to a growing number of studies to define and assess safety culture in a variety of complex, high-risk industries.

An unsafe culture is more likely to be involved in the causation of organizational rather than individual accidents. Safety cultures evolve gradually in response to local conditions, past events, the character of the leadership and the mood of the workforce. An ideal safety culture is the "engine" that drives the system toward the goal of sustaining the maximum resistance toward its operational hazards and risks, regardless of the leadership's personality or current commercial concerns.

According to Reason [41], several powerful factors act to push safety into the background of an organization's collective awareness, particularly if it possesses many elaborate barriers and safeguards. However, it is just these defenses-in-depth that render such systems especially vulnerable to adverse cultural influences. Organizations are also prey to external forces that make them either forget to be afraid or, even worse, avoid fear altogether. The penalties of such complacency can be seen in the recurrent accident patterns in which the same cultural drivers, along with the same uncorrected local traps, cause the same bad events to happen again and again.

An organization with a "safety culture" is one that gives appropriate priority to safety and realizes that safety has to be managed like other areas of the business. That culture is more than merely avoiding accidents or even reducing the number of accidents, although these are likely to be the most apparent measures of success. It is about doing the right thing at the right time in response to normal and emergency situations. The quality and effectiveness of that training will play a significant part in determining the attitude and performance – the professionalism. And the attitude adopted will, in turn, be shaped to a large degree by the "culture" of the company.

The key to achieving that safety culture is as follows:
–   Recognizing that accidents are preventable through establishing and following correct procedures and established best practices;
–   Constantly thinking about safety, creating "situational awareness" among employees.
–   Seeking continuous improvement, having a "questioning attitude" and being a learning community.
–   Not blaming people if they make mistakes, errors and even violations with good intentions. The problem only arises in cases of malintent and recidivism.

It is relatively unusual for new types of accidents to occur, and many of those that continue to occur are caused by unsafe acts. These errors, or more often violations of good practice or established rules, can be readily avoided. Those who make them are often well aware of the errors of their ways. They may have taken shortcuts they should not have taken. Most will have received training aimed at preventing them, but through a culture that is tolerant of the "calculated risk," they still occur.

Actually, if all employees followed all the existing safety procedures and rules in organizations and were risk-aware, almost no accidents would probably occur. This can be compared with traffic accidents: if all car drivers rigorously followed the driving laws and were attentive all the time, almost no car accidents would occur. Another comparison can be made with the Covid-19 pandemic rules: if all citizens strictly followed the rules of social distancing, washing hands with alcohol gel all the time, wearing face masks, etc., much less infections would be determined. People are people, and many don't like to follow rules or procedures despite the (indirect and uncertain) benefits for themselves and for the organization or society at large. They have to fully understand what the purpose is, what the consequence is if they don't follow the rule and why it is important for themselves, their coworkers, the organization and/or society.

Referring to the KPABC model of Section 2.6, there needs to be an emphasis on both the "knowledge, perception and attitude" of a person, which is called person-based safety, and the "behavior and consequences" of a person, which is called behavior-based safety. Behavior-based safety only lasts for a limited period of time, while person-based safety is much more long-lasting.

Following the Chernobyl accident in 1986, a lot more attention was paid to the term "corporate safety culture" and various definitions were proposed. Cooper [42] defines corporate safety culture as "that observable degree of effort by which all organizational members direct their attention and actions toward improving safety on a daily basis," thereby stressing that in a good safety culture, all members of an organization should deliver intentional efforts to continuously improve overall safety. Hale [43] refers to beliefs, values and perceptions of natural groups within an organization and the effect of these groups on values and norms. These values and norms will define how a company will handle its risk and risk control systems. Wiegmann et al. [44] try to capture all previous definitions of safety culture in the following elaborated formulation:

> Safety culture is the enduring value and priority placed on worker and public safety by everyone in every group at every level of an organization. It refers to the extent to which individuals and groups will commit to personal responsibility for safety, act to preserve, enhance and communicate safety concerns, strive to actively learn, adapt and modify (both individual and organizational) behavior based on lessons learned from mistakes, and be rewarded in a manner consistent with these values.

It is important to also note that the aspect of learning from mistakes is adopted in the latter definition. Mohamed [45] presents a very pragmatic definition by stating that a corporate safety culture is a mere subculture of the general organizational culture. He defines safety culture as *a sub-facet of organizational culture, which affects workers' attitudes and behavior in relation to an organization's on-going safety perfor-*

mance. In summary, a safety culture can be regarded and explained as the way that people behave and act (with respect to safety) in an organization when nobody is watching them.

Moreover, as already mentioned in the previous section, an important difference exists between *safety culture* and *safety climate.* Much like organizational culture and climate, Hale [43] states that safety culture is the whole of values and practices that are linked to the company in a strong, unobservable relation. These values and practices are stable in time and cannot easily be observed nor altered. The corporate safety climate is easier to observe. The explicit artifacts and values of a company are good examples of the safety climate components. Wiegmann et al. [44] add to this that the safety climate is a snapshot of the safety culture at one moment in time, and that the climate displays what is the perception of the culture by the members of an organization.

Safety culture is commonly viewed as an enduring characteristic of an organization that is reflected in its consistent way of dealing with safety issues. Safety climate is viewed as a temporary state of an organization that is subject to change depending on the features of the specific operational or economic circumstances. Therefore, just like personality researchers, safety researchers have attempted to identify key indicators of organizational safety culture and to develop methods for assessing the extent to which these key organizational features are consistent across time and situations.

Safety climate is the temporal state measure of safety culture, subject to commonalities among individual perceptions of the organization. It is, therefore, situationally based, refers to the perceived state of safety at a particular place at a particular time and is relatively unstable and subject to change, depending on the features of the current environment or prevailing conditions.

Safety climate is a psychological phenomenon, which is usually defined as the perception of the state of safety at a particular time. It is closely concerned with intangible issues such as situational and environmental factors. Hence, safety climate is a temporal phenomenon, a "snapshot" of safety culture, relatively unstable and subject to change.

Safety behavior presents a paradox to practitioners and researchers alike because, contrary to the assumption that self-preservation overrides other motives, careless behavior prevails during many routine jobs, making safe behavior an ongoing managerial challenge.

At present, there is no overall satisfying model for safety culture and climate or for security² culture and climate [46]. However, important features and capabilities that are essential for characterizing, elaborating and improving an organization's safety and security culture and its safety and security climate have been put forward

---

2 Remember that security is characterized with intentionality and indicates deliberate acts by humans to cause loss. Safety indicates accidental loss.

by various authors. Moreover, as Guldenmund [33] rightly puts it, when a given safety and security culture and climate have been assessed, the next question will certainly be – so what? The organization needs to make conclusions, and corrective actions have to be taken and carried out if required. This exercise can best be carried out by linking the risk scenarios and the objectives related to the risks with performance management indicators in a systematic approach. Section 3.6.3 presents a model that may be viewed and used to continuously improve the organization's safety and security culture or climate in order to achieve safety and security excellence and leadership.

3.6.3 The P2T model revisited and applied to safety and security culture

In Section 3.3.2, we proposed a three-dimensional model that can be applied as the first (observable) phase to develop an integrative safety and security culture model. With this model from Section 3.3.2, all observable safety and security culture aspects can be integrated and covered, as all observable elements concerning a good safety and security culture can be placed under one of the three dimensions. The suggested dimensions were people, procedures and technology, and the model was therefore referred to as the P2T model. The interplay between these three observable domains defines the present observable safety and security culture in any organization (see Fig. 3.25).

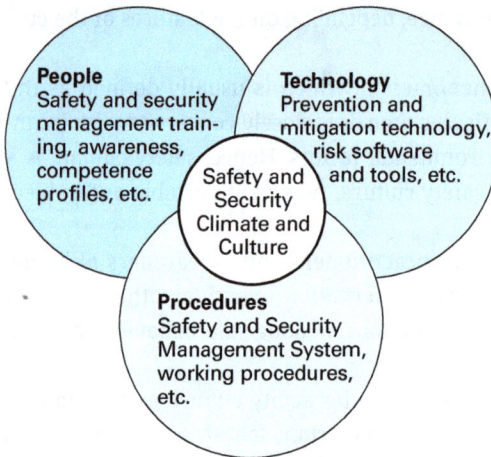

Fig. 3.25: Observable safety and security culture according to the P2T model.

We need little argument that the technological dimension is indispensable to ensure a good observable safety and security culture. With failing installations or equipment, a company inflicts a direct threat to its workers and surrounding people and buildings. Despite the large improvements in safety technology that have been gained in the past decades, security technology such as CCTV and biometric systems has not been applied to its full potential in many enterprises. Following the ALARP/A principle (as low as reasonably practicable/achievable), it should be noted that no risk can be reduced to zero without expenses that can be justified economically. That is why the technological dimension has to be designed in a way that the resulting risk lies between socially accepted boundaries. Governments will impose these bounds, but often organizations will surpass the required measures.

The second dimension (of the observable part of the safety and security culture), procedures, is being managed by a safety management system (in the case of safety) and a security management program (in the case of security). These management systems revise the existing procedures used to maintain a good observable safety culture and/or security culture. The term "procedures" can be interpreted very broadly. It concerns procedures to operate safely and securely; to safely store hazardous substances; to manage the competencies of employees; to manage emergency situations; to deter, detect, deny, delay and defend procedures in place, etc. Logically, the organizational structure and culture play a large role in this.

The third dimension influencing the observable safety and security culture is defined as people. Reason [12], Fuller and Vassie [3], CCPS [47] and many other researchers indicate that a majority of accidents and near-misses can be attributed to human error. According to some estimates, human error contributes to 90% of all accidents [48]. This number considers all possible sources of error, including frontline operating personnel, engineers and supervision. In the case of security, even all incidents are human-made. Another point for security is the necessity to understand the human element for threat and vulnerability analysis. Also, human errors can be made in analyzing, identifying and responding to security incidents, and this must be considered to minimize threats and decrease risk. That is why creating safety and security awareness among all employees is essential for a good safety and security culture, as well as providing proper training, providing safety and security incentives, creating a safety-driven and security-driven organizational community, enhancing the competencies of employees at all levels, etc.

It is evident that a good observable (and nonobservable!) safety culture depends on adequate and solid strategic management concerning risks. Section 3.7 discusses how strategic management should lead to continuous improvement, and Section 3.8 proposes and discusses a model to unify the principles of performance management and continuous improvement with the concepts of safety culture.

3.6.4 The Egg Aggregated Model (TEAM) of safety culture

While investigating literature regarding the observable aspects of safety, also sometimes called the engineering domain of safety, and the nonobservable aspects of safety, also sometimes called the perceptual and motivational domains of safety, it becomes obvious that all these aspects or factors are not isolated from one another. On the contrary, they are strongly related, influencing each other, and thus forming a cyclic framework. To illustrate the above, Fig. 3.26 uses the iceberg metaphor, showing how safety/security culture can and should be seen: only a part of the culture is visible, and the majority of the culture is hidden. The perceptual and motivational parts of the culture are very much present, but in a tacit way. Nonetheless, both the visible part and the nonvisible part should be understood, assessed and improved where possible.

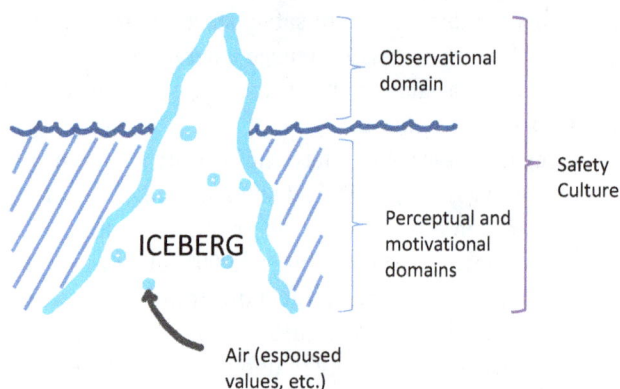

Figure 3.26: Observational (visible) domain and perceptual and motivational (nonvisible) domains of safety culture (and security culture).

From the previous discussion, it is obvious that a lot of research has been carried out on the subject of safety culture. This research has been conducted by a variety of scientific disciplines, for example, engineering, sociology, psychology, safety scientists and others. However, up until recently, there has never been an integrated and holistic overview of what a safety culture constitutes, and a vivid debate among scientists on the safety culture topic could be observed. Recently, a unifying model of safety culture was developed and proposed, called "The Egg Aggregated Model of safety culture," abbreviated TEAM, taking all aspects of safety science within an organization into consideration and explaining their position toward each other (see the study by [49]). Figure 3.27 illustrates TEAM of safety culture.

The egg displayed in Fig. 3.27 is composed of three different layers with distinct visibility, comparable with Guldenmund's [39] framework of safety culture. The observable factors are represented by the yolk. Most elements of these observable factors are cited by Guldenmund as particular manifestations of his outermost layer. The protein, a somewhat translucent mass that is harder to capture than the yolk, represents the perceptual and motivational factors sometimes grouped under the name of the "psychosocial factors for safety." The content of the protein is comparable with Guldenmund's middle layer, indicated by him as "espoused values/attitudes with a relatively explicit visibility." Beliefs, affective and cognitive processes, and self-control can be seen as the "air" present in the egg, that is, invisible though an essential part of a culture of an organization, just like the basic assumptions of Guldenmund's core level.

The safety culture of an organization can thus be conceptualized as three layers, which might be studied separately as the engineering domain of observable factors, the perceptual domain of safety climate and the psychological/motivational domain of intended behavior.

In summary, in the TEAM of safety culture, measurable factors and unmeasurable factors are mentioned. First, the unmeasurable factors involve beliefs, affective and cognitive processes, and self-control, also indicated by Guldenmund (2000) as being an essential part of the culture of an organization. These unmeasurable factors can be seen as the "air" present in an egg. Second, two types of measurable factors are present: observable measurable factors (represented by the yolk of the egg) and nonobservable measurable factors (represented by the protein of the egg). Three measurable domains thus constitute the safety culture of an organization, that is, the engineering domain of observable factors, the perceptual domain of safety climate and the psychological/motivational domain of intended behavior. It is obvious from previous research that the domains are linked, but there is no definite proof of causal relationships. Rather, it is clear that there is a loop-wise structure of the domains. This cyclic characteristic is so typical for safety thinking, thus also transpires in the TEAM of safety culture. The influence between the three egg domains is indicated by lightning arrows.

Further research is needed to translate the various elements that compose the TEAM of safety culture into measurable indicators. These indicators have to be measured by means of different research methods: (i) audits, inspections, document analyses and quantitative analyses for the engineering domain, (ii) quantitative analyses (questionnaires) for determining the shared perceptions on safety (or the so-called safety climate) and (iii) qualitative analyses (in-depth interviewing, focus groups, observations) to find out the individual and group human-related state, and the motivation of people, as regards safety. Hence, both quantitative and qualitative research techniques should be used to obtain a good idea of an organization's safety culture.

Remark that it is possible to define and elaborate on the security culture of an organization in a completely analogous way: three domains make up the constituting parts, that is, an observational domain with respect to security, a perceptual domain (the "security climate") and a motivational (intended behavior) domain. The first domain can be measured by (internal and external) audits and walking around the organization, document analyses, etc. The perceptual domain can be measured by using questionnaires for the employees on the topic of security (management commitment, security communication, etc.). The intended behavior domain can be measured by conducting interviews about the motivation of workers with respect to security practices, their knowledge and know-how on the topic, their security attitude and awareness, etc. (see Fig. 3.28).

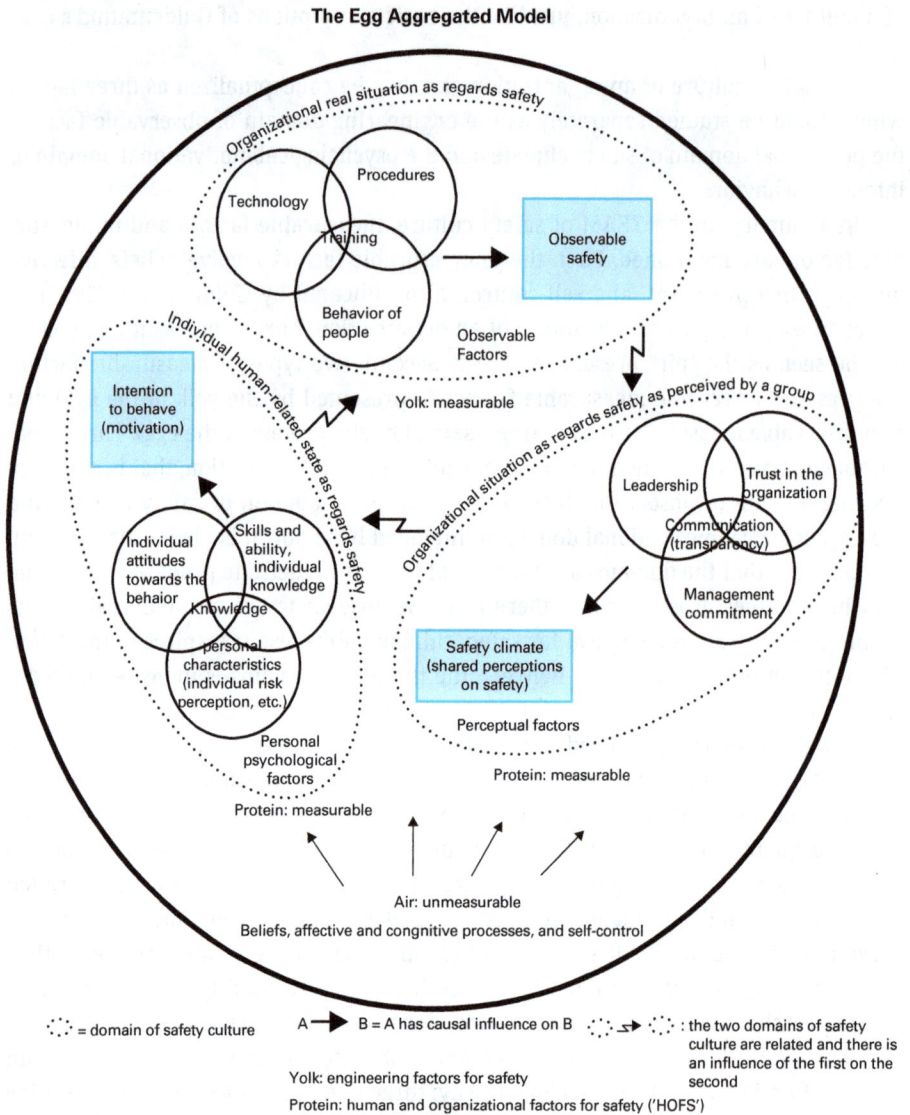

The Egg Aggregated Model

Fig. 3.27: The Egg Aggregated Model (TEAM) of safety culture (source: [50]).

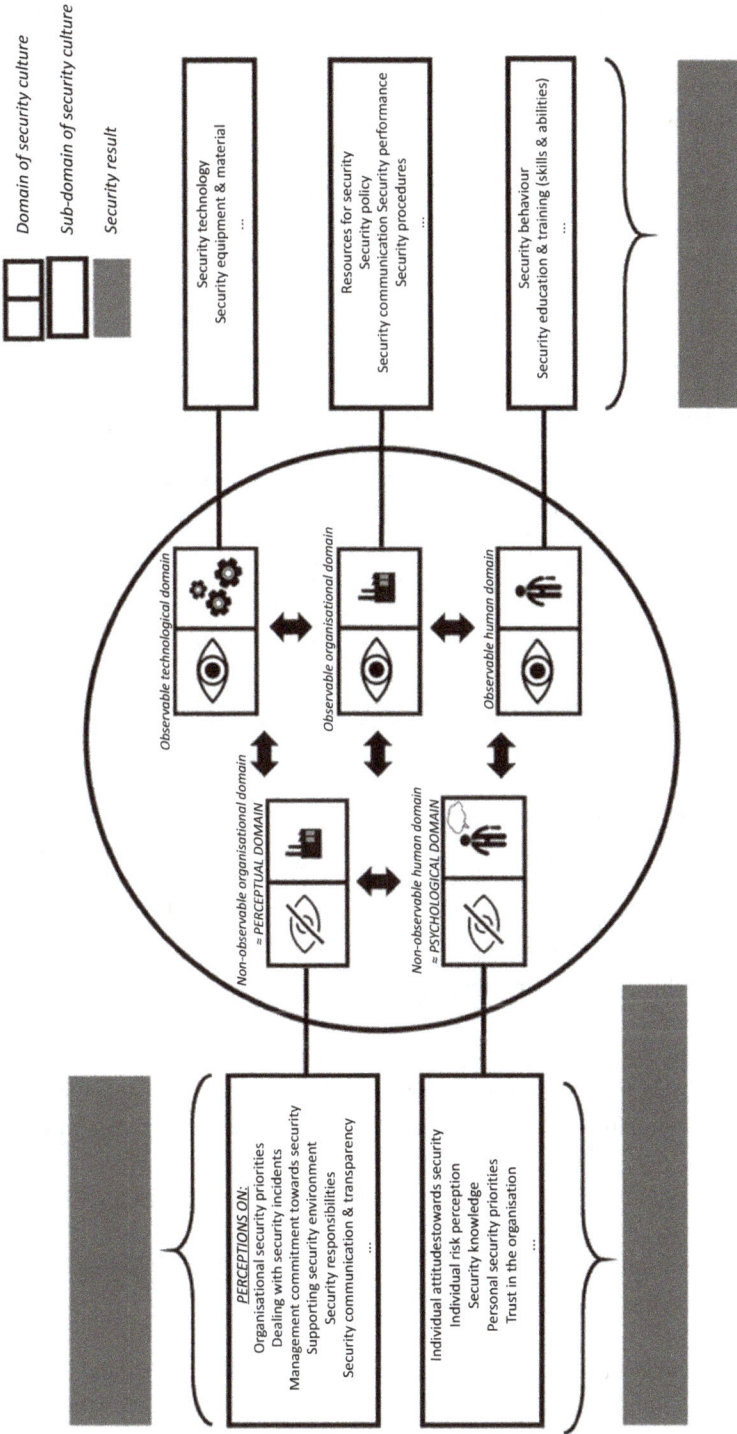

Fig. 3.28: The Security Culture model (based on the TEAM model).

Domain of security culture

Sub-domain of security culture

Security result

Security technology
Security equipment & material
...

Resources for security
Security policy
Security communication Security performance
Security procedures
...

Security behaviour
Security education & training (skills & abilities)
...

Observable technological domain

Observable organisational domain

Observable human domain

Non-observable organisational domain = PERCEPTUAL DOMAIN

Non-observable human domain = PSYCHOLOGICAL DOMAIN

PERCEPTIONS ON:
Organisational security priorities
Dealing with security incidents
Management commitment towards security
Supporting security environment
Security responsibilities
Security communication & transparency
...

Individual attitudes towards security
Individual risk perception
Security knowledge
Personal security priorities
Trust in the organisation
...

It needs finally to be said that a safety/security culture requires constant attention and constant labor to ensure its success. An (internal or external) measurement only provides an idea of the safety state (whether it is the real situation via an audit or the perceived situation via a questionnaire) at a certain point in time and thus does not give a true indication of the safety culture or the "safety DNA" of an organization. To have a more accurate picture of the safety culture of an organization, several research methods should be combined and strategic safety performance management should be established to ensure that there is continuous improvement over time.

One can look at the safety/security culture of an organization as an iceberg, where the top that can be seen from above the water represents the observational domain. But just like in the case of an iceberg, most of the ice is under water and cannot be seen. Nonetheless, this hidden part of the iceberg is as important as the visible part to assess the risk that the iceberg represents. The same is true for an organizational safety culture: the hidden, nonobservational part is as important as the auditable part and should also be measured if one desires an adequate understanding and insight into the safety/security culture. Climate and motivation should thus always also be measured, using a multimethod design composed of a variety of methods such as document analyses and audits, questionnaires, and interviews and focus groups.

3.7 Strategic management concerning risks and continuous improvement

Strategic management consists of five phases:
- The first phase determines the strategic vision of the organization. This vision makes it clear to the entire organization what the organization should look like and how it should evolve.
- The second phase consists of translating the strategic vision into clearly measurable objectives. This enables the company to measure if the desired results have been achieved.
- In the third phase, the organization should develop a strategy that makes it possible for the organization to reach its goals. This strategy should be specified for each functional domain within the organization, e.g., the safety domain and the security domain.
- In the fourth phase, the chosen strategy should be implemented in an efficient and effective way.
- Finally, in the fifth phase, the performance of the organization should be evaluated, and if necessary, changes should be implemented [50].

In order to execute the last phase of strategic management, a company should have a system of continuous improvement. One of the most widely used systems is the well-known *plan-do-check-act* loop of continuous improvement or the *Deming cycle*. The different steps of the Deming loop, which is also called the PDCA cycle, are below:

– Plan – develop a policy and determine the goals and processes necessary for achieving certain objectives, based on the risk analyses carried out and the subsequent action programs.

– Do – execute the actions and measures to realize the policy objectives; implement the processes (e.g., the risk management process).

– Check – monitor, measure and analyze the realization of the aims, targets, action programs, etc., and their effects by means of inspections, audits, etc. The result of these measurements and analyses is the definition of new corrective and/or preventive improvement actions that aim to improve organizational processes and products in relation to the policy and its objectives.

– Act – take measures to continuously improve process achievements; review and, eventually, revise company policy.

The Deming cycle describes the quality management principle of continuous improvement that needs to be applied to all aspects, processes and activities throughout the whole company [46]. The PDCA cycle can be illustrated in different ways, and it can be filled in for different types of activities. Figure 3.29 illustrates the basic philosophy of the PDCA cycle for risk management.

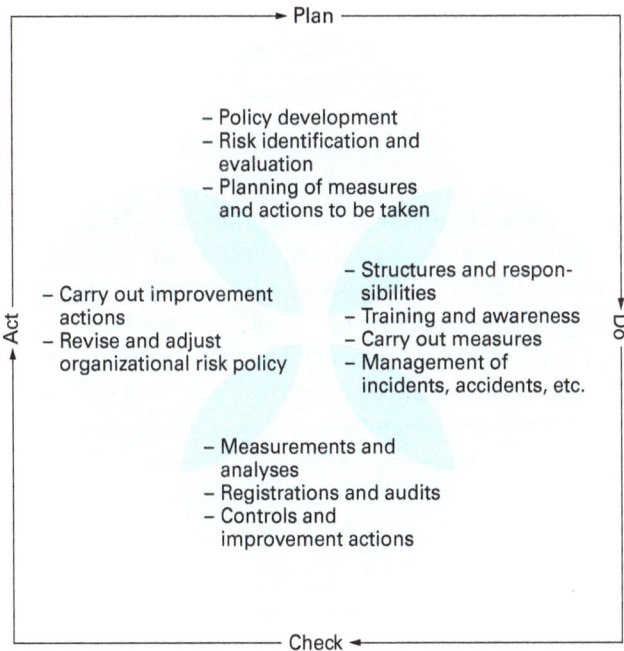

Fig. 3.29: Basic philosophy of PDCA applied to risk management.

3.8 The IDEAL S&S model

In current industrial settings, companies follow the *plan-do-check-act* loop of continuous improvement because of their acquired know-how of internationally accepted business standards such as the ISO9000 series and the ISO14000 series, addressing quality and environment management systems, respectively. The OHSAS18000 series – the international Occupational Health and Safety Management Assessment System specification that empowers an organization to control its occupational health and safety risks and improve its performance concerning those risks, as well as the ISO45000 series concerned with occupational risks – also use the PDCA cycle as a basic management concept and are often used to work out safety management systems.

The ISO norms and OHSAS are very well known throughout all industrial sectors and hence some degree of basic management standardization often already exists. Available risk management models mention performance management as a technique to measure safety performance using proactive/leading indicators or reactive/lagging indicators. The models fail to unambiguously recognize the strong link between performance management and organizational culture and climate.

Reniers et al. [46] present a model to integrate a safety and security culture with performance management, leading to company excellence achievement and leadership in safety as well as security. The suggested elements to this end are embedded in the Deming cycle. The proposed model – Improvement Diamond for Excellence Achievement and Leadership in Safety and Security (IDEAL S&S) – uses the P2T model (explained in Section 3.3.2) to visualize the observable part of a safety and security culture. Figure 3.30 illustrates the IDEAL S&S model.

In order to explain the IDEAL S&S model, a number of terms have to be defined. The IDEAL S&S model shows two fields of tension. The tension between optimal resources versus deployed resources makes up the first field of tension. The term "resources" should be interpreted very broadly, e.g., money, knowledge, installations, know-how and people. On the one hand, optimal resources are those that are necessary to reduce a certain risk component until it is considered – in some way or another – acceptable. They represent resources "as should be in ideal circumstances." On the other hand, deployed resources are/can be deployed in reality and thus in the real industrial setting of the plant – the situation "as is in real circumstances." In an ideally equilibrated situation, the optimal resources are equal to the deployed resources. If the level of optimal resources lies above the level of the deployed resources, a potentially hazardous situation can occur. Conversely, if the deployed resources surpass the optimal ones, a company has wasted resources because the risk level was already reduced below an acceptable level.

A second field of tension exists between short term and long term. Influencing a safety culture requires goals in the long term that also need to be translated into manageable short-term goals. Hence, achieving the long-term requirements for a company's safety and security culture is visualized in the (long-term) upper part of the model, whereas the safety and security climate (which can be seen as a short-term perceptual snapshot of the company's culture) is represented by the (short-term) lower part of the model.

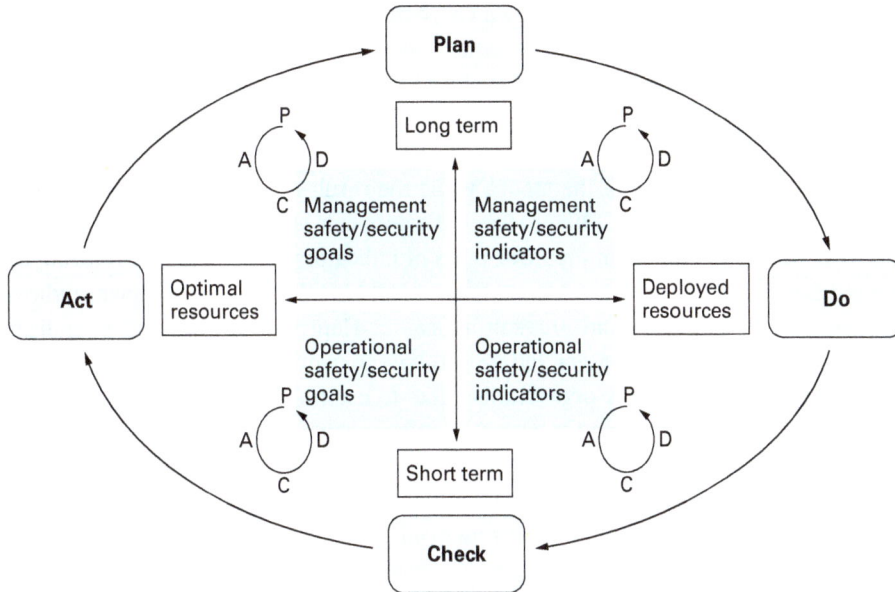

Fig. 3.30: IDEAL S&S model.

The IDEAL S&S model employs safety and security indicators. These indicators assign a qualitative or quantitative value to different safety and security culture aspects. The long-term deployed resources are situated between the plan phase and the do phase of the Deming cycle and can be steered by using management safety and security indicators. The short-term deployed resources are situated between the do phase and the check phase and can be steered by using operational (process and result) safety and security indicators. Further information concerning developing leading and lagging (safety and security) indicators can, for instance, be found in HSE [51] and parmenter [52].

Actually, three types of indicators exist: management indicators, process indicators and result indicators. Management indicators are proactive/leading indicators answering the question "With what means?" Hence, such indicators provide an idea of whether the conditions are present to achieve certain safety-related goals. Some examples are the percentage of yearly turnover used for safety measures or the percentage of 5-yearly company strategic goals related to health and safety. The second type of indicators, process indicators, is also proactive/leading by nature, and they tell you what to do to optimize (human, work, etc.) processes. They provide an answer to the question "How?" and they indicate whether the efforts to achieve a predefined goal are carried out according to plan. Some examples include the number of yearly risk analyses carried out over the number of risk analyses planned in the period of a year or the percentage of processes audited externally. The third and last type of indicators, the result indicators, is the only reactive/lagging indicators, and they tell you what was done in a perspective. They answer the (lagging) question "What has been achieved?" Result indicators indicate what was achieved and whether a predefined goal was reached. Some examples of such indicators are the number of first aids of contractors per 6 months or the yearly insurance premium for damages.

In relation to the IDEAL S&S model (Fig. 3.30), the management indicators can be linked to the "Plan" phase of the model, while the result indicators are related to the "Check" phase of the model. The process indicators can be linked to all phases of the model. A rule of thumb about the three types of indicators is the 10–80–10 rule: out of 100 indicators used in an organization, there should be about 10 management indicators, 80 process indicators and 10 result indicators. Hence, by far, most of the indicators should be process indicators, telling you how to continuously improve processes all the time. Note that many organizations like to use result indicators because they are simple to develop, use and interpret. However, such indicators are reactive/lagging, and therefore the wrong has already been done. In the case of result indicators, the only way to learn is by mistake, while the mistake could perhaps have been prevented if a process indicator had been employed.

The IDEAL S&S model also uses safety and security goals. The amounts of optimal resources have to ensure that these goals can be met. Management safety and security goals and operational safety and security goals, respectively, are quantitative or qualitative figures set to be achieved by company management or by business unit management for a management or a operational safety and security indicator. Management safety and security goals (e.g., with a yearly, 2-yearly or 5-yearly frequency) and operational safety and security objectives (e.g., with a weekly, monthly or 3-monthly frequency) are used to control, remediate and continuously improve the organization's safety and security achievements, which ultimately lead to company excellence in safety and security.

Remember the ISO31000 definition of risk, which is "the effect of uncertainty on achieving objectives"; one can see that the "objectives" are an important feature of every risk (scenario). Without an objective, there is no risk and no risk scenario. The approach of determining the (management, process and result) indicators starts with determining the risk scenarios (for every domain of the TEAM model of safety/security culture) and the accompanying objectives. Once the objectives are known (for dealing with some risks within one of the safety/security culture domains), these can be linked/verified with indicators, and the indicators can then be used to improve the safety/security culture of the organization.

Indicators and objectives themselves should be continuously planned, implemented, checked and adapted (if necessary) according to the Deming Wheel of Improvement.

Furthermore, all management process and result safety and security indicators should be worked out by using the three cultural domains, that is, the observational, perceptual and motivational domain. The risks related to the domain are further translated into different objectives and various safety and/or security indicators are linked to each of these objectives. For all indicators, a set of minimum organization-specific safety and/or security goals can be established by the user of the model. Based on the current literature (among others, HSE [51] and OECD Guidance on Safety Performance Indicators [53]), a list of possible safety and security objectives may be identified. The objectives from Tabs. 3.2 and 3.3, respectively, for the long term and the short term, are used for illustrative purposes. Note that the indicator examples given in Tabs. 3.2 and 3.3 are not categorized according to the three existing types of indicators, that is, management, process, and result indicators. The reader is invited to do the exercise and assign an indicator category to each of the mentioned indicator examples.

Based on a safety culture or security culture assesment, every company should thus define its own safety and security risk and objectives, which are used to develop management safety and security indicators as well as process and result safety and security indicators and goals. The objectives presented in Tabs. 3.2 and 3.3 should be regarded as illustrative guidance.

Tab. 3.2: Safety and security long-term objectives and indicator examples.

Long-term objectives	Nonexhaustive list of indicator examples[1]
Technology	
Software, tools, etc. for safety and security prevention, mitigation, emergency, etc. are used	Two-yearly budget available for safety software
Technology (other than software) for safety, security prevention, mitigation and emergency purposes is used	Five-yearly budget available to maintain installations according to the best available practices
Technological knowledge and know-how of chemical processes, products, installations, etc. are regarded as essential	Two-yearly budget for training/educating personnel on installation of state-of-the-art knowledge
Installations are safely and securely designed	Percentage of installations that comply with international norms (DIN norms, ISO norms, etc.) within 2 years

Tab 3.2 (continued)

Procedures

Existence of a company safety and security policy.	An external audit of the company's safety and security policy is carried out every 5 years
Compliance with safety and security legislation at all times	The number of legally prescribed safety procedures that are not fulfilled is <5% of all legally prescribed safety procedures; this is checked every 3 years
Likelihood and severity of potential accidents are reduced to an acceptable level using risk assessments and threat assessments	Every 5 years, the entire plant is checked by using security vulnerability assessments (i.e., at least every 5 years, an SVA is carried out for every installation within the plant)
Safety and security improvements are driven by learning	Number of incidents attributed to the same cause in 2 years
A well-functioning safety management system is put in place as well as a security management program	A 3-yearly internal audit of the security management program is carried out
Emergency preparedness and response procedures as well as business continuity plans are in place	The BCP is tested every 2 years
Internal and external audits are drivers for change management and continuous improvement of safety and security within the company	When an internal audit is performed, long-term recommendations for continuous improvement are required in the audit report
Documentation of procedures	A 4-yearly check is carried out by the security department whether all security procedures are written down, understandable, up-to-date and whether they can be easily consulted by its users

People

Involvement of top management in safety and security policy	Overall, 2-yearly budget is assigned to security activities
Involvement of employees in safety and security practices	Every 3 years, a security survey is organized among company personnel
Involvement with parties external to the company	Every 2 years, contractor safety achievements are discussed with the contractors
Employees are sufficiently competent concerning safety and security issues, and they have adequate experience/expertise when needed	A learning trajectory for employees exists within the company
There is a very open atmosphere regarding safety; all employees are well-informed and are free to express ideas, discontentment, etc.; employees are involved in the decision-making process	Score given to "corporate openness regarding safety" in a 2-yearly questionnaire

Tab 3.2 (continued)

People	
Safety and security are the number one priorities, and this is acknowledged by all employee.	Score employees receive during safety observations using 360° feedback reviews carried out every 3 years
There is mutual communication based on mutual trust concerning safety and security among all employees	Score given to "mutual communication as regards security topics" in a 2-yearly questionnaire
Employees are prepared for emergency situations	Percentage of executed improvement propositions within 2 years resulting from emergency plan exercises

[1]One indicator is given as an example per objectives. Source: [46].

Tab. 3.3: Safety and security short-term objectives and indicator examples (nonexhaustive).

Short-term objectives	Nonexhaustive list of indicator examples[1]
Technology	
Installations and chemical products and processes are regularly investigated for working as expected (safe and secure), and they are maintained wherever deemed necessary on a regular basis	Safety inspections are carried out at least every 6 months in every installation of the plant
Risk assessment software and threat assessment software are employed for risk studies, and they are regularly updated	Access and gate control: the number of daily controlled people
All technology (besides software) used within the company or business unit with regard to safety and security is regularly maintained and updated	Ratio of corrective/predictive maintenance per month
The workplace is safe and ergonomic to work in, and technological security measures (e.g., CCTV) are applied wherever necessary	Weekly score given to workplace safety and housekeeping
Procedures	
Company safety and security policy are clearly and unambiguously translated into operational safety and security goals per level and per business unit of the organization	The number of yearly improvement proposals as a result of an internal audit in one of the company's installations
Hazardous substances are stored properly	Daily housekeeping checklists are used for storing materials

Tab 3.3 (continued)

Procedures

Working procedures, safety procedures, security requirements, installation specifications, etc., are well documented	Percentage of standardization of security documentation, checked every 6 months
Efficient and adequate (user-friendly) procedures have been developed for the staff to follow	Percentage of procedures still leading to difficulties and incidents, evaluated per year
Company procedures and guidelines are in place regarding the frequency and the necessity/circumstances to use risk assessment or threat assessment software	A frequency of SVAs to be carried out per installation is determined, and the circumstantial conditions/approaches are described
Procedures are in place to comply with existing safety and security regulations and to follow-up and comply with new safety and/or security legislation	The degree to which existing security legislation is taken into account by company procedures is checked every 6 months
The procedural expectations regarding safety and security are understood by all employees, and everyone commits themselves to respect these procedural standards	The degree to which the working procedures of a business unit or installation are easy to understand and are followed is formally checked by the shift supervisor every 3 months (in addition to daily informal checks)
All company safety and security procedures, guidelines, working instructions, etc., are documented	Level of standardization of security documents (procedures, guidelines, work instructions, etc.)
Company procedures and guidelines are in place regarding the frequency and the necessity/circumstances for conducting internal and/or external audits	Number of scenarios (circumstances) for which the frequency of external audits is fixed per year
The company has internal and external emergency plans, as well as a business continuity plan.	The degree to which the external emergency plan is elaborated and tested for security situations (e.g., a terrorist attack) every year.

People

Competence schemes and profiles are maintained for all employees, and a learning trajectory exists within the company	Percentage of employees within an installation that has similar competencies (which leads to more flexibility within the work shift)
All employees are aware of all safety and security knowledge and know-how for carrying out their functions	Number of weekly visits by management to the work floor
Every person fully understands their safety and security responsibilities and acts appropriately	Number of monthly meetings where employees receive information and feedback about the importance of security

Tab 3.3 (continued)

People	
The company can ensure safe cooperation with – and operations of – contractors and outsiders	Levels of satisfaction (questionnaire scores) regarding cooperation with external partners after annual emergency exercises
360° feedback reviews and open communication initiatives are undertaken within the company	Daily operational staff meetings are held on safety
Emergency exercises and drills are regularly carried out	Every 3 months, a drill for security guards and dogs is held

[1]One indicator is given as an example per objectives. Source: [46].

3.8.1 Performance indicators

Obviously, there should be clear and unambiguous objectives that can be used to evaluate a company's safety/security culture and to continuously improve it. As mentioned above, the objectives are ideally linked with risks (scenarios) and fields within one of the three cultural domains that need improvement within the organization.

Indicators should be "SMART"ly formulated:
– Specific and clearly defined
– Measurable so that the performance of the indicator can be checked on a regular basis
– Achievable so that each indicator provides a target that is stretching but not so extreme that it is no longer motivational (the indicator needs to have sufficient support)
– Relevant to the organization and what it is aiming to achieve
– Time-bound in terms of (realistic) deadlines or timing for when each indicator will be achieved

Goals (and indicators) can be formulated in different ways: as an absolute number (target numbers), as a percentage (decrease of x%, satisfy x% of criteria, satisfy x% of a checklist, etc.), or as a relative position to a benchmark (higher than the national mean, lower than the mean of the industrial sector, lower than one's own performance of the past x years, etc.).

The different parts of performance management, as part of the organizational safety/security management system, are dimension, subdimension, indicator, objective, result and deviation.

Mazri et al. [54] indicate that certain basic information, technical data, organizational information and IT data are required for every indicator. Table 3.4 provides an overview of the information required to ensure adequate use of performance indicators. By keeping all this information for every indicator, a company memory approach is installed in the organization. Such a company memory is essential for the learning aspect of the organization and to ensure that the same mistakes and errors are not repeated as time goes by.

Moreover, as already mentioned, different types and levels of indicators exist. It is obvious that indicators should be defined for all three dimensions (and all their subdimensions). In Chapter 2, the risk sandglass was introduced, indicating the existence of positive as well as negative risks (the risks can, e.g., be determined by the combination of a SWOT analysis and more traditional risk analyses), and that positive risks should be maximized, while negative risks should be minimized. Thus, indicators should be elaborated for both positive and negative risks. Moreover, one should not be blind toward major catastrophes and toward risks characterized by extremely high uncertainties. Hence, indicators should be identified for the two types of risks (types I and II). Furthermore, different decision levels require different indicators: management, process and result. Note that management indicators are always long term, and result indicators are usually (but not necessarily) short term. Process indicators can be long term or short term. It should be obvious that indicators sometimes will be extremely hard to imagine and to think of or will simply not exist.

Tab. 3.4: Performance indicators: information table.

General information	
Short name	Unique codified name of the indicator
Long name	Detailed name of the indicator
Description and purpose	What does and doesn't the indicator measure? (What could possibly cause confusion?)
Source	Who issued this indicator?
References	Available reference documents concerning the indicator
Nature	Qualitative, semiquantitative or quantitative
Risk domains covered	Depending on the needs and the management systems implemented, a myriad of risk domains can be covered. For example, environmental risks, health and safety risks, security risks, operational risks, process risks, occupational risks, quality risks, ethical risks, etc., or any combinations thereof. Note that a unique indicator may be more or less relevant for several domains.
Technical information	
Formula and unit	With which formula was the indicator value calculated (if applicable)?
Target value	Target value (to reach a predefined performance)

Tab 3.4 (continued)

Technical information	
Minimal and maximal values	Describe the minimum and maximum limit values within which the indicator value may be considered as "acceptable." If the indicator value is outside these limit values, actions need to be taken.
Input data required	Information required to implement the formula described above (which led to the calculation of the indicator)
Frequency of measurement	What is the frequency with which this indicator should be measured (the periodicity of monitoring will influence the level of resources required)?
Related indicators	Indicators are part of a "network of indicators" monitoring different system components. The relationship(s) between the indicators should be mapped, and a list of additional indicators providing extra information on the indicator under consideration should be drafted.
Organizational information	
Indicator reference person (or owner)	A reference person in the organization should be assigned to each indicator. This person will be responsible for the quality of the whole process from data and information collection to interpretation and communication of the results.
Data provider(s) or registrator(s)	Person(s) need to be appointed to collect and deliver the required data/information (necessary input data).
Interpretation procedure	Person(s) need to be identified who are capable of, and who have the competence and the authority to, correctly interpret the measured indicator value and translate this value into knowledge and insights.
Communication procedure	Person(s) within and outside the organization who should be informed about the indicator results are to be identified. The method of communicating the results is to be determined.
Organizational information	
Relevance assessment procedure	The relevance of any indicator should be questioned at regular time intervals and according to a predefined procedure.
IT information	
Software availability	Existing software is listed that improves the use of the indicator or makes it easier.
Adequacy with existing/local information system	The configuration of existing software may facilitate the input of collected data, or it may complicate this process. This fact should be taken into account beforehand.

As mentioned above, the aim of performance indicators and performance management in relation to safety or security is obviously to improve the different domains of an organization's safety/security culture, for instance, the observable domain of it (i.e., people, procedures and technology). Proactive and reactive monitoring allows for taking measures in organizational processes, activities, attitudes, etc. Figure 3.31 visualizes the usefulness of performance management.

Based on the TEAM model, for every organizational risk, a three-dimensional representation can be suggested for all three domains of organizational safety and security culture. As an example, Fig. 3.32 shows a quick overview of the three individual subdomains (P, P, T) of the observable (engineering) domain of organizational culture, for the negative as well as the positive sides of risks.

This way, a quick overview is provided of the status of risks/scenarios, based on monitoring leading (for proactive monitoring) and lagging (for reactive monitoring) indicators. The levels at which actions are required are indicated by light blue = no actions needed, medium blue = actions needed but not urgently (e.g., within 4 months) and dark blue = immediate action required.

> **!** Performance management is a very powerful tool to systematically map the effectiveness with which every aim or goal (short-term or long-term) within the different safety and security dimensions and sub-dimensions of an organization is reached. It can also be used to prioritize actions, budget allocations, etc.

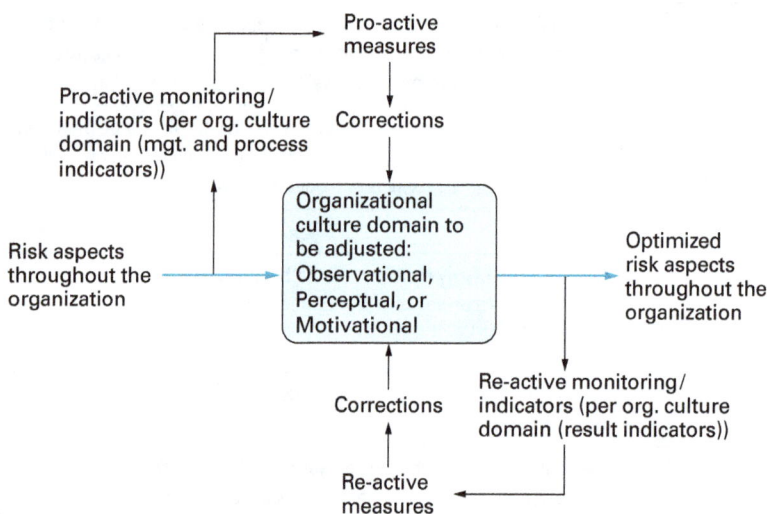

Fig. 3.31: Optimization of the risk aspects throughout the organization and performance management, with reprint permission from Die Keure [55].

Organizational risk aspects

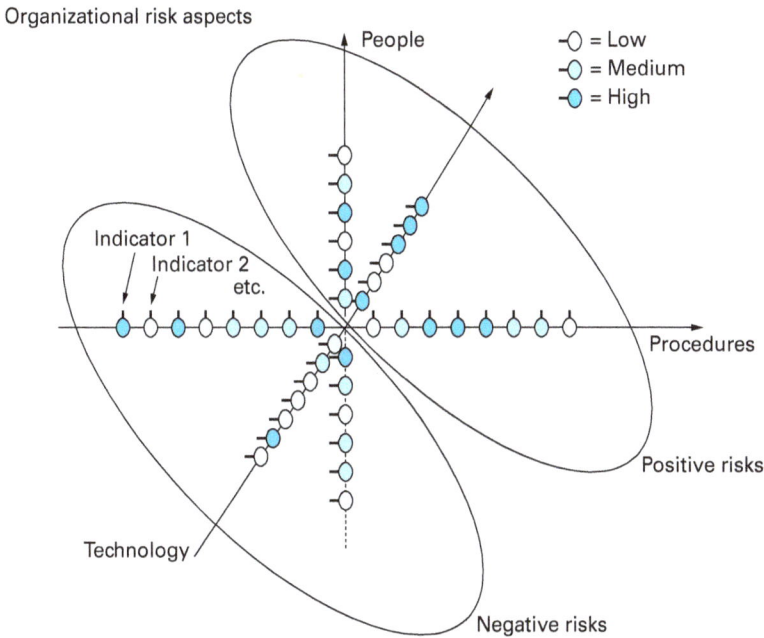

Fig. 3.32: Possible depiction for a quick visualization of the observable subdomains of people, procedures and technology, with reprint permission from Die Keure [55].

3.9 Continuous improvement of organizational culture

Based on social system theory, Wu et al. [56] studied the potential correlation between safety leadership, safety climate and safety performance. The results of the statistical analysis indicated that organizational leaders would do well to develop a strategy by which they improve the safety climates within their organizations, which will then have a positive effect on safety performance. We can therefore assume that employing the suggested IDEAL S&S model will lead to truly safer and more secure companies that will enhance organizational safety and security leadership.

In summary, a safety culture is like a human relationship: it needs constant attention and constant labor to ensure its success in the short term and especially in the long term. An (internal or external) audit only provides an idea of the safety climate at a certain point in time and does not give a true indication of the safety culture or the "safety DNA" of an organization. To have a more accurate picture, safety performance management should be established, and the three domains of safety culture should be assessed/measured on a regular basis, and the results should be compared with each other and analyzed to ensure that there is continuous improvement over time. Figure 3.33 shows how this may be achieved.

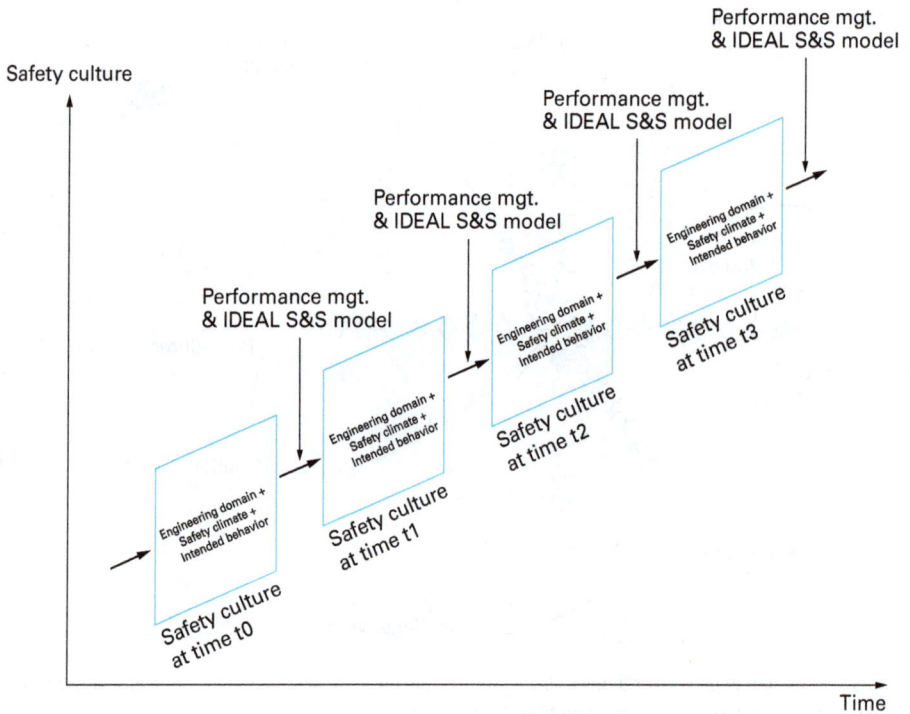

Fig. 3.33: Continuous improvement of organizational safety culture.

The way an organization deals with risks is also explained by the general attitude of the people working in the organization and its leadership. Different levels of maturity of safety culture can be discerned, from pathological over reactive, calculative and proactive to generative. The difference between the proactive level and the generative level is mainly due to attitude and trust (the motivational domain): in the generative level, people communicate freely about problems and trust each other within and across every layer of the organization. Things are done safely or not at all.

Nonetheless, most best-of-class companies today are situated at the proactive level, where they actively look for potential problems and adopt the attitude of "if you think safety is expensive, try an accident." However, they should really try achieving the generative level. The IDEAL S&S model can be employed to get there in a realistic way.

3.10 Link between Performance, Safety/Security and Risk

There exists a time-relationship between the fields of performance (management), safety/security (management) and risk (management), if we look from a bird's eye view. Performance is about the past, and trying to obtain info to influence the future.

Safety/security is about the present: it is a 'state' (real state or a state of mind about the perceived current situation) that continuously (every microsecond) evolves into a new 'state'. Risk is about the future and tries to come up with possible future scenarios (risks) to anticipate what the future could bring in certain contexts (situational circumstances). Figure 3.34 illustrates the link between performance, safety/security and risk.

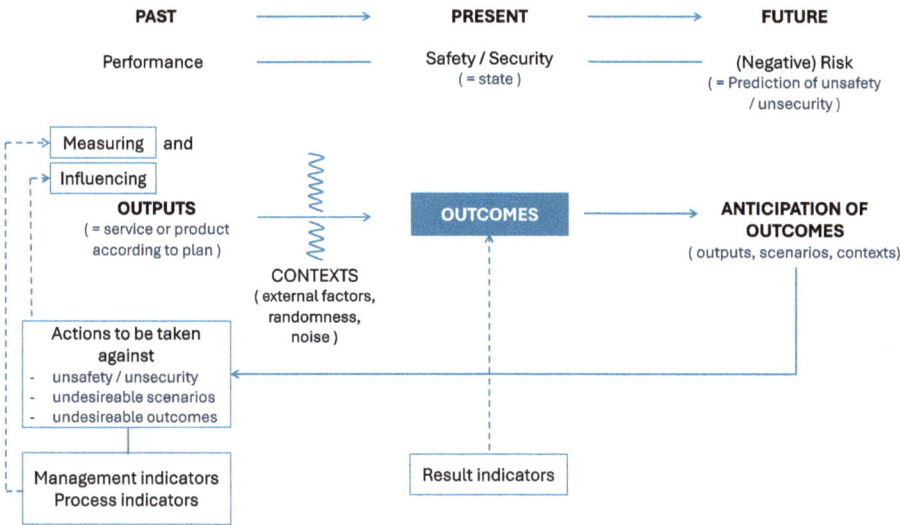

Figure 3.34: Time-relationship between Performance, Safety/Security and Risk.

3.11 High reliability organizations and systemic risks

3.11.1 Systems thinking

The different subsequent work process steps need to be understood as well as the existing links between them. However, it is also very important to gain insights into the "logic behind the system." Such insights are essential to be able to interpret existing causalities at the level of organizational working processes. Some *general insights into systems thinking* are given hereafter. It should be noted that gaining insights within any organization requires "trial and error" procedures within the organization.

3.11.1.1 Reaction time or retardant effect
Every safety measure, taken based on risk management, has a certain "reaction time" – it takes a certain amount of time before the effects of a measure become apparent. It is important to know, or at least to have an idea of, this reaction time in order to avoid

taking new measures too quickly. Hence, a long-term vision, while taking safety and health measures and interpreting the results, needs to be supported by the insights into the working of the system and the long-term effects of measures on the system.

Senge [57] illustrates the idea by using the metaphor of a hot-water faucet. If one does not take into account the fact that 10 s are needed for the water to become hot, one might further open the hot water faucet, so that by the time the water becomes warm, it might be so hot that it hurts, leading the operator to turn the tap toward cold water; with the result of having cold water instead of hot water, and time is lost. Eventually, after a certain period of trial and error, the desired water temperature may be achieved, but a lot of time has been lost, dangerous situations may have occurred and the work process is all but optimal.

3.11.1.2 Law of communicating vessels

The law of communicating vessels simply states that in physics, matters are often linked to one another, and that in some way, they have an impact on each other. This is no different for socioeconomic systems such as organizations. Every measure or change within a system leads to changes and shifts within other parts of the system, and these domino-changes need to be identified and, if necessary, additional measures need to be taken.

The whole system at once, instead of different parts of the system, needs to be considered, and the relationships between the parts of the system need to be taken into account when making risk decisions. Hence, awareness of possible unexpected changes and continuous vigilance always need to be present.

3.11.1.3 Nonlinear causalities

Linear causality is very hard to find in real life and in real industrial practice: almost never is an accident the result of one cause; on the contrary, an accident is usually (almost always) caused by the concurrence of circumstances and a variety of factors. One factor itself often does not suffice to cause an accident. Also, if another order of events had happened, then the consequences of an accident might have been completely different. Moreover, cause and consequence are regularly not closely linked in space and time. Nonetheless, the urge to "think in linear causalities" by humans is very strong, and for risk managers, it certainly has also been so in the past.

Instead of viewing reality as a static picture and, based on this picture, taking preventive measures, risk managers need to discern change patterns, looking at positive and negative feedback loops between events, and, based on this improved perception of reality, take health and safety measures.

3.11.1.4 Long-term vision

Research indicates that business failures more often originate from poor adaptation of a business to slowly emerging threats, instead of being caused by sudden threats [57]. Insidious, latent (long-term) problems usually do not receive the attention they deserve, while, on the contrary, short-term failures leading to problems often do.

Because analyses are limited in space and time, gradually emerging failures and problems are much more difficult to detect. Increasing space and time while analyzing a certain part of reality improves the perception of this piece of reality. An improved perception of reality leads to better decisions.

3.11.1.5 Systems thinking conclusions

All previously mentioned "laws of systems thinking" indicate the need for insights into a system, thereby considering all relevant system parts, their relationships and their interdependencies. Often, "easy solutions" are sought, looking for symptoms instead of underlying causes and structures. This way, the "visible problem" is solved for a short time, cannot be seen anymore, and all seem to be well. Of course, in reality, the problem is still present and it will reappear later in time, often with more persistence and possibly somewhere else in the system. Also, the problem has often become that of someone else, who finds it harder to solve the problem because of the superficial actions taken previously.

Such nonsystems thinking behavior is strongly present in people, and knowledge and reasoning are therefore necessary to implement true systems thinking behavior in organizations. This can be illustrated by a fireman who is considered a hero when extinguishing fire at the risk of his own life, but who is considered a stickler when implementing fire safety regulations [58].

3.11.2 Normal accident theory and high reliability theory

Two schools of thought exist on how to prevent major accidents: the high reliability theory (HRT) and the normal accident theory (NAT). HRT believes that with intelligent organizational design and management techniques, a company can compensate for weaknesses within the organization and guarantee accident-free operations. NAT, on the contrary, believes that major accidents are inevitable and suggests that complex organizations make decisions in ways that are different from the rational models used by the HRT theorists.

According to the HRT, four aspects lead to zero-accident safety: leadership that prioritizes safety objectives as an organizational goal, high levels of human and nonhuman redundancy, an organizational setting of high reliability with decentralized authority and a line-level culture of reliability and training, and an approach to trial-and-error learning where organizations learn from internal and external experiences.

According to NAT theorists, however, complex organizations may work hard to maintain safety and reliability, but despite all efforts, major accidents will be a "normal" result or an integral characteristic of the system. NAT theorists provide some causes for this: different individuals at different levels of an organization may hold conflicting goals, there may be important communication problems, the processes may not be fully understood by the employees, essential risk knowledge may have left the company with retirement, etc.

In any case, the levels of vulnerability and resiliency of organizations are very important in order to decide how to deal with prevention, mitigation, emergency management, etc., within an organization. Furthermore, the type of activity of an organization has an impact on its levels of resiliency and vulnerability. Perrow [59] discerns two important dimensions in organizations: "interactions" and "coupling."

Interactions can be linear or complex. Linear interactions may be complicated, but they are always clear and visible to some extent. Complex interactions, on the other hand, can be unexpected or even incomprehensible. Ideally, systems should be made as linear as possible; this way, an incident and its possible effects become more predictable. However, this is not always possible in the complex industrial world of today. Systems subject to complex interactions are not always more dangerous or hazardous, and they are characterized by low predictability. Complex interactions require decentralization to be able to adequately cope with problems.

Some organizations or systems (within organizations) are tightly organized according to fixed procedures and with strong interdependence. Perrow [59] calls this "tightly coupled systems." In this type of organization, taking care of errors, disturbances and failures is built in, without many possible ways to deviate from suggested scenarios and solutions. There is little room for improvisation in case an accident is unfolding, and centralization of command is needed. In other "loosely coupled" organizations, there is much more possibility to improvise and to use alternative solutions.

NAT theory, elaborated by Perrow [59, 60], explains "system accidents" by using both these dimensions and their implications (such as nonlinearities, reinforcing causalities, complexities and interactions, and strong structural links ("tight coupling"). Organizations can be divided into four quadrants, as illustrated in Fig. 3.35. The nuclear industry and, to a lesser extent, the chemical industry are examples of organizations belonging to complex, tightly coupled working environments. These industrial sectors are in a quadrant where, on the one hand, centralization is required due to the low room for improvisation when things go wrong (tight coupling), and on the other hand, decentralization is needed due to the low predictability of things going wrong. Hence, problems will keep arising in such environments according to NAT theorists.

Insights gained from the NAT can also be applicable to noncomplex working environments. The theory boils down to people (regardless of the environment in which they work) understanding and appreciating the possibility of minor faults and failures interacting in unexpected ways, and, through the existence of structural relationships between system components, leading to a cascade of faults/failures, eventually leading to an accident. The higher the complexity of a system, the more it depends on coincidence and unexpected events, from both a positive and a negative point of view. A failure in a linear system (e.g., an assembly line) can usually be anticipated, and it is understandable and visible. Such a failure can be avoided by taking the correct preventive measures, being correctly executed and maintained. Complex interactive systems can be subject to different failures, of which each of the failures separately may not cause any problems because the correct preventive measures are in place. However, it may be the case that because of unexpected and incomprehensible interactions between different simultaneous failures of the complex system, the separate preventive measures are bypassed or nullified. If the system, in addition to being complex, is tightly linked as well, the failures may get out of control, and an avalanche of failures may arise, resulting in a major system failure and a subsequent major accident.

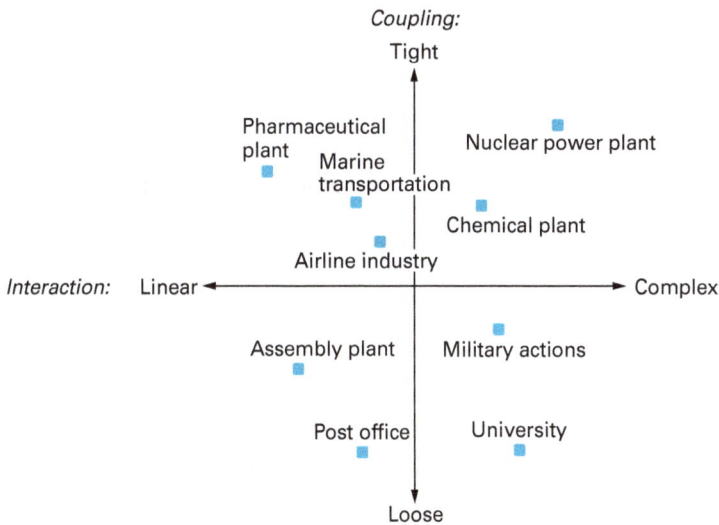

Fig. 3.35: Interaction versus coupling for different industrial sectors (inspired by Zanders [61]).

Conversely to the NAT theory, HRT theorists maintain that organizations can be designed and managed to achieve near-perfect reliability and safety records. NAT theorists maintain that, in these same organizations, major "system" accidents are inevitable because of "system faults." The difference in perception between the theorists of HRT and NAT boils down to how much influence each one believes risk management has on the performance of the organization.

In this regard, the difficulty of having a technological disaster should be stressed. For example, chemical plants with catastrophic potential have been around for more than a century, yet there have been few disasters that claimed thousands of lives. There has, e.g., only been one Bhopal. Nonetheless, the US EPA estimated in 1989 that in the previous 25 years, there were 17 releases of toxic chemicals in volumes and levels of toxicity exceeding those that killed many thousands of lives in Bhopal. Mostly because of sheer luck, only five people were killed in these accidents [62]. As Perrow [63] indicates, the reason is the flip side of NAT: just as it takes the right combination of failures to defeat all the safety devices, so does it take the right combination of circumstances to produce a true disaster.

3.11.3 High reliability organization principles

Organizations capable of gaining and sustaining high reliability levels are called "high reliability organizations" or HROs. Despite the fact that HROs operate hazardous activities within a high-risk environment, they succeed in achieving excellent health and safety figures. Hence, they identify and correct risks very efficiently and effectively. Some examples of organizations having no choice but to function reliably are nuclear aircraft carriers, air traffic control systems, aircraft operations systems, emergency medical treatment teams, nuclear power generation plants, continuous processing firms, wildland firefighting crews, etc.

A typical characteristic of HROs is collective mindfulness. Hopkins [64] also indicates that HROs organize themselves in such a way that they are better able to notice the unexpected in the making and halt its development. Hence, collective mindfulness in HROs implies a certain approach to organizing themselves. Collective mindfulness can only be achieved by individuals developing two characteristics: (i) situational awareness and (ii) a questioning attitude.

Situational awareness could be described as a personal trait that leads an individual to be fully devoted to professional knowledge regarding risks, brutally honest self-assessment with respect to safety and risk behavior and thinking, continuous improvement and intellectual integrity. Situational awareness is, however, very difficult, if not impossible, to reach. It is a mental model, a mindset, that individuals need to strive for. The questioning attitude, as explained by Digenonimo and Koonce (2016), is a thought process that employs a person's fundamental knowledge to critically evaluate the processes, ideas or operations and tasks that his/her organization uses and/or are carried out. Some pitfalls where one needs to be careful are a lack of ownership, a lack of training, a lack of conscientiousness and a lack of communication, among others.

Five key principles are used by HROs to achieve such mindful and reliable organizations (see also [65]). The first three principles mainly relate to anticipation, or the ability with which organizations can cope with unexpected events. Anticipation concerns disruptions, simplifications and execution and requires means of detecting small clues and indications, with the potential to result in large, disruptive events. Of course, such organizations should also be able to decrease, diminish or stop the consequences of (a chain of) unwanted events. Anticipation implies the ability to imagine new, noncontrollable situations, which are based on little differences with well-known and controllable situations. HROs take this into account in principles 1, 2 and 3.

Whereas the first three principles relate to proaction, the fourth and fifth focus on reaction. It is evident that if unexpected events happen despite all precautions taken, the consequences of these events need to be mitigated. HROs take this into account in principles 4 and 5.

3.11.3.1 HRO principle 1: targeted at disturbances

This principle asserts that HROs are very actively, and in a proactive manner, looking for failures, disturbances, deviations, inconsistencies, etc., because they realize that these phenomena can escalate into larger problems and system failures. They achieve this goal by urging all employees to report (without a blame culture) mistakes, errors, failures, near-misses, etc. HROs are also very much aware that a long period of time without any incidents or accidents may lead to employee complacency and may thus further lead to less risk awareness and less collective mindfulness, eventually leading to accidents. Hence, HROs rigorously ensure that such **complacency is avoided** at all times.

3.11.3.2 HRO principle 2: reluctant for simplification

When people – or organizations – receive information or data, there is a natural tendency to simplify or reduce it. Parts of the information considered non-important or irrelevant are omitted. Evidently, information that may be perceived as irrelevant might, in fact, be very relevant in order to avoid incidents or accidents. HROs will therefore question the knowledge they possess from different perspectives and at all times. This way, the organizations try to discover "blind spots" or phenomena that are hard to perceive. To this end, extra personnel (as a type of human **redundancy**) are used to gather information.

3.11.3.3 HRO principle 3: sensitive toward implementation

HROs strive for continuous attention toward real-time information. All employees (from frontline workers to top management) should be very well informed about all organizational processes, not only about the process or task they are responsible for. They should also be informed about the ways that organizational processes may fail and how to control or repair such failures.

To this end, an organizational **culture of trust** between and among all employees is an absolute must. A working environment in which employees are afraid to provide certain information, e.g., to report incidents, will result in an organization lacking information, and in which efficient working is impossible. An "engineering culture," in which quantitative data/information is much more appreciated than qualitative knowledge/information, should also be avoided. HROs do not distinguish between qualitative and quantitative information.

HROs are also sensitive toward routines and routine-wise handling. Routines can be dangerous when they lead to mindlessness and distraction. By implementing job rotation and/or task rotation intelligently, HROs try to prevent such routine-wise handling.

Furthermore, HROs view near-misses and incidents as **opportunities to learn**. The failures that go hand-in-hand with the near-misses always reveal potential (otherwise hidden) hazards; hence, such failures serve as an opportunity to avoid future similarly caused incidents.

3.11.3.4 HRO principle 4: devoted to resiliency

HROs define "resiliency" as "the capacity of a system to retain its function and structure, regardless of internal and external changes." The system's flexibility allows it to keep on functioning, even when certain system parts do not function as required anymore. An approach to ensure this is that employees organize themselves into **ad hoc networks when unexpected events happen**. These can be regarded as temporary informal networks capable of supplying the required expertise to solve the problems. When the problems have disappeared or are solved, the network ceases to exist.

3.11.3.5 HRO principle 5: respectful for expertise

Most organizations are characterized by a hierarchical structure with a hierarchical power structure, at least to some degree. This is also the case for HROs. However, in HROs, the power structure is no longer valid in unexpected situations in which certain expertise is required. The **decision process and the power are transferred** from those at the top of the hierarchy (in normal situations) **toward those with the most expertise** regarding certain topics (**in exceptional situations**).

Two theories exist regarding accidents in large-scale systems with catastrophic potential: HRT and NAT. HRT believes that organizations can learn from operating and regulatory mistakes, put safety first and empower lower levels, thereby making risky items quite safe. NAT suggests that, no matter how hard organizations try, there will be serious accidents because of the interactive complexity and tight coupling of most risky systems.

3.11.4 Risk and reliability

The safety of an organization, especially a complex, tightly coupled organization, depends on a variety of factors, evidently starting with the good design of processes, equipment, installations, etc. As a process, installation or equipment, no matter how well designed, cannot operate indefinitely without intervention, the degree of safety depends on the maintenance procedures and on actions intended to keep the organization safe. Reliability engineering, as a part of risk engineering, is concerned with analyzing the *reliability, availability, and maintainability* (RAM) of a safety system. Several issues related to performance must be considered: hardware and software failures, human errors, incorrect operating procedures and also the interactions between these. But what are the differences between those terms? Definitions are given by Sutton [66]:

- The reliability of a component or a system is the probability that it will perform a required function without failure under stated conditions for a stated period of time.
- The availability of a repairable system is the fraction of time that it is able to perform a required function under stated conditions.
- The maintainability of a failed component or system is the probability that it will be returned to its operable condition in a stated period of time under stated conditions and using prescribed procedures and resources.

The difference between reliability and availability arises because reliability does not account for the possibility that a given system can be repaired after its failure. This indicates that the reliability as a function of time t, noted as $R(t)$, predicts the time t until the system has undergone its first failure (thus reliability refers to the *first* system failure), whereas the system may have failed in the past but has been repaired so that it is operational at time t with predicted availability $A(t)$. Reliability can also be seen as the complement of the failure probability $F(t)$, hence $R(t) = 1 - F(t)$.

A system with redundant subsystems can exhibit subsystem failures without system failure. For an availability analysis, the ongoing repair actions continue. Availability is thus used for systems, and reliability is employed for components of those systems.

Maintainability is the ability of a system component to be restored to a state in which it can perform its intended function when maintenance is performed under prescribed procedures. It involves actions typically performed according to procedures established by the manufacturer of the component.

For further information on reliability engineering, including probability theory, Boolean algebra, failure rates, probability distributions, mean time between failures, mean time to failure, mean time to repair and mean downtime, we refer to specialized literature such as Sutton [66], Smith [67], Lee and McCormick [68] and Zio [69].

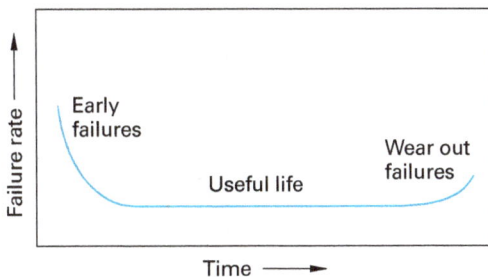

Fig. 3.36: Bathtub curve.

One important and much-quoted concept from reliability theory is the "bathtub distribution." The bathtub curve seeks to describe the variation of the failure rate of components during their life. Figure 3.36 illustrates this generalized relationship.

From a conceptual point of view, the bathtub curve is interesting as it clearly demonstrates the usefulness of RAM for safety purposes. The failures shown in the first part of the curve, where the failure rate is decreasing, are called "early failures" or "infant mortality failures." They are usually related to manufacturing and quality assessment, e.g., connections, joints, dirt, impurities, cracks, insulation or coating flaws, incorrect adjustment and incorrect positioning. The middle portion is referred to as the "useful life," and it is assumed that failures exhibit a constant failure rate, i.e., they occur at random. The failures from this region are usually assumed to be stress-related. The latter part of the curve describes the wear-out failures, and it is assumed that the failure rate increases as the wear-out mechanisms accelerate. Such failures result from corrosion, oxidation, breakdown of insulation, atomic migration, friction wear, fatigue, etc.

3.12 Accident reporting

This section deals with the question of how and in what ways incident and accident figures can be used as reactive (lagging) indicators to measure safety effectiveness within an organization. It is an important topic because many organizations still use this type of performance indication to have an idea of how well – or how bad – they are performing regarding safety. Several types of metrics are used. If we discern, as the Bird pyramid (Fig. 3.5) suggests, serious accidents from minor injuries, different metrics for different kinds of accidents are used.

For serious accidents, which we may define as occurrences that result in a fatality, permanent disability, or time lost from work of one day/shift or more, usually the lost-time injury frequency rate (LTIFR) is used. The LTIFR is the number of LTIs per million hours worked, calculated using this equation:

$$\text{LTIFR} = \frac{\text{Number of lost-time injuries over the accounting period} \times 1{,}000{,}000}{\text{Total number of hours worked in accounting period}}$$

Hence, the LTIFR is how many LTI occurred over a specified period per 1,000,000 h (or some other number; 100,000 is also often used) worked in that period. Mostly, the accounting period is chosen to be 1 year. By counting the number of hours worked, rather than the number of employees, e.g., discrepancies that may be caused in the incidence rate calculation by part-time workers and overtime are avoided. However, this metric using the employees instead of the hours worked – the LTI incidence rate (LTIIR) – is also used in many organizations. To calculate the LTIIR, which is the number of LTIs per 100 (or whatever figure you want) employees, the following equation is used:

$$LTIIR = \frac{\text{Number of lost-time injuries over the accounting period} \times 100}{\text{Average number of employees in accounting period}}$$

Next, the severity rate takes into account the severity per accident. Depending on how this is expressed, you will need at least the information from above and the number of work days lost over the year. Often, the severity rate is expressed as an average by simply dividing the number of days lost by the number of LTIs. Another way of calculating the severity rate, the lost-time injury severity rate (LTISR), is using the following equation (the figure 1,000,000 may be replaced by any other figure; it just tells us that the LTISR in this case is expressed per million hours worked):

$$LTISR = \frac{\text{Number of work days lost over the accounting period} \times 1,000,000}{\text{Total number of hours worked in accounting period}}$$

Also, the medical treatment injury frequency rate is often measured. This frequency rate measures how often medical treatment injuries are occurring. It is expressed as the number of medical treatment injuries per million hours worked:

$$MTIFR = \frac{\text{Number of medical treatment injuries over the accounting period} \times 1,000,000}{\text{Total number of hours worked in accounting period}}$$

Finally, the total recordable injury frequency rate measures the frequency of recordable injuries, i.e., the total number of fatalities, LTIs, medical treatment injuries and restricted work injuries occurring per million hours worked:

$$TRIFR = \frac{\begin{array}{c}\text{Number of recordable injuries (fatalities + lost-time injuries}\\ \text{+ medical treatment injuries + restricted work injuries)}\\ \text{over the accounting period} \times 1,000,000\end{array}}{\text{Total number of hours worked in accounting period}}$$

A regularly used fatality accident measure is the so-called FAR. The FAR reports the number of fatalities based on 1,000 employees working their entire lifetime. The employees are assumed to work a total of 50 years. Hence, this measure is based on 10^8 working hours:

$$FAR = \frac{\text{Number of fatalities} \times 10^8}{\begin{array}{c}\text{Total number of hours worked in}\\ \text{accounting period}\end{array}}$$

For more information regarding accident reporting, see [70].

3.13 Conclusions

On the one hand, risk management principles cannot be summarized easily or in simple terms: the various ways to handle and manage risks are diverse and focused on the type and the characteristics of the risks at hand. On the other hand, although risk is an abstract concept, our natural human understanding of "risk" and especially of decision-making related to risk is pretty sophisticated. As an example, a predictable or expected loss is not the same as an unexpected loss or a catastrophic loss; we all intuitively understand this. We also understand that different risks require different approaches and different actions. We even have our own ideas about what approach would be suitable for which circumstances. But these ideas are not always right, and in fact, a lot of our intuitive thoughts about risks cannot be trusted. Dealing with risks in our era is something that should be learned and that requires experience and expertise. In this chapter, different ways and models to deal with different types of risks are presented. Although rather generic and theoretical, the models are easy to understand, and they can be applied on an operational level quite easily. Most importantly, they provide profound and valuable insights into the abstract world of risks, incidents and accidents and the abstract and concrete requirements to deal with them.

Furthermore, for decades, there has been discussion among safety scientists about what constitutes a safety culture, and how it should be conceptually viewed. In this chapter, an integrative conceptual framework for safety culture is presented, composed of three domains: the engineering domain, the perceptual domain and the intended behavior domain. Every domain further consists of safety dimensions/factors and can be measured. Safety culture can thus be steered by monitoring its three domains. The domains can, in turn, be monitored by measuring their constituting dimensions/factors. Since the dimensions/factors are interrelated and influence one another, they should not be viewed in an individual, analytical, linear way, but rather in a holistic, systemic, cyclic way.

References

[1]   Bernstein, P.L. (1998). Against the Gods. The Remarkable Story of Risk. New York: John Wiley & Sons, Inc.
[2]   Vaughan, E.J. (1997). Risk Management. New York: John Wiley and Sons.
[3]   Fuller, C.W., Vassie, L.H. (2004). Health and Safety Management. Principles and Best Practice. Essex, UK: Prentice Hall.
[4]   ISO 31000:2009 Risk Management Standard – Principles and Guidelines, International Organization for Standardization, 2009.
[5]   Heinrich, H.W. (1950). Industrial Accident Prevention. 3rd edn. New York: McGrawHill Book Company.
[6]   Bird, E., Germain, G.L. (1985). Practical Loss Control Leadership, the Conservation of People, Property, Process and Profits. Loganville, GA: Institute Publishing.
[7]   James, B., Fullman, P. (1994). Construction Safety, Security and Loss Prevention. New York: Wiley Interscience.

[8] Krause. (2011). New Findings on Serious Injuries and Fatalities. Ojai: California, USA: BST (Behavioural Science Technology).

[9] Hopkins, A. (2010). Failure to Learn. The BP Texas City Refinery Disaster. Sydney, Australia: CCH Australia Ltd.

[10] OGP. (2005). Human Factors. London: International Association of Oil and Gas Producers.

[11] Reniers, G., Dullaert, W. (2007). Gaining and Sustaining Site-integrated Safety and Security in Chemical Clusters. Zelzate, Belgium: Nautilus Academic Books.

[12] Reason, J.T. (1997). Managing the Risks of Organizational Accidents. Aldershot, UK: Ashgate Publishing Limited.

[13] Heinrich, H.W. (1950). Industrial Accident Prevention. 3rd edn. New York: McGrawHill Book Company.

[14] Heinrich, H.W. (1959). Industrial Accident Prevention. 4th edn. New York: McGrawHill Book Company.

[15] Heinrich, H.W., Petersen, D., Roos, N. (1980). Industrial Accident Prevention. 5th edn. New York: Mc GrawHill Book Company.

[16] Powell, R.L. (1996) Process Safety and Control Systems Integrity. International Conference and Workshop on Process Safety Management and Inherently Safer Processes, October 8–11,1996, Orlando, Florida, 227–241. New York: American Institute for Chemical Engineers.

[17] Reniers, G.L.L. (2010). Multi-plant Safety and Security Management in the Chemical and Process Industries. Weinheim, Germany: Wiley-VCH.

[18] CCPS, Center for Chemical Process Safety. (1996). Inherently Safer Chemical Processes. A Life Cycle Approach. New York: American Institute of Chemical Engineers.

[19] International Standard IEC-61511 (2003) 1st edn.

[20] Gardner, R.J., Reyne, M.R. (1994). Selection of Safety Interlock Integrity Levels. Wilmington, DE: Dupont Engineering.

[21] CCPS, Center for Chemical Process Safety. (2003). Guidelines for Analyzing and Managing the Security Vulnerabilities of Fixed Chemical Sites. New York: American Institute of Chemical Engineers.

[22] Vincoli, J.W. (1994). Basic Guide to Accident Investigation and Loss Control. New York: John Wiley & Sons, Inc.

[23] Ale, B.J.M. (2009). Risk: An Introduction. The Concepts of Risk, Danger and Chance. Abingdon, UK: Routledge.

[24] CCPS, Center for Chemical Process Safety. (2000). Guidelines for Chemical Process Quantitative Risk Analysis. 2nd edn. New York: American Institute of Chemical Engineers.

[25] Ale, B.J.M. (2005). Living with risk: A management question. Reliab. Eng. Syst: Safe 90: 196–205.

[26] Reniers, G., Ale, B.J.M., Dullaert, W., Foubert, B. (2006). Decision support systems for major accident prevention in the chemical process industry: A developers' survey. J. Loss Prev. Process Ind. 19: 604–662.

[27] Pasman H. (2015). Risk Analysis and Control for Industrial Processes – Gas, Oil and Chemicals. A System Perspective for Assessing and Avoiding Low-probability High-consequence Events. Oxford, UK: Butterworth-Heinemann.

[28] OHSAS(18001:2007). (2007). Occupational Health and Safety Management Systems – Requirements. London, UK: British Standardisation Institute.

[29] Le Ray, J. (2010). Gérer Les Risques: Pourquoi? Comment? Saint-Denis-La Plaine. France: Afnor.

[30] Schein, E.H. (2010). Organizational Culture and Leadership. San Francisco, CA: Jossey-Bass.

[31] Irani, Z., Beskese, A., Love, P. (2004). Total quality management and corporate culture: Constructs of organizational excellence. Technovation 24: 643–650.

[32] Guldenmund, F.W. (2007). The use of questionnaires in safety culture – an evaluation. Safety Sci. 45 (6): 723–743.

[33] Guldenmund, F.W. (2010). Understanding and Exploring Safety Culture. Oisterwijk, The Netherlands: Boxpress.

[34] Zohar, D. (1980). Safety climate in industrial organizations: Theoretical and applied implications. J. Appl. Psychol. 65: 96–102.

[35] Niskanen, T. (1994). Safety climate in the road administration. Safety Sci 17: 237–255.

[36] Coyle, I.R., Sleeman, S.D., Adams, N. (1995). Safety Climate. J. Safety Res. 26: 247–254.

[37] Diaz, R.I., Cabrera, D.D. (1997). Safety climate and attitude as evaluation measures of organizational safety. Accident Anal. Prev. 29: 643–650.

[38] Seo, D.-C., Torabi, M.R., Blair, E.H., Ellis, N.T. (2004). A cross-validation of safety climate scale using confirmatory factor analytic approach. J. Safety Res. 35: 427–445.

[39] Guldenmund, F.W. (2000). The nature of safety culture: A review of theory and research. Safety Sci 34: 215–257.

[40] Van Leeuwen, M. (2006). De Veiligheids barometer, dissertation, University of Twente.

[41] Reason, J. (1998). Achieving a safe culture: Theory and practice. Work Stress 12: 293–306.

[42] Cooper, M. (2000). Towards a model of safety culture. Safety Sci 36: 111–136.

[43] Hale, A.R. (2000). Culture's confusions. Safety Sci 34: 1–14.

[44] Wiegmann, D.A., Zhang, H., Von Thaden, T.L., Sharma, G., Mitchell, A.A. (2002) A Synthesis of Safety Culture and Safety Climate Research. Technical Report ARL-02-3/FAA-02-2 for the Federal Aviation Administration, Atlantic City International Airport, NY.

[45] Mohamed, S. (2003). Scorecard approach to benchmarking organizational safety culture in construction. J. Construct. Eng. Manag. 129: 80–88.

[46] Reniers, G., Cremer, K., Buytaert, J. (2011). Continuously and simultaneously optimizing an organization's safety and security culture and climate: The improvement diamond for excellence achievement and leadership in safety and security (IDEAL S&S). J. Clean. Prod. 19: 1239–1249.

[47] CCPS, Center for Chemical Process Safety. (2007). Human Factors Methods for Improving Performance in the Process Industries. Hoboken, NJ: American Institute of Chemical Engineers.

[48] Kletz, T.A. (2001). An Engineer's View of Human Error. 3rd edn. Rugby, UK: Institute of Chemical Engineers.

[49] Thompson, A.A., Strickland, A., Gamble, J.E. (2007). Crafting and Executing Strategy – Text and Readings. New York: McGrawHill/Irwin.

[50] Vierendeels, G., Reniers G.L.L., Van Nunen K., Ponnet K. (2016) An integrative conceptual framework for safety culture: The Egg Aggregated Model (TEAM) of safety culture; forthcoming.

[51] HSE, Health and Safety Executive. (2006). Developing Process Safety Indicators. A Step-by-step Guide for Chemical and Major Hazard Industries. Sudbury, UK: HSE Books.

[52] Parmenter, D. (2007). Key Performance Indicators. Developing, Implementing, and Using Winning KPIs. Hoboken, NJ: John Wiley & Sons, Inc.

[53] OECD. (2003). Guidance on Safety Performance Indicators. Paris, France: OECD Environment, Health and Safety Publications.

[54] Mazri, C., Jovanovic, A., Balos, D. (2012). Descriptive model of indicators for environment, health and safety management. Chem. Eng. Trans. 26: 465–470.

[55] Van Heuverswyn, K., Reniers, G. (2012). Performance Management 4: Van Safetymanagement Naar Performant Welzijnsmanagement. Brugge, Belgium: Die Keure.

[56] Wu, T.C., Chen, C.H., Li, C.C. A correlation among safety leadership, safety climate and safety performance. J. Loss Prevent. Proc. 21: 307–318.

[57] Senge, P. (1992). De Vijfde Discipline, De Kunst En De Praktijk Van De Lerende Organizatie. Schiedam, The Netherlands: Scriptum Management.

[58] Bryan, B., Goodman, M., Schaveling, J. (2006). Systeemdenken, Ontdekken Van Onze Organizatiepatronen. Den Haag, The Netherlands: Sdu Uitgevers.

[59] Perrow, C. (1984). Normal Accidents. Living with High Risk Systems. New York: Basic Books.

[60] Perrow, C. (1999). Normal Accidents. Living with High-tech Risk Technologies. Princeton, NJ: Princeton University Press.

[61] Zanders, A. (2008). Crisis Management. Organizaties Bij Crises En Calamiteiten. Bussum, The Netherlands: Uitgeverij Coutinho.

[62] Shabecoff, P. (1989). Bhopal Disaster Rivals 17 in US. New York Times, 30 April.

[63] Perrow, C. (2006). The limits of safety: The enhancement of a theory of accidents. In: Key Readings in Crisis Management. Systems and Structures for Prevention and Recovery, Smith & Elliott, Editors. Abingdon, UK: Routledge.

[64] Hopkins, A. (2005). Safety, Culture and Risk. The Organizational Causes of Disasters. Sydney, Australia: CCH Australia Ltd.

[65] Weick, K.E., Sutcliffe, K.M. (2007). Managing the Unexpected. Resilient Performance in an Age of Uncertainty. 2nd edn. San Francisco, CA: John Wiley & Sons, Inc.

[66] Sutton, I. (2010). Process Risk and Reliability Management. Operational Integrity Management. Oxford, UK: Elsevier.

[67] Smith, D.J. (2011). Reliability, Maintainability and Risk. Practical Methods for Engineers. Oxford, UK: Butterworth-Heinemann.

[68] Lee, J.C., McCormick, N.J. (2011). Risk and Safety Analysis of Nuclear Systems. Hoboken, NJ: John Wiley and Sons.

[69] Zio, E. (2012). An Introduction to the Basics of Reliability and Risk Analysis. Singapore: World Scientific Publishing.

[70] Health and Safety Executive (HSE). (2009). HSE Event Injury Illness Classification Guide. Sudbury, UK: HSE Books.

4 Risk diagnostic and analysis

4.1 Introduction to risk assessment techniques

A risk assessment is an important step in protecting workers and businesses as well as for complying with the law. It helps to focus on the risks that really matter at the workplace – the ones with the potential to cause real harm. Currently, over a hundred risk analysis techniques are available in the literature. Most of them identify initiating events (causes), consequences, safeguards and recommendations. The main difference between the methods is in the way they approach the identification of causes or consequences. Empirical research revealed that the four most commonly used techniques in the chemical and process industries are hazard and operability studies (HAZOP), failure mode and effects analysis (FMEA) or failure mode, effects and criticality analysis (FMECA), what-if analysis and the risk matrix [1]. Other techniques that are mainly used in the process industries are event tree analysis, fault tree analysis (FTA), human reliability analysis and check lists. Some of them will be discussed here, but for more information, we refer to Groso et al. [2]. Preliminary hazard analysis (PHA) is also presented as a relatively simple and inexpensive technique that still provides meaningful results. In other industries, other techniques will be more popular, for instance, HACCP in the food industry, FMEA in the manufacturing industry and checklists in the construction industry.

Through the use of systematic methods, risk analysis aims to objectify the risks incurred by a system (globally speaking). In this sense, it only serves as a starting point and as a support for the decision-making process relative to the acceptable level of risk, which is risk management. The latter is based on a wider range of criteria consisting of subjective factors. As mentioned in the previous chapter, risk management is the political (decision-making) process to deal with risks, while risk analysis and risk assessment are only technical approaches to have a good notion of risks. Hence:

$$\text{Risk management} = \text{risk analysis} + \text{risk assessment} + \text{risk communication}$$

$$+ \text{business continuity planning} + \text{preventive training}$$

$$+ \text{incident analysis} + \text{influencing risk perception} + \ldots$$

All kinds of risk analysis techniques exist, both for technical and human-related assessments. The context of a particular risk analysis discussed in this book can be defined by two main situations:
- Cases involving a technical object, e.g., a plane, vehicle or machine.
- Cases involving more complex systems, e.g., industrial plants, agricultural and urban installations. These systems also include machines and other technical objects, but in this case, they will be closely linked with their environment.

https://doi.org/10.1515/9783111493633-004

The methods and tools for risk analysis differ for the two situations:
- In the first case, one mainly uses classical tools for safety engineering (e.g., preliminary risk analysis and fault or event tree analysis).
- In the second case, these tools only allow an individual analysis of the system compartments, mainly technical objects, and it would be necessary to refer to other methods that are capable of carrying out systemic analyses.

Well-known available tools can be classified into two categories:
- Semiempirical tools such as PHA, which eventually lead to the development of grids derived from experimental results, FMEA and FMECA, HAZOP and functional analysis.
- Logical tools such as tree charts (FTA) and cause–consequence charts and logical models such as Markov chains (relations between probabilities and partial differential equations) or Petri networks.

All these tools allow approaches through mathematical calculations. The implementation of these tools can present certain difficulties, as most of them originate from the reliability analysis of objects or object elements and thus are not fully adequate for complex risk analysis. Furthermore, their implementation requires information that does not come from the tools themselves.

Therefore, tools to model complex sociotechnical systems, such as functional resonance analysis method (FRAM) and STAMP, are also needed to increase the knowledge of the system in which the system compartment that is being investigated operates. For when a system is designed, there is, from the start, a need to know how it will function and what parameters will have what kind of influence. Nonetheless, knowledge is always insufficient, and virtually all classical risk analyses are conducted in a state of relative ignorance of the full behavior of the system [34].

The reader should realize that risk analyses will mostly reveal the hazards and the risks that may lead to accidents. Most – almost all – accidents happen following known risks within a company. However, either the risks were not believed to be credible, or the preventive and protective measures were inadequate (incomplete or absent) or the risk level was higher (or has become higher) than assessed. It should be noted (and warned) that the limitation of a risk analysis is not only constrained by the imagination of the person carrying it out but also by the self-censorship of this person, determining what he/she believes to be "credible" or "incredible" risk.

4.1.1 Inductive and deductive approaches

Two main approaches to risk analysis exist: inductive and deductive, also called, respectively, bottom-up and top-down:

–   The deductive methodologies analyze the causes of an adverse event (accident) by answering the question, "How is it that this event may occur (search for causes)?"
–   The inductive methodologies analyze the consequences of failure (initiating event) and answer the question, "What adverse events can result in (search for consequences)?"

Inductive analyses are headed in the direction of an accidental process, while the deductive analyses back up to this process (see Fig. 4.1).

There is also a complementary method that classifies risk analysis methodologies depending on the purpose fields of the research:
–   The prospective analyses use a preventive approach. They can improve the system in its design phase and thus form part of the philosophy "prevention rather than cure."
–   Conversely, retrospective studies investigate accident scenarios that have occurred to find the causes and improve the system afterward.

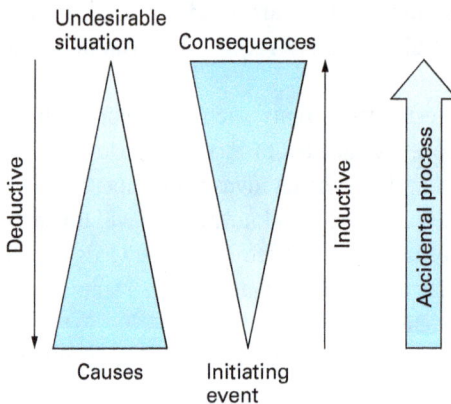

Fig. 4.1: Inductive and deductive methods.

Prospective analyses rely on inductive methods, while retrospective analyses almost always use deductive approaches.

4.1.2 General methods for risk analysis

In this chapter, selected methods of risk analysis will be briefly presented to cover the field of risk analysis. None of them is fully adequate for every situation, but one needs to use the method best suited to each study. The techniques can be classified into three distinct groups: basic methods, static methods and dynamic methods.

Among the well-known basic methods, there are
– Functional analysis
– Preliminary risk analysis
– FMECA
– HAZOP study

These basic methods are often used during preliminary stages and proceed toward a first global and high-level analysis. Depending on the complexity of the system, a more detailed analysis may or may not be required. Most basic methods are inductive (cause → effect) and do not require much more than a pencil and a table chart; there are no underlying mathematical models involved. These methods are indispensable for understanding how the system works and for correctly identifying the risks, regardless of the type of system and the nature of the study.

The most common static methods are:
– Reliability block diagram
– FTA
– Event tree analysis (ETA)

These methods, compared to the previous basic ones, allow an analysis from a structural (topological) point of view. This is obtained through Boolean mathematical models, which are static as they cannot model temporal effects on the system. To simplify, we can say that this approach only allows a logical representation of the system at a certain time and does not take into account changes in the system over time. Among the cited methods, FTA should be mentioned in particular because it is the only deductive (effect → cause) tool in reliability engineering.

The most well-known dynamic methods are
– Markov chains
– Stochastic Petri nets
– Bayesian (belief) networks

These methods, which now must include formal languages such as AltaRica, were developed to take into account the temporal and compartmental effects that cannot be appropriately covered by static models. While the Markov approach is analytical, the stochastic Petri net method requires calculations through Monte Carlo simulation. Bayesian belief networks will be developed later in this chapter.

The above-mentioned risk analyses are all analytic methods and, hence, fit for looking at a part of a sociotechnical system (at best) and not at the complex sociotechnical system as a whole. To conduct a so-called "systemic" risk analysis, and thus have an idea of the risks from a sociotechnical system's perspective, other risk analysis techniques are required, such as the FRAM method (see later in this chapter) and the STAMP/STPA technique (methods can be used both retrospectively and prospectively) or Accimap (retrospective technique).

A question that is immediately raised by the reader is probably: "Why are there so many methods?" (Tab. 4.1 provides a non-exhaustive overview of risk analysis methods for engineering projects.)

Among the several existing methods, it is critical to select the one best suited to the study being carried out. An inappropriate selection can result in incorrect results or a significant loss of time.

The choice of method mainly depends on
– the nature and requirements of the study;
– the amount of knowledge about the system;
– availability of quantitative data; and
– availability of time and resources.

The key questions to be asked before choosing a method should be:
– What are the aims and what is the scope of the analysis?
– Is the analysis prospective or retrospective?
– Is the analysis specific (linked to an event, failure, etc.) or does it consider the system as a whole?
– What is the required depth of analysis?
– How much time and resources are available?
– How well is the system known and which data are available?

> **!** There are no universal methods. The most suitable method should be selected for each situation, and neither is there a specific guideline for selecting a method, rather the situation should be globally studied in order to determine the appropriate method.

Furthermore, certain methods are more adequate than others depending on the phase of the project. Aims, as well as the required knowledge and data, vary according to the phase. The following table illustrates the possible choice of suitable methods in engineering projects depending on the project phase and specifies the aims, at what point in the process the risk analysis should be used, and which documents are required as support.

Table 4.2 summarizes the main characteristics of different risk analysis techniques. Not all of them will be developed in this chapter, but this will provide some information about the pros and cons of the mentioned methods.

Furthermore, risk analysis techniques can also be divided into the following two categories: deterministic methods and probabilistic methods.

> **!** A deterministic method assumes that an unwanted event takes place and the physical effects and the damage of the event are calculated (hence, the event's probability of occurrence = 1 in such methods). On the contrary, a probabilistic approach takes both the likelihood and the outcome of an unwanted event into consideration.

Tab. 4.1: Suitable methods for engineering projects.

Project phase	Objectives	When	Required documentation	Analysis method
Conception	– Selection process – Identify unacceptable risks – Input to the design process – Identification of changes to reduce risks	Design evaluation	– Basic documentation	– PHA – Functional analysis – What-if – Brainstorming – Checklist – Project FMECA
Preliminary	– Identify the hazards associated with the process	– Design of the process – Flowcharts completed	– PFD* – Control flowchart – PID* – Process description	– What-if – Checklist – FMECA – HAZOP – FTA
Detailed engineering	– Identify hazards – Identify exploitation difficulties – Provide information for operation processes, design modifications, activation and maintenance	– Detailed engineering	– Final PFD* – Final control logic – Final PID* – Process description – Supplier drawings – Operation information	– Checklist – HAZOP – FMECA
Construction	– Verify conformity to plans and estimates during construction and equipment installation	– During construction	– PID* – Isometric design – Mechanical design – Norms and specifications – Supplier drawings	– Workplace inspection – Leak testing – X-ray sealing inspection – Nondestructive testing

Tab. 4.1 (continued)

Project phase	Objectives	When	Required documentation	Analysis method
Equipment handover	– Verify conformity of installation upon completion	End of construction	– PID* – Isometric design – Mechanical design – Norms and specifications – Supplier drawings	– Physical audit
Preoperation	– Verify that equipment and processes function as described	After equipment handover	– List of preoperational tests – Functional description – Supplier information	– Dynamic tests with inert substances and controlled quantities of reactive substances
Before				
Start-up	– Verify that the production system is safe before introducing chemicals	Before operational tests	– Risk analysis – HAZOP report – Training – List of deficiencies	– Plant inspection – Checklist
Preproduction	– Verify functionality of production system	After preactivation review	– Report of preactivation review – Corrective actions	– Dynamic tests with chemical substances

*PID, piping and instrumentation diagram; PFD, process flow diagram.

> FMECA, failure mode, effect and criticality analysis; HAZOP, hazard and operability studies; ETA, event tree analysis; FTA, fault tree analysis; RADM, risk assessment decision matrix; PHA, preliminary hazard analysis; HRA, human reliability analysis.

A further distinction can be made according to the way in which a method is used:
– A qualitative approach: likelihood and consequences are treated purely in a qualitative way.
– A quantitative approach: both likelihood and consequences are fully quantified.
– A semiquantitative approach (or semiqualitative approach): both likelihood and consequences are quantified to a certain extent (within certain predefined limits).

Tab. 4.2: Summary of the main characteristics of the risk analysis techniques.

Techniques	Procedure	Advantages	Disadvantages
FMECA	Examine whether components or processes can have some failures	Good for equipment, mechanic systems	Little attention given to human factors Does not estimate the cost of failure
HAZOP	Use the nodes of industrial plants to search for deviations from the designed intent	Improve chemical process and operability	Time-consuming Experienced team leader required
ETA	Structuring causes back to the consequences	Quantitative with graphic tool Good for technology performing	Cannot analyze multiples failures
FTA	Structuring consequences back to the causes	Reveals the main causes of failure Give graphical view	Problem of reliability when data are minimized
RADM	Combining probability and severity of hazard. Determining a risk priority number	Graphical tool: Good relationship between probability and severity ranking risks	Inadequate if there are many risks Cannot be used to deduce causes and consequences
PHA	Ask questions about potential failure and fault	Prioritize recommendations	Cannot be used to find details concerning a hazard
What-if	Checks for potential hazards by posing "What-if" questions	Very fast in searching for consequences	Cannot determine causes Very basic
Checklist	Use a list of hazards to record consequences and safety actions	Useful to have an overview of the hazards list	Much time required to find a list of hazards
HRA	Evaluates human–machine interface and carries out task analysis	Can help reduce human errors by improving performance shaping factors	Much time is required if there are many personnel

Methods used according to a purely qualitative approach are generally less complex and are based on the use of arbitrarily definable evaluation standards. Likelihood and potential consequences are described in detail. Such methods are rather simple, easy-to-use and flexible in their use and can be applied to nearly all situations. The disadvantage consists in their dependence on subjective impressions and perceptions and in the fact that not all elements are taken into account.

Methods used following a quantitative approach try to structure events and situations in a systematic manner. A variety of scenarios and cause–consequence events are analyzed, and the relevant parameters are identified. For every cause–consequence event, the likelihood (in the form of a frequency or a probability) and the consequences are quantitatively determined. The result of such a study thus strongly depends on the reliability of the values and data used and the validity of the models used for the method. Such an approach is generally much more complex and more time-consuming than the qualitative approach.

Semiquantitative approaches use detailed descriptions of likelihood and consequences and assign values to them according to the definitions provided. However, compared with a quantitative approach, the values are indicative and have much less of a statistical background.

In Tab. 4.3, an overview is given of the advantages and disadvantages of quantitative and qualitative approaches. Semiquantitative approaches are situated in between.

Tab. 4.3: Comparison between quantitative and qualitative risk analysis approaches.

	Consequence-based	Likelihood-based	Risk-based
Qualitative approach	Estimation of the consequences (no calculations)	Estimation of the likelihood (nonnumerical basis)	Estimation of consequences and likelihood
	Advantages: – Quick – Easy-to-use to compare certain parameters, e.g., type of product Disadvantages: – Likelihood is not taken into account – Provides no idea of the total risk, only an indication of the potential consequences – Cannot be applied to all applications/situations – Possibly not all relevant factors are considered	Advantages: – Quick – Easy-to-use to compare certain parameters, e.g., the number of hazardous transports Disadvantages: – Consequences are not taken into account – Provides no idea of the total risk, only an indication of the potential likelihood – Cannot be applied to all applications/situations – Possibly not all relevant factors are considered	Advantages: – Both consequences and likelihood are taken into account; thus the risk can be compared between specific parameters Disadvantages: – Indicative consideration of risk is usually based on a subjective opinion (expert judgment). This entails the possibility that estimation by other experts may lead to different results. – Cannot be applied to all applications/situations – Mainly for "rough" analyses, and not for detailed analyses

Tab. 4.3 (continued)

	Consequence-based	Likelihood-based	Risk-based
	– Provides only an estimation of the consequences, usually based on a subjective opinion (expert judgment) – this entails the possibility that estimation by other experts may lead to different results	– Provides only an estimation of the likelihood, usually based on a subjective opinion (expert judgment). This entails the possibility that estimation by other experts may lead to different results	
Quantitative approach	– Calculation of the consequences	– Calculation of the likelihood, based on causal and historical data and statistics	– Both the consequences are calculated, and the likelihood is estimated ($\rightarrow$ QRA, see also Section 4.9)
	Advantages: – Provides a scientific idea of the potential consequences based on objective parameters – Results are less dependent on the person carrying out the risk analysis – Can more easily be used to suggest or use acceptance criteria Disadvantages: – More time-consuming than (semi)qualitative method – Certain software and modeling required – Requires experience to handle the software and the models. – Likelihood is not taken into account	Advantages: – Provides a scientific idea of the potential likelihood based on objective parameters – Results are less dependent on the person carrying out the risk analysis – Can be more easily used to suggest or use acceptance criteria Disadvantages: – More time-consuming than in the case of a (semi)qualitative approach – More data and analysis required than in the case of a (semi) qualitative approach – More knowledge and experience required than in the case of a (semi)qualitative approach	Advantages: – Provides the most accurate picture of the risks – May be used to use acceptance criteria (if applied in an adequate way) Disadvantages: – Most time-consuming approach – Certain software and models required – Experience required to handle software and models – A lot of background data and detailed information required

Tab. 4.3 (continued)

	Consequence-based	Likelihood-based	Risk-based
	– Provides no picture of the total risk, only a calculation of the potential consequences – Detailed data required, such as detailed environmental factors, detailed weather conditions, topography and soil condition to obtain an accurate calculation		

4.1.3 General procedure

All risk analyses follow the same general procedure, typically in an iterative process that continues until the system being assessed reaches an accepted/acceptable level of risk. In practice, a risk analysis rarely goes beyond two or three iterations, except for systems that require exceptionally high safety and reliability, such as nuclear power plants or spacecrafts.

The different steps are summarized in Fig. 4.2. The essential starting point is the definition of the system. At this moment, a qualitative risk analysis can be undertaken. This analysis cannot always be quantified by figures or probabilities, often because of the lack of reliable and pertinent data. However, whenever it is useful and feasible, the system will be quantified. The final question that remains to be asked is: "Is the risk acceptable?" If yes, the analysis is complete, and if not, it becomes necessary to modify the concept or the design and repeat the analysis. This is the iteration of the process.

> For both economic and efficiency reasons, it is generally favorable to integrate risk analysis at the project design stage (prospective analysis) (the measures are called "inherent" or "design-based" safety measures) because measures issued afterward ("add-on" safety measures) are often more expensive and mainly palliative (like a bandage on a wooden leg).

Figure 4.3 illustrates the evolution of costs linked to technical and operational modifications imposed by the results of risk analysis as well as their efficiency over time and the development of the project.

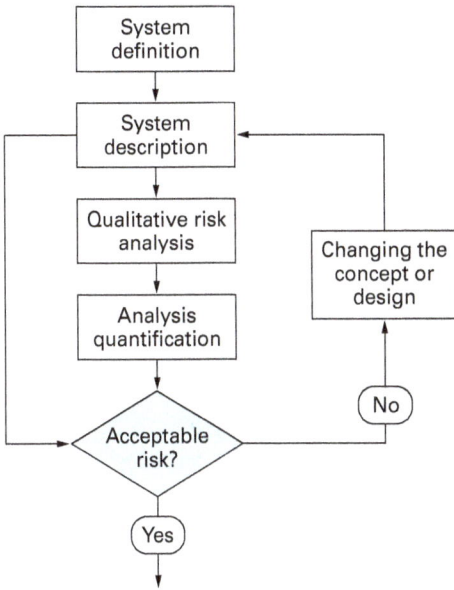

Fig. 4.2: General procedure of risk analysis.

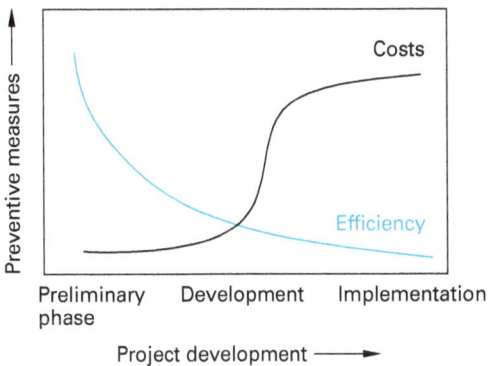

Fig. 4.3: Cost evolution depending on the development phases.

4.1.4 General process for all analysis techniques

Risk assessment is an important step in protecting workers and businesses as well as for complying with the law. It helps to focus on the risks that really matter in the workplace – the ones with the highest potential to cause real harm, whether they are type I or II risks. The general framework to assess risk follows five steps:

- Step 1: Identify the hazards.
- Step 2: Decide who or what might be harmed and how.
- Step 3: Evaluate the risks and decide on precautions.
- Step 4: Record your findings and implement them.
- Step 5: Review your assessment and update it if necessary.

One of the crucial parts is the risk analysis procedure on which the decision-making process will be based. We can draft a global procedure that is valid for most available risk analysis methods. Some homemade techniques might also follow these rules; however, if they are not openly accessible, it is impossible to include them. The general risk analysis procedure could be divided into nine parts:

1. Definition of the system
   - Objective(s) and scope of the study, definition of the system to be studied, identify the elements to be analyzed and subdivide complex processes.
2. Team selection
   - Choose experts according to the process. Important factors are multidisciplinary, expertise and availability.
   - Designate a secretary (generally the future user) or a moderator who will record the identified risks, causes, corrective measures, unsolved problems, etc.
3. Information gathering
   - Collect all the necessary information before the analysis (products and equipment properties and descriptions, operating procedures, technical drawings, process and flow diagrams, schemes, general drawings, process manuals, heat and mass flows, emergency procedures, weather conditions, environment, topography, human reliability, etc.).
   - Identify intended use.
4. Perform the analysis with the adequately chosen method.
   - Identify and list the elements to assess, make risk analysis meetings, save the results of the analysis in table form or an appropriate document, control the evaluation table by a system engineer and follow the methodology without introducing "feelings" statements.
5. Recommend corrective actions and an action plan:
   - Define preventive and corrective solutions.
   - Recommend actions to reduce unacceptable risks.
   - Assign responsibility and schedule for corrective actions.
6. Monitor the implementation of the solution.
   - Regularly monitor the implementation of corrective measures.
   - Update the analysis in case of major changes.
7. Record hazards
   - Record identified hazards in the safety quality assurance system (if any).
   - Establish record-keeping.

–   Establish documents of the complete analysis with diagrams, drawings, tables and processes.
–   Update information according to the completion of corrective measures.
8.   Forecast to update the system
–   It must evolve to reflect changes in raw materials, formulation (recipe), market, habits or consumer demands, new hazards, scientific information or inefficiency.
–   It must provide, at the outset, why, when and how the system will be reviewed.
9.   Continuous monitoring and follow-up
–   Once the analysis is completed, the story does not stop there. As time is a factor of change, iterations of the procedure must be performed when (sometimes minor, certainly major) changes happen.

4.2 SWOT

SWOT stands for strengths, weaknesses, opportunities and threats. Strictly speaking, it is not a risk analysis technique. It is a widely used framework for organizing and using data and information gained from a situation analysis. It encompasses both internal and external environments. It is one of the most effective tools in the analysis of environmental data and information. However, it allows for assessing strategic risks as the other techniques do not.

A SWOT analysis systematically and methodically identifies internal strengths and weaknesses, aligning them with external opportunities or threats. This technique is used in business or strategic planning to summarize the main elements of strategic environments, which are classified into three categories:
–   Internal environment
–   Industry environment
–   Macroenvironment

Sometimes the last two are combined and are called "the external environment." The method of SWOT analysis is to take the information from an environmental analysis and separate it into internal (strengths and weaknesses) and external issues (opportunities and threats). Once this is completed, a SWOT analysis determines what may assist the firm in accomplishing its objectives and what obstacles must be overcome or minimized to achieve desired results.

Factors affecting an organization can usually be classified as shown in Fig. 4.4:

SWOT is very often represented as a matrix to allow comparison of all the sensitive aspects in one shot (Fig. 4.5).

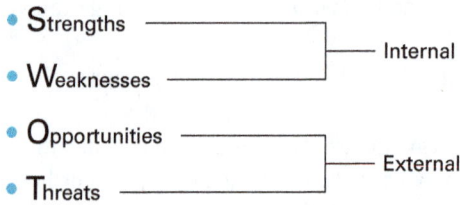

Fig. 4.4: Internal and external environment.

Effectively, one has to match each component with another. For example, match the internal strengths with external opportunities and list the resulting strengths/opportunities strategies in the matrix chart. The four strategy types are:

–   S–O strategies pursue opportunities that match the company's strengths. These are the best strategies to employ, but many firms are not in a position to do so. Companies will generally pursue one or several of the other three strategies first to be able to apply S–O strategies.
–   W–O strategies overcome weaknesses to pursue opportunities. Match internal weaknesses with external opportunities and list the resulting W–O strategies.
–   S–T strategies identify ways that the company can use its strengths to reduce its vulnerability to external threats. Match internal strengths with external threats and list the resulting S–T strategies.
–   W–T strategies establish a defensive plan to prevent the firm's weaknesses from making it susceptible to external threats. Match the internal weaknesses with external threats and record the resulting W–T strategies.

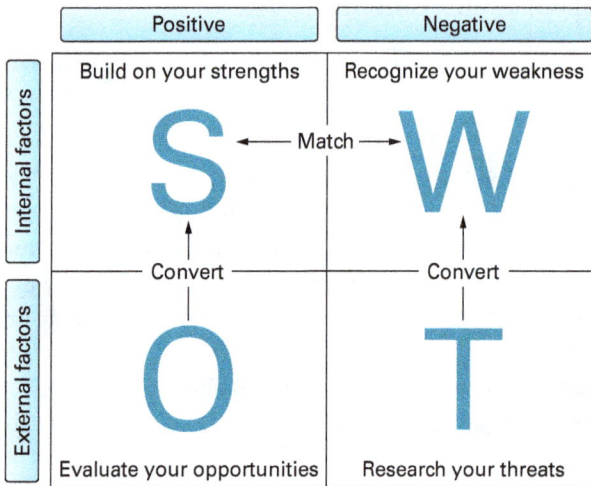

Fig. 4.5: The SWOT matrix.

In order to fill this matrix, several questions must be answered related to the four aspects of SWOT, as depicted in Fig. 4.6 (some illustrative questions are presented).

As stated in the beginning, a SWOT analysis will summarize your strategic analysis. It will be an integral part of the four-step planning process as illustrated in Fig. 4.7.

> In conclusion, a SWOT analysis is a useful technique for understanding your strengths and weaknesses and for identifying both the opportunities open to you and the threats you face. Used in a business context, a SWOT analysis helps you carve a sustainable niche in your market. Used in a personal context, it helps you develop your career in a way that takes best advantage of your talents, abilities and opportunities.

	Positive	Negative
Internal factors	• Which strengths are unique to the team? • What are we good at doing? • What are the things that had gone well?	• What should be done better in the future? • What knowledge do we lack? • Which skills do we lack? • What system do we need to change?
External factors	• What are the key success enablers? • Which additional services can we offer? • What new market should we investigate?	• Barriers to progress • What are the possible impacts of what competitors are doing? • Which regulatory issue might cause us concern?

Fig. 4.6: The SWOT sample questions.

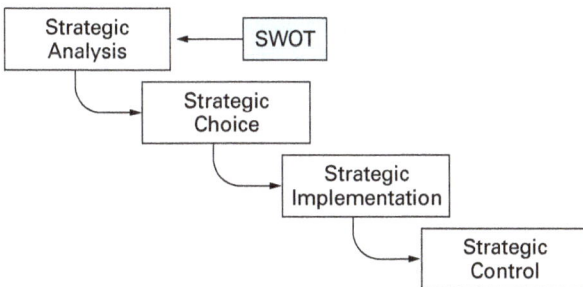

Strategic Analysis ← SWOT
↓
Strategic Choice
↓
Strategic Implementation
↓
Strategic Control

Fig. 4.7: A link to strategy.

4.3 Preliminary hazard analysis

Preliminary hazard analysis (PHA) is a relatively simple technique to implement, and it allows for a quick identification of the main risks of a system. PHA was instituted and promulgated by the developers of the US Air Force standard practice for system safety (MIL-STD-882) since 1969 [3]. PHA is often used in the early life of the processes (conceptual phase or research and development phase) [4, 5] to affect the design for safety as early as possible. The technique is a safety analysis tool for identifying hazards, their associated causal factors, effects, level of risk and mitigating design measures when detailed design information is not available. To perform the PHA analysis, the system safety analyst must have three information inputs: design knowledge, hazard knowledge and a preliminary hazard list (collection of identified hazards). The output of a PHA includes identified and suspected hazards, hazard causal factors, the resulting mishap effect, mishap risk, safety critical function, and top-level mishaps. The advantages of PHA are that it is easily and quickly performed; it is comparatively inexpensive in providing meaningful results; it is a methodological analysis technique; most of the system hazards are identified, and an indication of system risk is provided. While there are no major disadvantages to the PHA, there is sometimes an (improper) tendency to have it as the only applied analysis technique [3].

A PHA is applied in two ways:

1. Alone, as risk analysis for systems with simple or easily identifiable hazards and not complex accidental processes.
2. In combination with other methods, a PHA is seen as a preliminary risk study to prepare the complex or poorly defined case. In this sense, it is mainly used in the early design phase of a project.

PHA is an identification and analysis technique of hazard frequency that can be used in preliminary conception phases in order to identify hazards and evaluate their criticality. However, usually, PHA is not limited to the risk evaluation phase but also gives a certain directive in order to master these risks, making it a management method rather than a risk analysis.

PHA aims at listing all sources of hazards present in the system: hazardous materials, equipment or components that could present a hazard, and risky processes or procedures. Each of these elements will be assigned to one or more adverse events, possible causes, as well as potential compensatory measures. Results are generally presented in a table. It requires, as a first step, identifying the hazardous parts of the installation/ process. These harmful elements refer most often to

– hazardous substances and preparations, whether in the form of raw materials, finished goods, utilities, etc.;
– hazardous equipment such as storage facilities, reception areas, shipping, reactors and energy supply systems (e.g., boilers); and
– hazardous operations associated with the process.

An example of a PHA applied to a storage tank is presented in Tab. 4.4.

The main advantage of the PHA is to enable a relatively quick review of hazardous situations in the facilities. Relatively economical in terms of time and resources, it does not require a very detailed level of description of the studied system (generally implemented at the design stage). This method is a simplified form of the FMEA, but it is very limited in failures propagation and consequences of multiple failures. Only the direct consequences of failure are known. Moreover, this method does not require modeling of the system, making it difficult to ensure a systematic approach, and therefore it is based on the knowledge of the team conducting the analysis. PHA must therefore really be considered as a preliminary method or a sufficient method for systems simple enough to be appropriately studied without modeling.

PHA is a qualitative analysis that is performed to
- identify all potential hazards and accidental events that may lead to an accident;
- rank the identified accidental events according to their severity;
- identify required hazard controls and follow-up actions.

Tab. 4.4: Example of PHA analysis on the storage of flammable gas under pressure.

Element	Hazard	Hazardous event	Causes	Consequences	Measures
Tank	Heat stress (fire outside the tank)	Explosion of the tank and important release of gases	Presence of combustible elements near the tank	– Fire – Property damage – Casualties	– Changing logistics storage – Move individual hazards away
Tank	Mechanical impact against the shell of the tank	Gas release	– Accident with a crossing vehicle – Intentional damage	– Fire – Property damage – Casualties	– Inspection program – Continuous monitoring of air quality
Tank	Weakening of the tank shell	Gas release, explosion of the tank and important release of gases	– Corrosion – Fatigue (crack) – Not well-sized tank (does not withstand the pressure imposed)	– Fire – Property damage – Casualties	– Inspection program – Verification of design
Valve	Unexpected opening	Gas release	– Valve or control system failed – Error during routine maintenance	Gas release	– Inspection program – Continuous monitoring of air quality

4.4 Checklist

> **!** Several variants of PHA are used, sometimes under different names, such as rapid risk ranking or hazard identification (HAZID).

A checklist analysis uses a written list of items or procedural steps to verify the status of a system [5]. Traditional checklists vary widely in their level of detail and are frequently used to indicate compliance with standards and practices. The checklist analysis approach is user-friendly and can be applied at any stage in the lifetime of the process. Even though it will not replace a detailed risk analysis process, a checklist is a good starting point and a highly cost-effective method for common hazards.

The checklist is the easiest to implement but remains very limited in what it can offer. Either it is too generic and could be difficult to apply to the system, or it could be considered appropriate to the system, which implies in this case that hazards have been previously identified.

The checklist uses a standard form containing a series of specific questions related to the potential hazards of the considered process.

4.4.1 Methodology

The checklist involves the use of a pre-established form containing a series of specific questions related to the potential hazards of the considered process. A control list, or checklist, is also an operation involving the methodical verification of the necessary steps that are required for the process to run with maximum safety. An example would be the use of such a procedure in aviation. It becomes a safety and security procedure that methodically verifies whether the plane is ready to undergo the next phase of flight. Such operations are usually carried out vocally or by ticking a written procedure list. The principle of hazard determination and action planning with the use of a checklist is presented in Fig. 4.8.

Its principle is based on the operating procedure structured in numbered steps, each corresponding to a simple operation. Every step of the process is evaluated in terms of failures that could occur and possible deviations from the normal operating procedure. This widely used method can be adapted to many situations. Several organizations have developed checklists for different activities (e.g., Suva in Switzerland, www.suva.ch).

The methodology takes place in three steps:
1. Each element for a given step, or whose failure represents a risk, will be marked with a cross, indicating a potential hazard.

Identify hazards	1	What are the risks to **health and safety** at the workplace?
		The checklists can help you **identify hazards**.
Take the **necessary** measures	2	Plan and implement the **appropriate safety** measures.
		Checklists and other publications provide appropriate safety measures.
Act methodically	3	**It is important to act methodically!**
		Proceed methodically to ensure safety in the long-term business.

Fig. 4.8: Principle of hazard identification and action planning using a checklist.

2. Each marked hazard will be defined and described in detail as a script, so as to assess its likelihood of occurrence and severity.
3. Measures will be implemented in order to reduce risks of intolerable situations.

4.4.2 Example

Let us assume that we have divided a process into several steps, each identified with a reference number. In the table (see Tabs. 4.5 and 4.6) we identify by a cross mark the process steps that are sensitive to the different failure possibilities.

4.4.2.1 Step 1a: critical difference, effect of energy failures

In the example displayed in Tab. 4.5, point 12 (whether it refers to a stage, a unit operation or a process step) is liable to risk in case of electricity or compressed air regulation failures.

We then move forward by implementing the same procedure, but this time it is initiated by possible deviations from the operating procedure. The question to be answered is, "Which process steps might be affected when an operating parameter deviates from a desired value or function?" One can observe in Tab. 4.6 that step no. 18 is sensitive to ventilation as no. 75 is liable to risk when the set temperature is no longer what is expected.

Tab. 4.5: Process steps sensitive to energy failures.

Unit operation/step number	12	43	54	68	75
Electricity	*				*
Water		*			
Vapor			*		
Brine/ice					
Nitrogen					
Compressed air (regulation)	*				
Compressed air (command)					
Vacuum				*	
Ventilation			*		
Absorption					

4.4.2.2 Step 1b: critical difference, deviation from the operating procedure

Tab. 4.6: Process steps sensitive to operating procedure deviations.

Unit operation/step number	14	18	55	74	75
Cleaning	*				
Plant inspection					
Discharge					
Equipment ventilation		*			
Charges, dosage					
Quantity, flow rate	*				
Operation succession			*		
Speed of addition	*				
Product mismatch					
Electrostatic charges					
Temperature					*
Pressure	*				
pH				*	
Heating/cooling					
Speed of agitation					
Reaction with coolant					
Catalyst, inhibitor					
Connecting lines, valves					
Process interruption			*		

4.4.2.3 Step 2: establish the risk catalog

For the identified critical steps (defined by the hazard description) one may evaluate the risk by identifying the likelihood of occurrence and the severity. Then, mitigation measures have to be identified and proposed to reduce the risk. The first row of the table (see Tab. 4.7) helps to identify the process, product, plant building, the author, date and revision number (when successive analysis was performed). In the example, we note that step 75 reveals a hazard that may lead to the overheating of the reactor R-321. It has been marked "low" for the occurrence and "high" for the severity. The following mitigation measure was proposed: setting a temperature control with a high alarm level and setting the stoppage of the steam feed F-32. The authors of this example then evaluate both occurrence and severity as "low" when talking about residual risk (the risk after corrective measures have been taken).

4.4.2.4 Step 3: risk mitigation

Measures to reduce the risk for each hazard will be clearly defined so that their descriptions can be used as specifications. We then evaluate the progress by assessing the residual risk.

Tab. 4.7: Extract of an example of a risk catalog.

Product:	No. identity:		Process:		
Plant:	Bldg:		Date:		
Author:				Revision:	
Hazard description (causes, consequences)	Evaluation		Description of (technical) risk mitigation measures	Evaluation residual risk	
	O	S		O	S
Overheating of reactor R-321 at step 75 (synthesis step)	F	H	Temperature control, alarm high, cutting feed stream F-32	F	F
No more control system at step 12, loading the reactor, in the absence of electric current	F	H	UPS system for automatic control	F	M
Failure of ventilation at step 18, product loading TX24	M	H	No technical measures possible	M	H

O, likelihood of occurrence; S, severity; F, low; M, medium; H, high.

4.4.3 Conclusion

Although this qualitative method is limited in discovering hazards that have not already been identified, it can still prove to be efficient when applied to a new system with similar functions to an existing one for which the checklist was established. Furthermore, the checklist can be easily used by relatively inexperienced personnel, provided that the list has been established by experts. In this case, the checklist allows the user, e.g., to confirm that the process design does not exhibit safety vulnerabilities. The checklist is not adapted to new technologies whose hazards are insufficiently recognized, identified and understood.

> ❗ Checklist analysis is a systematic evaluation against pre-established criteria in the form of one or more checklists. It is a systematic approach built on the historical knowledge included in checklist questions. It is used for high-level or detailed analysis including root cause analysis. It is applicable to any activity or system including equipment issues and human factors issues. It is generally performed by an individual trained to understand the checklist questions. It generates qualitative lists of conformance and nonconformance determinations, with recommendations for correcting nonconformances. The quality of evaluation is determined primarily by the experience of people creating the checklists and the training of the checklist users.

4.5 HAZOP

The HAZOP study was developed by Imperial Chemical Industries (ICI) in the late 1960s, and the earliest work from ICI Mond Division in the northwest of England was published in 1968 [6]. The HAZOP system as we know it today was published by Lawley from ICI Petrochemicals Division in the northeast of England in 1974. The method is now standardized through the IEC 61882:2016 standard. After the Flixborough disaster in 1974 [7], this technique became widely used.

HAZOP is an organized, methodological technique for analyzing hazards and operational concerns of a system, often used in chemical industries [8–12]. According to HAZOP, normal and standard operations are safe, and hazards occur only when there is a deviation from the normal operation [3].

The standard CEI 61882 defines the objectives of the original HAZOP method as follows:

- Identification of potential hazards in the system. The hazard can be limited to the immediate proximity of the system or spread its effects beyond the system such as environmental hazards.
- Identification of potential exploitation problems of the system, particularly their causes and identification of functional perturbations and deviations in production liable to lead to the production of noncomplying products.

With the introduction of the directive Seveso II in Europe and new requirements regarding the prevention of industrial risks, the original HAZOP method became insufficient for the analysis of major risks. A phase of risk evaluation was added to the original method, and the previously purely qualitative HAZOP method became semiquantitative, contributing to the improvement of the knowledge of risk and thus the safety of facilities.

Input	HAZOP process	Output

• P&ID/PFD
• Process
• Team leader
• Team members
• Directives,
  methodology
	1. Establish HAZOP plan
	2. Select team members
	3. Identify system elements
	4. Choose guide/key words
	5. Perform the analysis
	6. Documenting the HAZOP
		• Hazards
		• Risks
		• Correctives
		  measures/actions

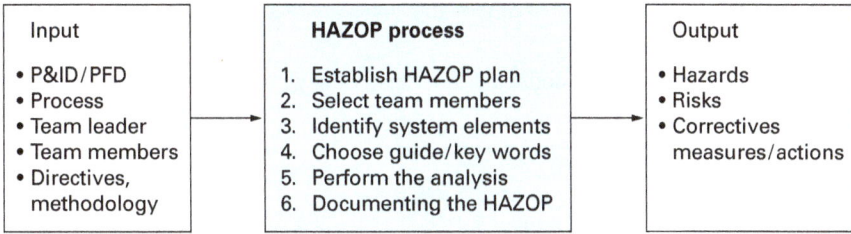

Fig. 4.9: Principle of the HAZOP process.

4.5.1 HAZOP inputs and outputs

HAZOP's inputs are design data, guidelines, and process descriptions (see Fig. 4.9). In order to perform systematic searches for conceivable departures from the design intent [normally presented in a piping and instrumentation diagram (P&ID)], the technique uses guide keywords (*more, none, less*, etc.) combined with process/system conditions such as speed, flow, pressure and temperature. P&IDs are scrutinized vessel by vessel, pipe by pipe, to ensure that all potentially hazardous situations have been taken into account. Once the hazard resulting from potential deviation in design operation is identified, a search backward is performed to find possible causes and forward to find possible consequences (middle-bottom-up technique).

HAZOP's outputs are, therefore, hazard, cause, consequence and corrective measures (see Fig. 4.9).

4.5.2 HAZOP process

HAZOP allows, in an inductive process, for the systematic analysis of the deviations of the plant or process parameters and the prediction of their potential consequences. This method is particularly useful for the examination of thermohydraulic systems, for which parameters such as speed, temperature, pressure, level and concentration are particularly important for the safety of the installation.

By its nature, this method requires particular consideration of schemes and plans of fluid flow patterns, or P&ID, and process flow diagrams (PFD). However, the method can be applied and elaborated for virtually any system.

Based on a detailed description of the process, each part of the facility will be analyzed for possible deviations from normal operation. Causes, effects or consequences, and remedies are sought. It requires a team of experts, each with special skills and competencies, and a moderator or facilitator controlling the method.

The method could be expressed in five steps (see also Fig. 4.10):

1. Subdivision of the process
   - Divide the process into nodes with their auxiliary instruments.
2. Expected function
   - Scope of the node: intention description.
3. Search for deviations
   - Each node and ancillary facility will be reviewed in light of possible deviations from the targets using the guide's word.
   - Deviation = guide/keyword + parameter.
   - Identify the possible causes and consequences of those deviations.
   - Repeat the search for deviations at each node. The process then stops when all the nodes and subunits have been studied.
4. Risk assessment
   - Assessing the likelihood of occurrence and severity.
   - Comprehensive analysis of all conceivable realistic deviations.
5. Measures
   - For consequences deemed too serious, measures must be proposed to eliminate the cause or to reduce the severity or impact.

In step 3, for each constituent part of the system considered (line or mesh), a deviation generation (conceptual) is performed systematically by the conjunction of guide/keywords (7 technical + 4 temporal), see Tab. 4.8, and parameters associated with the system studied (see Tab. 4.9). The parameters commonly encountered are temperature, pressure, flow, concentration and also the time or the operations to perform.

How to conduct the analysis?

1. Initially, choose a line of process. It generally covers equipment and connections, all performing a function in the process identified in the functional description.
2. Choose an operating parameter.
3. Identify a guide/keyword and generate a deviation.
4. Verify that the deviation is credible. If yes, proceed to step 5; otherwise, return to step 3.
5. Identify the causes and potential consequences of this deviation.
6. Examine ways to detect this drift as well as those provided to prevent its occurrence or mitigate its effects.
7. Propose, where appropriate, recommendations and improvements (see Tab. 4.10).
8. Choose a new guide/keyword for the same parameter and return to step 3.
9. When all guides/keywords have been considered, choose another operating parameter and go back to step 2.
10. When all operating phases have been studied, choose another process line and go back to step 1.

4.5.3 Example

Results of the analysis are often represented in Tab. 4.11, indicating the function or the phase evaluated (0), the keyword used (1), the observed deviation (2), identification of the possible causes (3), the evaluation of the occurrence and severity (4 and 6), the consequences (5), the corrective measures to be applied (7), the re-evaluation of the occurrence and probability (8 and 9), who is responsible for applying the measures (10) and the deadline for implementation (11).

Let us take an example that deals with the water supply to a cooling system. The results of the analysis are presented in Tab. 4.12, revealing that the parameters of fluid and electricity are the most critical. We intentionally used another representation, indicating that there is no universal result table. It is important that all information is present.

4.5.4 Conclusions

The HAZOP method is a structured and systematic approach that focuses on simple parameter deviations within a system, thereby avoiding the need – unlike FMEA – to consider every possible failure mode of each system component. However, HAZOP has limitations, such as difficulty in analyzing events arising from combinations of multiple simultaneous failures. Additionally, assigning a suitable guideword or keyword to a specific system section can be challenging, complicating the thorough identification of potential causes of deviations. The systems analyzed are often made up of interconnected parts, meaning that a deviation in one line or segment can impact adjacent sections. Although it is theoretically possible to trace the effects of a deviation across system parts, this process can quickly become complex. Finally, HAZOP also struggles to effectively address events caused by multiple simultaneous failures.

HAZOP has several advantages. It is easily learned and clearly structured. It provides rigor for focusing on system elements and hazards; it is a team effort with many viewpoints. The introduction of modeling could help, e.g., by taking into account the discrete components in the system; their connection and behavior over time [13, 14].

HAZOP has also some inconveniences:
- HAZOP analysis focuses on single events rather than combinations of possible events.
- It is based on guide words, allowing us to overlook some hazards not directly related to them.
- HAZOP analysis training is essential for optimal results, especially for the facilitator.
- HAZOP analysis can be time-consuming and thus expensive [3]. It is not uncommon for over 25 man-days' work to be necessary to perform a HAZOP study. Of course, the time depends on the scope (e.g., complexity of an installation) of a study.

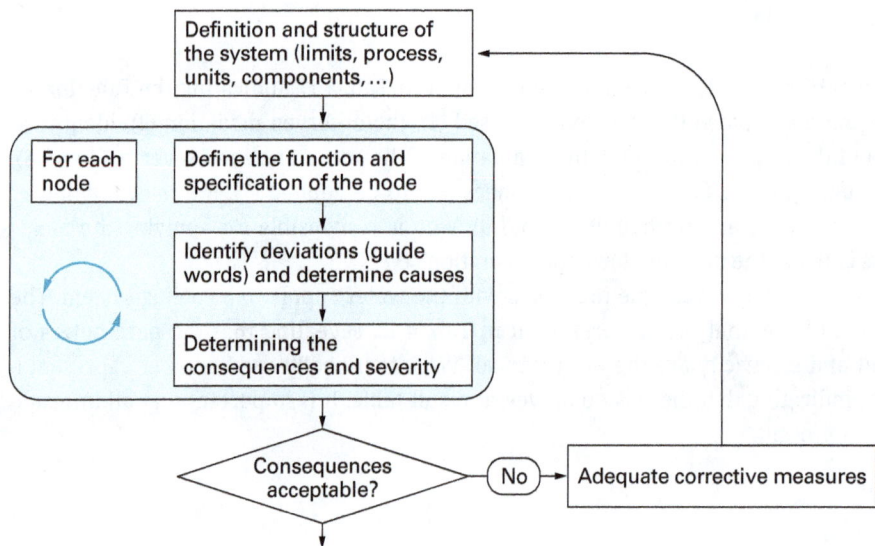

Fig. 4.10: Methodology of HAZOP.

Tab. 4.8: HAZOP guidewords or keywords.

Keywords	Signification	Commentary	Examples
No or not	No part of the intention is fulfilled	The purpose or function is not fulfilled at all, not even partially	– No agitation – No flow
More	Overrun or increase quantitatively	Refers to the quantities and properties (T, P), but also activities (heating, reaction)	– Higher temperature – Too much product
Less	Insufficient or quantitative reduction		– Lower flow rate than expected – Less agitation
As well as	Qualitative increase	– The intent (design and procedure) is performed with additional activity – Concomitant adverse effect	Heating started at the same time as the addition of reagent A
Part of	Qualitative modification/ diminution	Only part of the intention is realized	Only part of the reagent is added
Reverse	The logical opposite of the intention	Reversal of the activity or sequence	– Liquid flows in the opposite direction – It heats instead of cooling

Tab. 4.8 (continued)

Keywords	Signification	Commentary	Examples
Other than	Total substitution	Result different from that of intention	Reagent A is loaded in place of B
Earlier than	On the time clock	The action takes place before or after a defined time	We started heating 15 min before the deadline
Later than	On the time clock		The reaction, taking place over 2 h, has crossed the deadline by 1 h and 45 min
Before	On the sequence or order	Action is taken before or after the defined sequence	A was loaded before B
Later	On the sequence or order		It was cooled after stirring

Tab. 4.9: Example of operating parameters.

Measurable physical quantities		Operations		Actions	Functions–situations
Temperature	pH	Loading	Control	Start-up	Protection
Pressure	Intensity	Dilution	Separation	Sampling	Utility default
Level	Speed	Heating	Cooling	Stop	Freezing
Flow rate	Frequency	Stirring	Transfer	Isolate	Spill
Concentration	Amount	Mixing	Maintenance	Purge	Earthquake
Contamination	Time	Reaction	Corrosion	Close	Malevolence

Tab. 4.10: Example of safety barriers.

Safety barriers		Definition	Example
Technical	Passive safety devices	Unitary elements aim to fulfill a safety function without an external energy supply from the system to which they belong and without the involvement of any mechanical system	– Holding tank/tray – Rupture disk
	Active safety devices	Items not passively designed to perform a safety function without an external energy supply system to which they belong	– Safety valve – Excess flow valve

Tab. 4.10 (continued)

Safety barriers		Definition	Example
	Safety instrumented systems	Combination of sensors, processing units, and terminal elements aim to fulfill a function or a subsafety function	– Measuring elements that control a valve or switch power
Organizational		Human activities (operations) that do not involve technical safety barriers to prevent the occurrence of an accident	– Emergency plan – Containment
Systems with manual action		Interface between a technical barrier and human activity to carry out a safety function	– Pressing an emergency button – Low-flow alarm, followed by manual closing of a safety valve

Tab. 4.11: Example of HAZOP analysis table.

Phase, function: (detailed description, notated)			
Guide words:	M4: as well as	M8: later than	Level of P and G: (L)ow, (M)iddle, (H)igh
(examples)	M5: part of	M9: earlier than	P1, G1 occurrence and severity before measures
M1: no or not	M6: reverse	M10: before	
M2: less	M7: other than	M11: later	
M3: more			P2, G2, after measures

Guide word	Deviation	Possible cause	P1	Consequences	G1	Measures	P2	G2	Who	When
1	2	3	4	5	6	7	8	9	10	11

! Hazard and operability analysis is a structured and systematic technique for system examination and risk management. In particular, it is often used as a technique for identifying potential hazards in a system and identifying operability problems that are likely to lead to nonconforming products. It is based on a theory that assumes risk events are caused by deviations from design or operating intentions. Identification of such deviations is facilitated by using sets of "guide words" as a systematic list of deviation perspectives.

What-if, an inductive method similar to HAZOP (although much less systematic and more intuitive), is a brainstorming approach in which a group of experienced people familiar with the subject process raise the question "What-if?" instead of using keywords when examining the P&ID and voice concerns about possible undesired events.

4.6 FMECA

The FMEA was developed by the US military in 1949 and further encouraged in the 1960s in the aerospace industry. The Ford Motor Company reintroduced it in the late 1970s for safety and regulatory considerations and used it effectively for production and design improvement. Nowadays [15–18], this method is often used in industries producing machinery, motorcars, mechanical and electronic components [19]. A more detailed version of FMEA is called FMECA, which adds prioritization of actions to be taken based on a risk score. The method is now standardized through the IEC 60812:2018 standard.

The terms used in FMECA are as follows (see also Chapter 3):
- Reliability: The ability to present no default for a specified period under specified conditions.
- Availability: The ability to provide a given function under given conditions at a given time.
- Maintainability: The ability to return to service within a given period under given conditions.
- Safety: The ability to present no hazard to people, property and the environment.
- Failure: The termination of the ability of an entity to perform a required function. An element or state action no longer fulfills its original function, whether expected or anticipated.
- Failure mode: A potential failure mode describes the manner in which a product or process could fail in fulfilling its primary function.
- Failure cause: The process or mechanism responsible for the initiation of a failure mode.
- Failure consequence: The result that the failure mode induces on the operation or function.
- Detection: An assessment of the likelihood that the controls (design and process) will detect the failure cause or the failure itself.
- Control: Controls (design and process) are the mechanisms preventing the cause of a failure from occurring.

A failure occurs between a cause and an effect. A single cause may have multiple effects. A combination of causes could lead to single or multiple effects. Causes can themselves have causes, and effects may have subsequent following effects (see Fig. 4.11).

FMEA is an inductive method based on the study of elementary failures of composed systems in order to deduce what they can result in and, therefore, what situations can be expected due to these failures. FMECA includes an additional evaluation, which studies the severity of these situations. It includes identifying and evaluating the impact of elementary failures on the corresponding system, functions and environment. FMEA and FMECA are so commonly used and well-known that they have

Tab. 4.12: Partial results of a HAZOP analysis.

No.	Object	Function	Parameter	Key word	Consequence	Cause	Hazard	Risk P/G	Recommendation	Risk 2 P/G	Who	When
1	Line	Bring water to the system	Fluid	No	Loss of cooling in the pump	Line rupture	Damaged pump	LM	–	LM	–	–
2				More	Pressure rise in the line	No pressure regulation	Line rupture	LM	Add a safety valve to the loop	LL	TJ	2012.10
3				Less	Not enough cooling capacity in the pump	Leakage at the pipe and fittings	Damaged pump	MH	Periodic control of fittings	LM	TJ	Bimonthly
4	Electrical supply	Supply power to motor M23	Electricity	No	Loss of power to the pump	Short circuit, power failure	Loss of control	LH	Backup power	LL	JK	2013.01

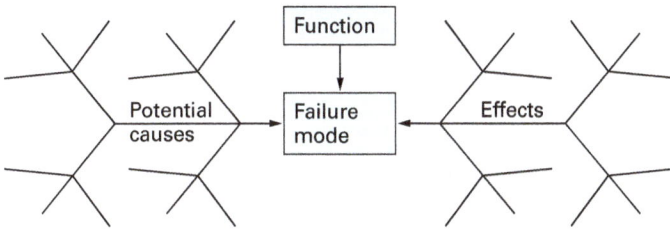

Fig. 4.11: Cause–effect failure.

practically become the safety symbol of functional safety. As FMEA is included in FMECA, we will refer to both as FMECA in this chapter.

4.6.1 FMECA inputs and outputs

FMECA is a tool for inductive and quantitative analysis based on the study of possible failure modes of components and functions constituting a system. It is a powerful method suitable for complex or innovative systems with substantial implementation. It is well-adapted to systematically and thoroughly identify all undesirable situations that may lead to individual failures.

FMEA evaluates the effects of potential failure modes of subsystems, assemblies, components and functions with design, functional diagrams and failure knowledge as input. Outputs of the technique are failure modes, consequences, reliability prediction, hazards and risks, and a critical item list (see also Fig. 4.12).

4.6.2 FMECA process

The global FMECA process is performed according to the following procedures: (i) system definition, (ii) FMECA plan, (iii) team selection, (iv) collecting information, (v) perform FMECA, (vi) define corrective mitigation measures, (vii) monitor the implementation, (viii) register hazards and (ix) analysis documentation.

The implementation of the FMECA methodology follows three steps:
1. Develop a hierarchical model of the system in question.
2. Identify failure modes associated with each base unit model.
3. Identify the propagation of failures through the model and determine the final consequences.

The analysis and evaluation from the functional diagram of failure modes can be summarized as follows (see Fig. 4.13):

Input	FMECA process	Output
• Construction and operating drawings • Potential failures • Types of failure mode • Frequency of failures	1. Assess the process, the concept 2. Identify potential failure modes 3. Evaluate the effect of each identified failure mode 4. Documenting the analysis on a table	• Failure mode • Consequences • System reliability • Hazard and risks • List of criticalities

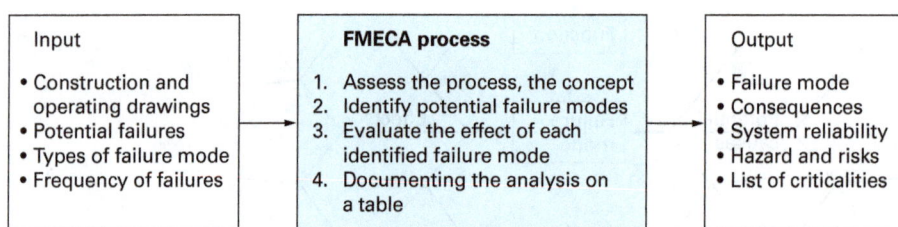

Fig. 4.12: Principle of the FMECA process.

- Look systematically for potential failure modes of components
- Accurately describe the effect on the client
- List the possible causes of the failure mode
- List the validation of systems planned
- Evaluate the modes and calculate the criticality index
- Prioritize modes

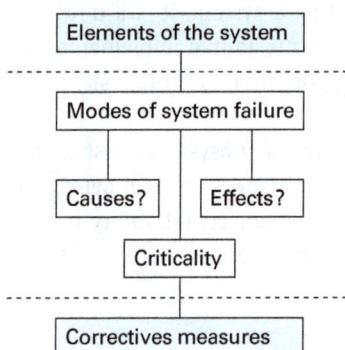

Fig. 4.13: Principle of the FMECA process.

4.6.2.1 Step 1: elaboration of the hierarchical model, functional analysis

It is important to note that the construction of the hierarchical structure is defined top-down, while the propagation of failures is achieved as inductive bottom-up (Fig. 4.14).

This first step is, therefore, the decomposition of the system into smaller elements (subsystems). What should we obtain at this stage?

- The decomposed elements must be simple in order to completely identify their failure modes.
- The group must consist of an expert for each element who is capable of describing its nominal function and its failure modes in detail.
- The whole system must be covered by the decomposition.

It is not important for the decomposition to be completely separated, but it is preferable for the various levels to be close enough to ensure that each component of the system is present in the decomposition at least once (ideally only once). The level of decomposition is the most important decision in this step. The main purpose is to understand why and how the studied product or process operates:
–   Product decomposition into simple elements or subsets.
–   The organizational system arranges all the elements of the system into different hierarchical levels from the top.

As an illustration, the decomposition of a grinder is shown in Fig. 4.15.

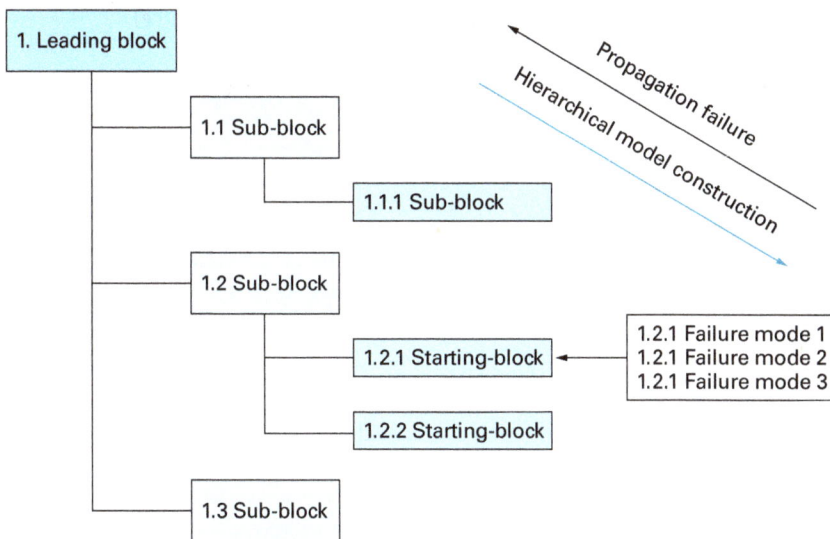

Fig. 4.14: Hierarchical model.

This first step, therefore, results in a list of elementary functions or a list of components of elementary subsystems containing each function, subsystem or component affected by the different failure modes it may encounter. One must not overlook the fact that the failure modes of the components do not only depend on the component itself but also on the conditions in which they are implemented. This is the reason why the real experience of the enterprise or profession is always more valuable than data available in other places around the world. It is also important not to forget, neglect or discard certain failure modes because of their apparent weak frequency or probability of occurrence.

4.6.2.2 Step 2: failure mode determination

The second step involves imagining and describing what would happen to the system when a studied failure mode appears. This is the stage where one may regret not having sufficiently decomposed the system. If each failure mode leads to a cascade of effects depending on various parameters (e.g., depending on the working stage), the procedure would be difficult to carry out, and it may be better to go one step back, define several FMECAs according to the same parameters and re-evaluate them one by one.

There are five generic failure modes in Fig. 4.16, ranging from function loss, undesired operation, refusal to stop or to start, to downgraded operation.

The following must be identified for each failure mode:
–   Its causes (weighted based on probability or likelihood of occurrence).
–   Its effects (weighted based on severity).
–   Measures to counter or limit the effects of the failure (weighted based on the probability of nondetection).

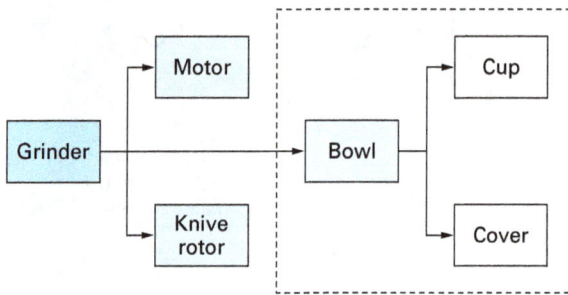

Fig. 4.15: Functional analysis of a grinder.

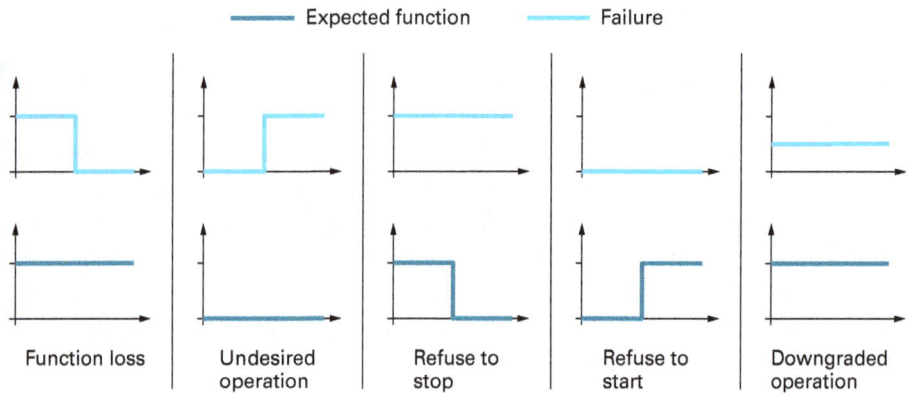

Fig. 4.16: Generic failure modes (top: the failure and bottom: the expected function).

One must describe the effects seen from outside of the system, the effects on the ac-
complishment of the function in the log, as described by an external functional analysis; and also the effects on the system's safety or the environment that are not documented anywhere else. The concept is then based on the following questions:

– What can fail?
– How does it fail?
– How frequently will it fail?
– What are the effects of the failure?
– What are the reliability/safety consequences of the failure?

FMECA is a qualitative and quantitative method based on five questions (Fig. 4.17):

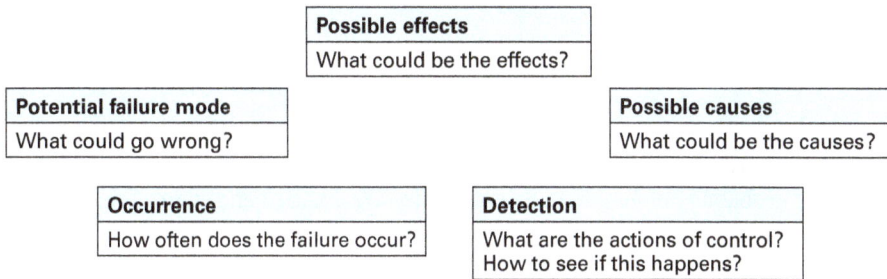

Possible effects
What could be the effects?

Potential failure mode		Possible causes
What could go wrong?		What could be the causes?

Occurrence	Detection
How often does the failure occur?	What are the actions of control? How to see if this happens?

Fig. 4.17: Five key questions of the FMECA analysis.

4.6.2.3 Step 3: the criticality determination

The criticality or risk priority number is an indicator of the importance of the identified impacts, and therefore a synthesis of several parameters that signify the importance of failure modes. The choice of parameters depends on the subjects being analyzed and the results being sought. It is important to explain what the criticality is
composed of (see Fig. 4.18). A common approach is to undertake a synthesis of the
severity of the consequences, the occurrence of the failure mode and its possibility of
detection as follows:

$$\text{Criticality} = \text{Severity} \times \text{occurrence} \times \text{nondetection}$$

This is the synthesis of questioning, and it allows for a prioritization of concerns. The risk
priority index is used to help identify the greatest risks, leading to corrective action:

– The severity of the consequences is based on a scale of 4, 5 or 10 levels. It is a type
  of measure of the consequences of various possible scenarios, weighed by their
  conditional probability once the initial failure takes place.
– The occurrence of the failure mode is an evaluation of the number of cases per
  unit of time. It is linearly represented on a scale containing as many levels as the
  severity scale.

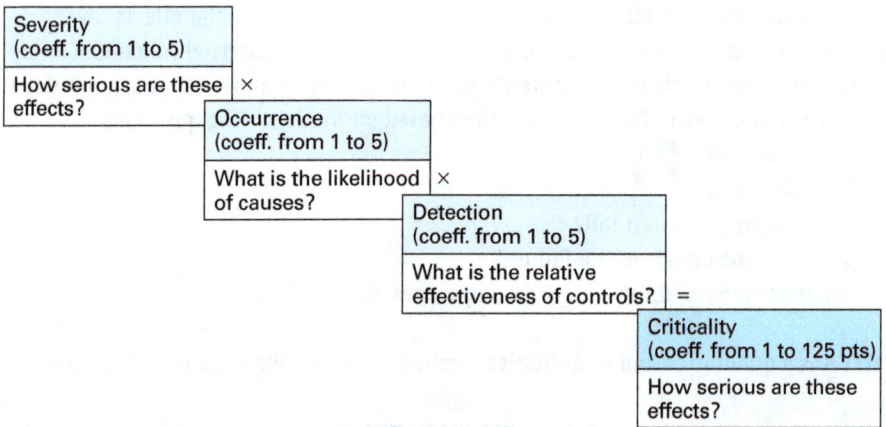

Fig. 4.18: Calculation of the criticality index.

- The detectability or nondetection is represented by a coefficient on a scale of 2, 3, 5, etc., levels reflecting the severity of the consequences in case of detection, weighed by the probability of detection. Failure modes are detected and stopped through:
- Controls
- Measurements
- Calculations/modeling
- Procedures
- Training

The main questions asked after the criticality evaluation can be summarized in Fig. 4.19.

4.6.3 Example

Results of the analysis are often summarized in Tab. 4.13, similar to the HAZOP risk analysis, where each corrective measure is described with a list of all defects on which it has an impact (mitigation of occurrence or consequences, including, when possible, negative impacts).

The following topics have to be determined: a reference number (1), the function or activity evaluated (2), the identified hazards (3), the potential damages or injuries (4), the estimation of the occurrence, severity, nondetection, and calculation of the criticality index (5–8), the corrective measures to be applied (9), the re-evaluation of the criticality (10–13), who is responsible for applying the measures and the deadline for implementation (14).

The following example in Tab. 4.14 depicts an extract of an FMECA analysis for a chemical lab.

FMEA and FMECA are methodologies designed to identify potential failure modes for a product or process, to assess the risk associated with those failure modes, to rank the issues in terms of importance
and to identify and carry out corrective actions to address the most serious concerns.

4.6.4 Conclusions

The analysis of failure modes and their effects, with or without criticality, is an approach based on logic and common sense. Considering that no system is infallible,
this analysis includes identifying, describing and evaluating risks that result from failures. It is a pertinent method whenever the failure modes of components and the internal functioning of the system are or can be known and understood. There are limitations because of the method itself (not suited to represent the dynamics of a system,
the temporal dimension and logical combinations) and the information available (like
all methods of safety engineering, the existing and available information and knowledge is worked with rather than new information or knowledge being created).

While FMECA is a relevant tool for safety engineering, it does not provide a simultaneous vision of possible failures and their consequences (e.g., if two failures occur
at the same time on two subsystems, what is the consequence on the system as a
whole?). Taking aeronautics as an example, we know that airplane crashes are rarely
linked to only one failure; there are generally multiple failures occurring simultaneously.

Furthermore, FMECA as a tool must not become an end in itself. It is common for
far-fetched risks to be unnecessarily associated with FMECA (e.g., someone could
break their leg skiing), or for the problems noted in the FMECA to be considered as
solved problems.

FMECA has several advantages: it is a detailed, rigorous method, providing reliability prediction, automated (commercial software) and is relatively inexpensive.
The main disadvantage is that it is not designed for hazards unrelated to failure
modes (hazards related to high voltage, radiation, etc.). It is limited to external inter-
ference and influences, where little attention is given to human factors and lacks consideration of combined failures, even though, for some systems, this issue has already
been addressed [20]. An estimate of failure cost is often missing, even though this
shortage can be overcome [21].

The FMECA is a simple methodology applicable to many facilities:

- The FMECA provides a systematic and methodical approach.
- It needs to evaluate all possible failure modes for each component of the system.
- It is poorly suited to identifying the consequences of multiple failures.
- It nonetheless highlights the specific points that should receive further study.

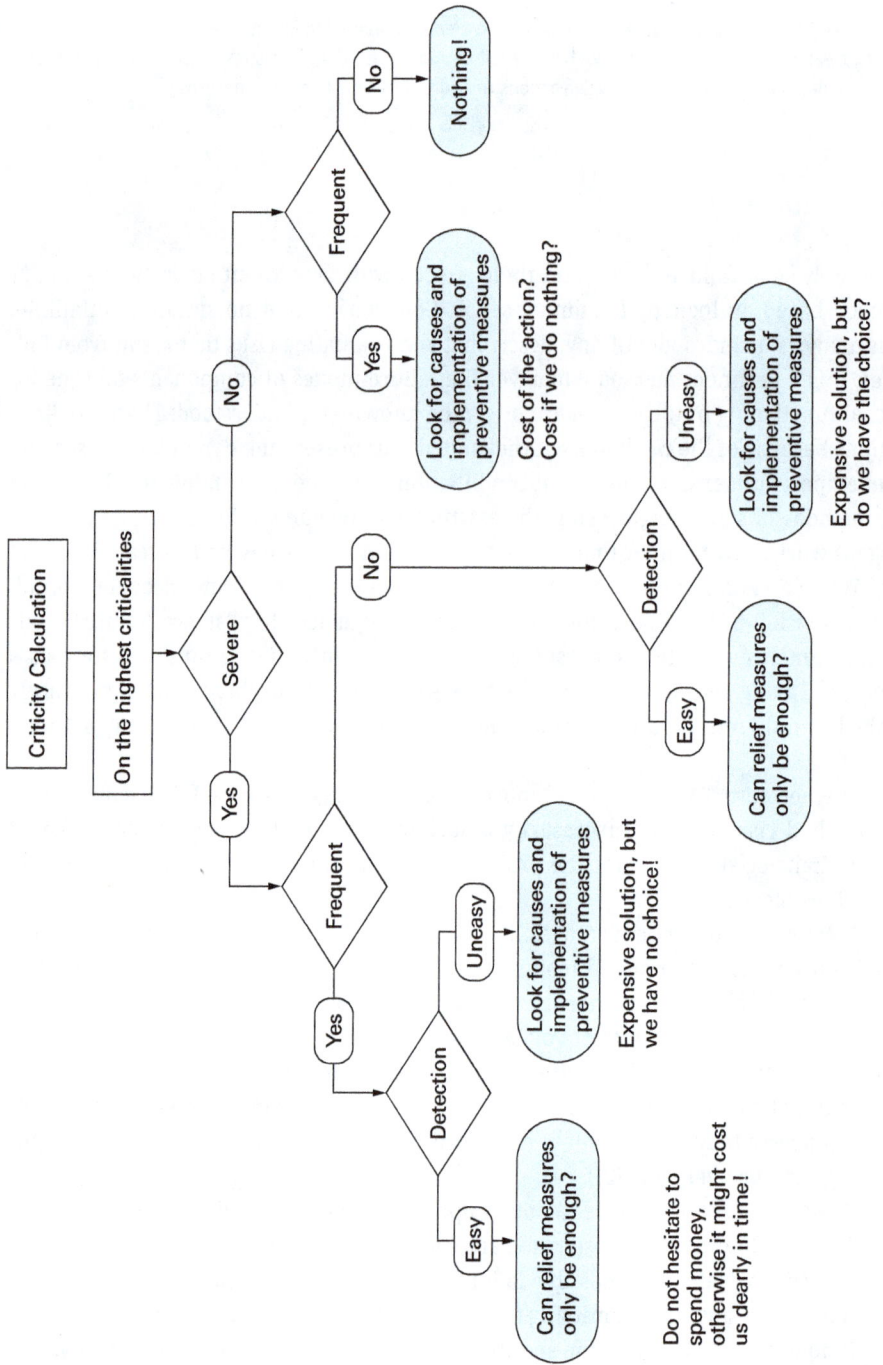

Fig. 4.19: Criticality calculation diagram.

Tab. 4.13: Example of an FMECA table.

Process		Occupational safety		Estimation				Corrective measures		Estimation			
No.	Function/ activity	Hazards	Damage/ injury	F	G	D	C	Measures	F	G	D	C	Date/visa
1	2	3		4	5	6	7	8	9	10	11	12	13	14

F: frequency; G: severity; D: nondetection; C: criticality index.

Tab. 4.14: Example of FMECA analysis.

Process		Occupational safety		Estimation			
No.	Function/activity	Hazards	Damage/injury	F	G	D	C
1	Gas bottle handling (>10 L)	Fall of the cylinder	Foot crush up	3	3	5	45
2		Valve rupture after shock	Person crush up	1	5	5	25
3	Handling chemicals	Spillage on person	Contamination	3	2	5	30
4		Spillage	Contamination	4	2	5	40
5	Cryogenics handling	Spillage on person	Burn	2	4	5	40
6		Asphyxia	Death	1	5	5	25

Corrective measures		Estimation				
Measures		F	G	D	C	Date/visa
Transportation chariot		1	3	5	15	2013.10/JM
Transportation only with protective cap		1	2	5	10	2012.07/TJ
Transportation with tray on chariot		1	2	5	10	2012.07/TJ
Transportation with tray on chariot		1	2	5	10	2012.07/TJ
Chariot ad hoc + gloves, goggles, lab coat		1	1	5	5	2012.11/JM
Oxygen detectors		1	2	1	2	2013.05/JB

There are several FMECAs:
- FMECA process (identify risks of the production process).
- FMECA product (identify risks induced by the concept).
- FMECA production (identify risks linked to production facilities).
- FMECA service.
- FMECA procedure.
- FMECA sustainability.

4.7 Fault tree analysis and event tree analysis

4.7.1 Fault tree analysis

FTA was developed at Bell Labs for the US Air Force in 1961 to evaluate and assess the safety of missile guidance [22]. It is an analytical and deductive method [23] that derives the causes of initiating an undesired event, the top event. Used in several plants such as nuclear, chemical and aeronautic, FTA provides both qualitative and quantitative results [24]. The method is now standardized through the IEC 61025:2006 standard.

As input, good knowledge of design, personal training, equipment and accident history are required. FTA involves logical paths (cut-sets) and Boolean algebra to determine the causes of failure (see Fig. 4.20). It is a combination of failures and their probability of occurrence, applying logic gates (OR gates and AND gates). An event occurs when the gate output changes state. The output of FTA consists of the fault diagram, which exhibits the root causes of the accident at the bottom. If the fault tree is evaluated, then the frequency of the top event is also an outcome.

Some research groups have performed FTA with probabilistic risk analysis methods [25]. The technique is suitable for the investigation and management of multiple cases of failures, reliability and maintainability.

The general process to build the FTA is:
1. Set the top event (failure or an accident) to be analyzed. This event should be specific enough to avoid creating an overly tree, which can quickly happen.
2. Define successively, and in a top-down approach, the direct causes of each event. All causes (events entering the logic gate) are necessary and sufficient for the occurrence of the consequence (event exiting the logic gate).
3. Each time a new level is built, it should be directly linked to higher levels by logic gates, respecting Boolean algebra (see Fig. 4.20).
4. Finish the process of tree building when the development of events is no longer feasible or desirable.
5. Assign a likelihood of occurrence to each basic event.
6. Calculate, using Boolean algebra, the likelihood of occurrence of the top event.
7. Identify vulnerabilities in highlighting the critical path (cut sets), i.e., the combination of the most likely basic events leading to the occurrence of the top event.

8. Propose measures that reduce the occurrence of the top event, giving priority to the most likely critical paths or addressing the common causes (basic events involved in several critical paths).

In fault trees, certain specific symbols are employed to represent the various events and the logic in the tree. A "basic event" is represented in an FTA by a circular shape. A basic event corresponds to a basic fault that does not need to be further developed into more basic events. An "intermediate event" is represented by a rectangular shape. An "underdeveloped event," that is, an event for which the development into basic events is not performed due to, for instance, insufficient information or assumed insignificance of the event, is represented by a diamond shape.

Furthermore, symbols are also used for the AND and OR gates, which are used (see [26]):

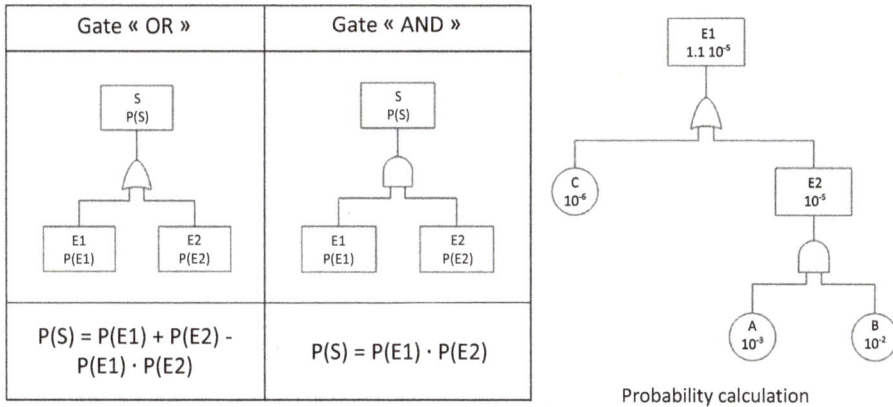

Fig. 4.20: Boolean logic gates and probability calculation.

- An AND gate means that all the incoming events have to be true for the outgoing event to be true.
- An OR gate means that one of the incoming events has to be true for the outgoing event to be true.
- An exclusive OR means that the outgoing event is true when only one of the incoming events is true, but not when more than one of the incoming events is true.
- A mutually exclusive OR means that the incoming events need to be different from each other; thus, if one is false, the other needs to be true. They cannot both occur at the same time. The outgoing event is, therefore, always true.

For the logic in the tree, the following Boolean laws apply:

$$\text{Associative law:} \quad (A.B).C = A.(B.C)$$
$$(A + B) + C = A + (B + C)$$
$$\text{Commutative law:} \quad A.B = B.A$$
$$A + B = B + A$$
$$\text{Distributive law:} \quad A.(B + C) = A.B + A.C$$
$$A + (B.C) = (A + B).(A + C)$$
$$\text{Idempotent law:} \quad A.A = A$$
$$A + A = A$$

In addition, the following Boolean simplification rules can also be formulated:

$$A.0 = 0$$
$$A.1 = A$$
$$A + 0 = A$$
$$A + 1 = 1$$

The probabilities that the outgoing event is true can now be calculated using the following formulae:

$$\text{AND gate:} \quad\quad\quad\quad\quad\quad\quad\quad\quad P(A \text{ and } B) = P(A).P(B)$$
$$\text{OR gate (independent events):} \quad P(A \text{ or } B) = P(A) + P(B) - P(A).P(B)$$
$$\text{Exclusive OR gate:} \quad\quad\quad\quad\quad P(A \text{ or } B) = P(A) + P(B) - 2.P(A).P(B)$$
$$\text{Mutually exclusive OR gate:} \quad P(A \text{ or } B) = P(A) + P(B)$$

In fault trees used for safety-related problems, usually "mutually exclusive OR gates" are assumed. The probability calculation according to the FTA procedure is described in a simple example in Fig. 4.20. The steps and details of the calculation are as follows:

Known data: Probability "P" of A to happen is: $P(A) = 10^{-3}$, for B it is 10^{-2} and for C it is 10^{-6}

Calculation of the probability for event E2 (logical gate AND) leads to

$$P(E2) = P(A) \times P(B) = 10^{-3} \times 10^{-2} = 10^{-5}$$

Calculation of the probability for event E1 (logical gate OR) leads to

$$P(E1) = P(C) + P(E2) = 10^{-6} + 10^{-5} = 1.1 \times 10^{-5}$$

When more than two gates are present, the probability calculations are defined as shown in Fig. 4.21, which provides solutions for cases with three gates as well as for n gates.

Gate « OR »	Gate « AND »
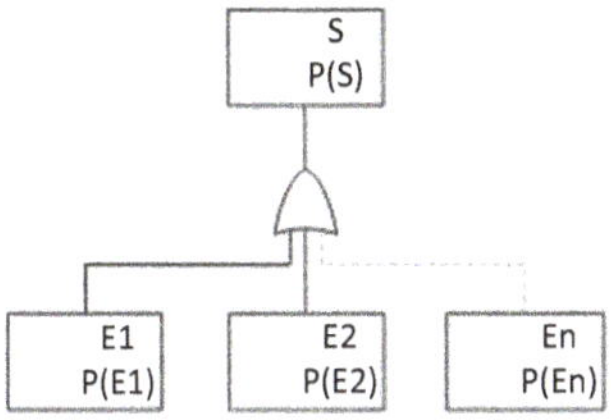 S P(S) / E1 P(E1), E2 P(E2), En P(En)	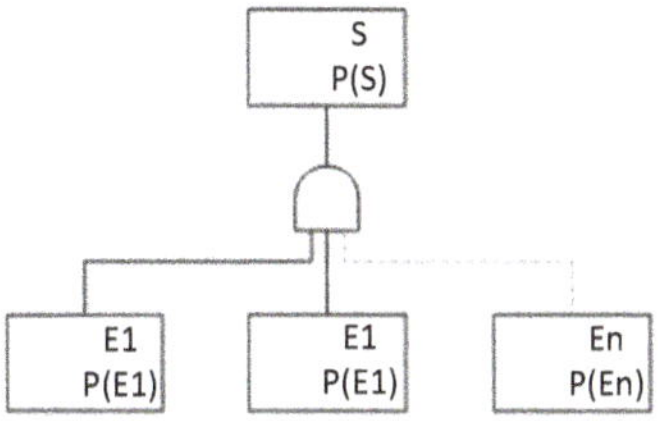 S P(S) / E1 P(E1), E1 P(E1), En P(En)
$P(S) = \sum (singles) - \sum(pairs) + \sum(triples)$ $- \sum(fours) + \sum(fives) - \sum(sixes) + \cdots$	$P(S) = \prod_{1}^{n} P(E_n)$
$P(S) = P(E1) + P(E2) + P(E3) -$ $P(E1) \cdot P(E2) - P(E2) \cdot P(E3) - P(E3) \cdot$ $P(E1) + P(E1) \cdot P(E2) \cdot P(E3)$	$P(S) = P(E1) \cdot P(E2) \cdot P(E3)$

Fig. 4.21: Boolean logic gates and probability calculation for 3 and n gates.

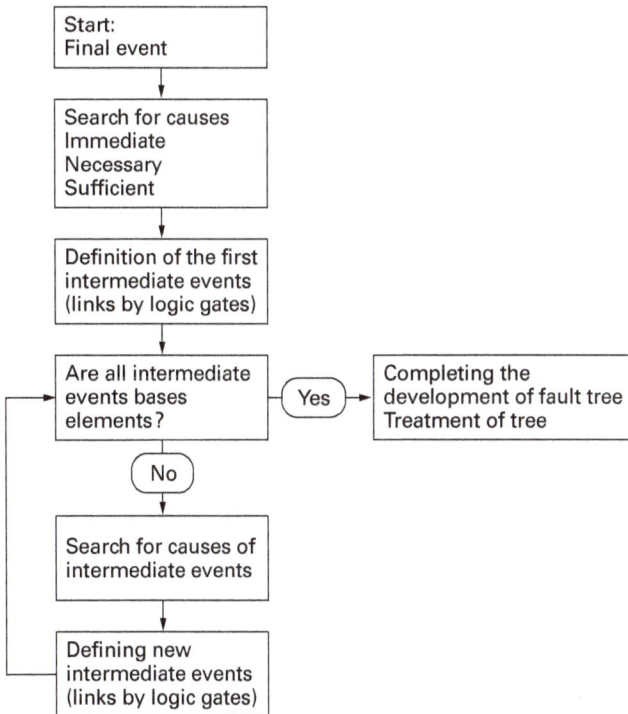

Fig. 4.22: FTA tree building.

The summary of tree building is depicted in Fig. 4.22, where the most important rule to remember is that the basic events of a fault tree should be strictly independent.

An example is given in Fig. 4.23 for the main event "a train passes at red signal." The tree is built according to the above method, also indicating the probability calculations.

FTA allows for considering combinations of events that can ultimately lead to a feared event and indeed represents a deductive method. It allows a good fit with the analysis of past incidents, indicating that major accidents reported result most often from a combination of several events that alone could not cause such accidents. Moreover, by estimating the likelihood of occurrence of the events leading to the final event, it will provide criteria for determining priorities for the prevention of potential accidents.

The FTA approach is not suitable for the risk analysis of a complete system, including a number of adverse events that may be very large. The quantification of FTA is extremely powerful, but it is often very difficult to implement because the rate of occurrence of basic events is often unknown. (All basic events without exception must be quantified in order for the top event occurrence to be calculated.) Seemingly simple, this technique can easily lead the beginner to build a tree completely wrong without knowing it, delivering results and analysis that are potentially dangerously false.

> **!** FTAs are logic block diagrams that display the state of a system (top event) in terms of the states of its components (basic events). An FTA is built top-down and in terms of events rather than blocks. It uses a graphic "model" of the pathways within a system that can lead to a foreseeable, undesirable loss event (or a failure). The pathways interconnect contributory events and conditions using standard logic symbols (AND, OR, etc.).

4.7.2 Event tree analysis

ETA was first used in the nuclear industry around 1974 to assess the risks of nuclear power plants using light water. It uses an inductive approach to determine the consequences of an undesirable event via a graphical tool [4]. The method is now standardized through the IEC 62502:2010 standard. The consequence of an event follows a series of paths to which probabilities are given. The technique can be used for qualitative as well as quantitative reliability and risk analysis. The required inputs are design knowledge, accident history or similar scenarios. ETA processes by identifying an accident and pivotal event, building event tree diagrams and evaluating the risks. The output of an ETA can be summarized as mishap outcome and risk probability, causal sources and safety requirements. The main disadvantage is that it cannot study multiple failures on the same initiating event. Multiple ETAs require multiple initiating events.

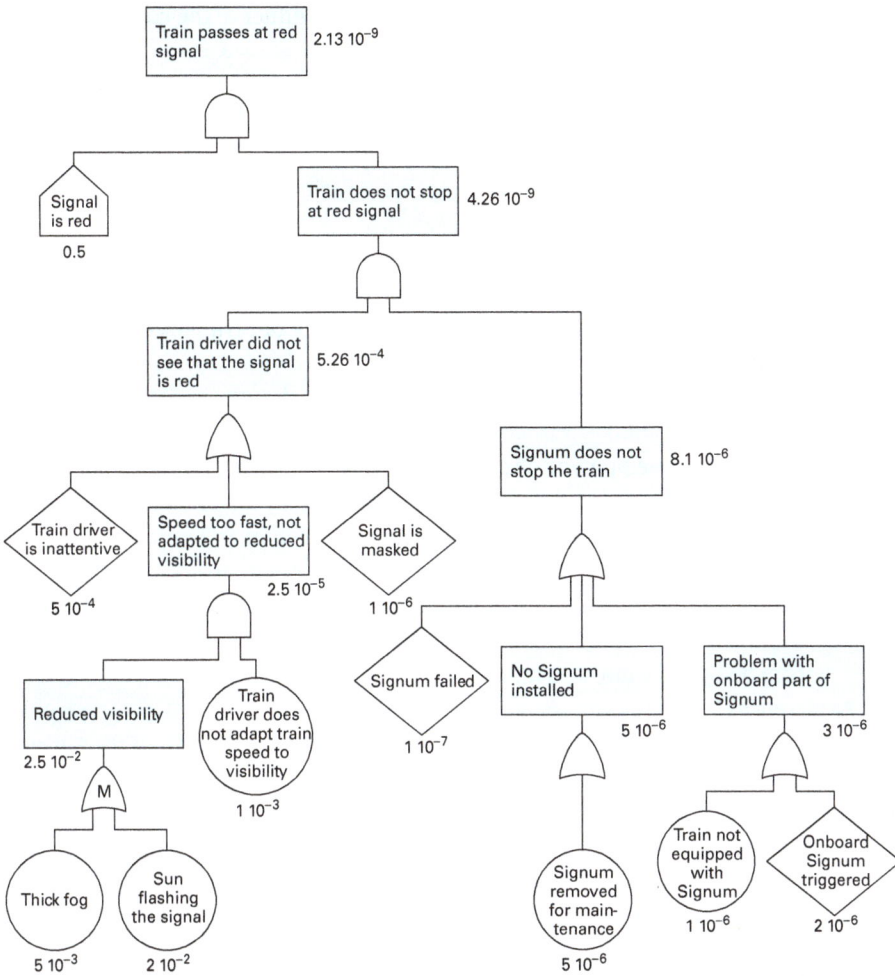

Fig. 4.23: Example of an FTA tree including probability calculations.

The ETA method is exploratory in that it allows for identifying long-term consequences that may be unknown. The methodology addresses both desirable and undesirable consequences, as it allows for the inclusion in the analysis of the responses of safety systems (alarm, automatic shutdown, etc.) and operating staff. The initiating event itself may be accidental (failure, incident) or perfectly anticipated (e.g., particular operation).

The method is quantitative as it allows, provided that the likelihood of occurrence of the initiating event and failure rates of various safety systems are known, for calculating the occurrence of each of the final consequences, e.g., of each scenario.

The general process to build the ETA is

Select an initiating event.

Identify safety functions or system players that can influence the course of events caused by the initiating event.

Build the tree successively considering possible responses to the different actors.

The build process stops once all the elements which, by their reaction influencing the spread of the consequences, have been considered.

Determine the final consequence of each scenario (e.g., in terms of material, human or environmental).

Assign a probability to each branch of the tree.

Calculate by simply multiplying the likelihood of occurrence for each scenario.

Propose adapted mitigation measures if the criticality of certain scenarios is considered too high.

The different steps are:

1. Starting and final points
   - The logical development of an event tree is to answer the question, "What happens if . . .?" This leads to a tree because usually we answer, "It depends: if . . . then . . . else . . ." It is this alternative that results in branches.
2. Identification of safety functions
   - Safety features must be provided as barriers in response to the initiating event. They generally aim to prevent, as far as possible, the initiating event from being the cause of a major accident.
3. Recurrence
   - Having posed an alternative, the question is repeated for each of the alternatives. It develops and branches out to explore the consequences.
4. Representing
   - Traditionally, the initial event is located on the left, and the tree grows from left to right. The upper branch of each alternative represents the desired operation, and the lower branch represents the undesired operation.
5. Likelihood, tree exploitation
   - A quantitative aspect could be added to the event tree, indicating the probabilities of events taken into account. The occurrence of each alternative is indicated in the upper branch, (the other being its complement to 1).

Figure 4.24a develops an ETA tree for an originating event, which is the failure of a cooling system, based on the abovementioned method.

The same tree could be represented in a more compact manner as depicted in Fig 4.24b.

The main limitation of the event tree is that it can only consider a single initiating event, and it is therefore not possible to identify scenarios for multiple initiating events. Furthermore, although more powerful than the FMECA approach in integrating the system responses, the ETA approach makes it difficult to conduct a systemic study because there is no systematic search for initiating events.

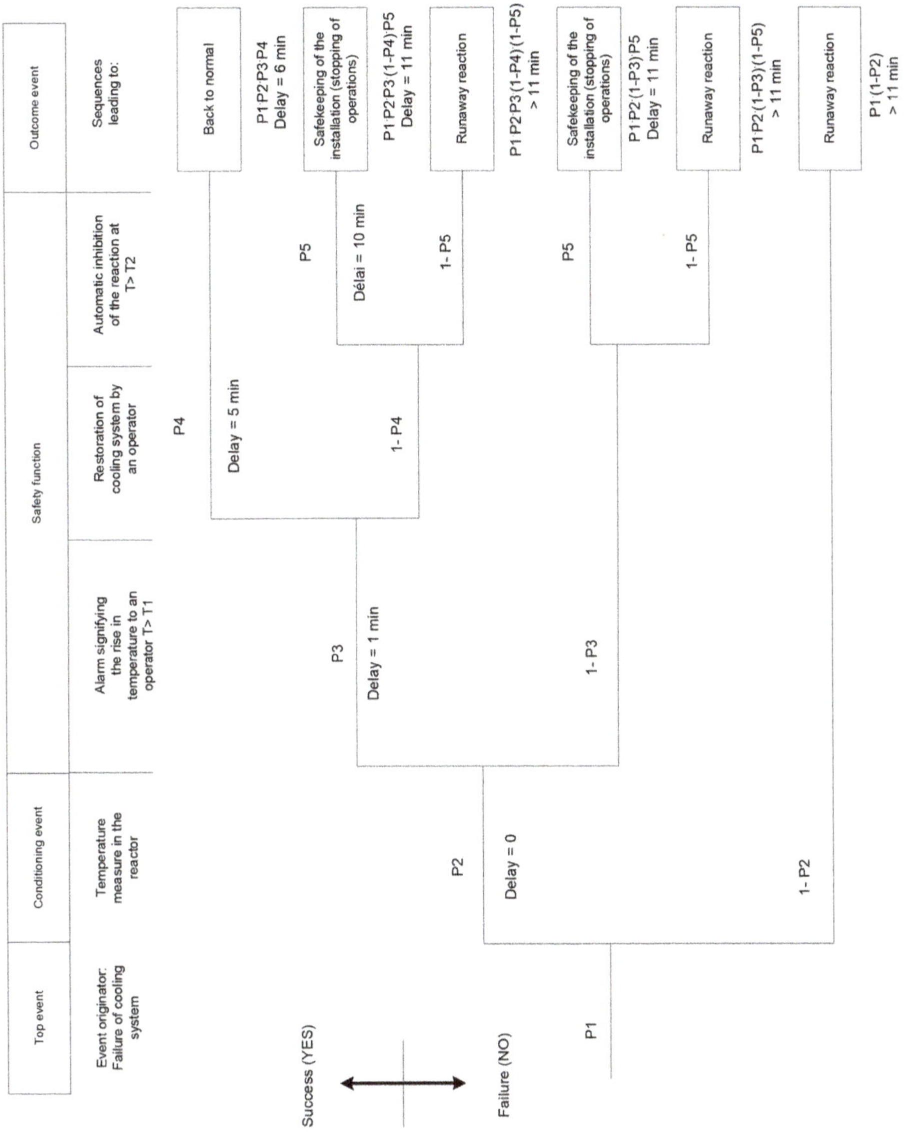

Fig. 4.24a: Example of an ETA tree.

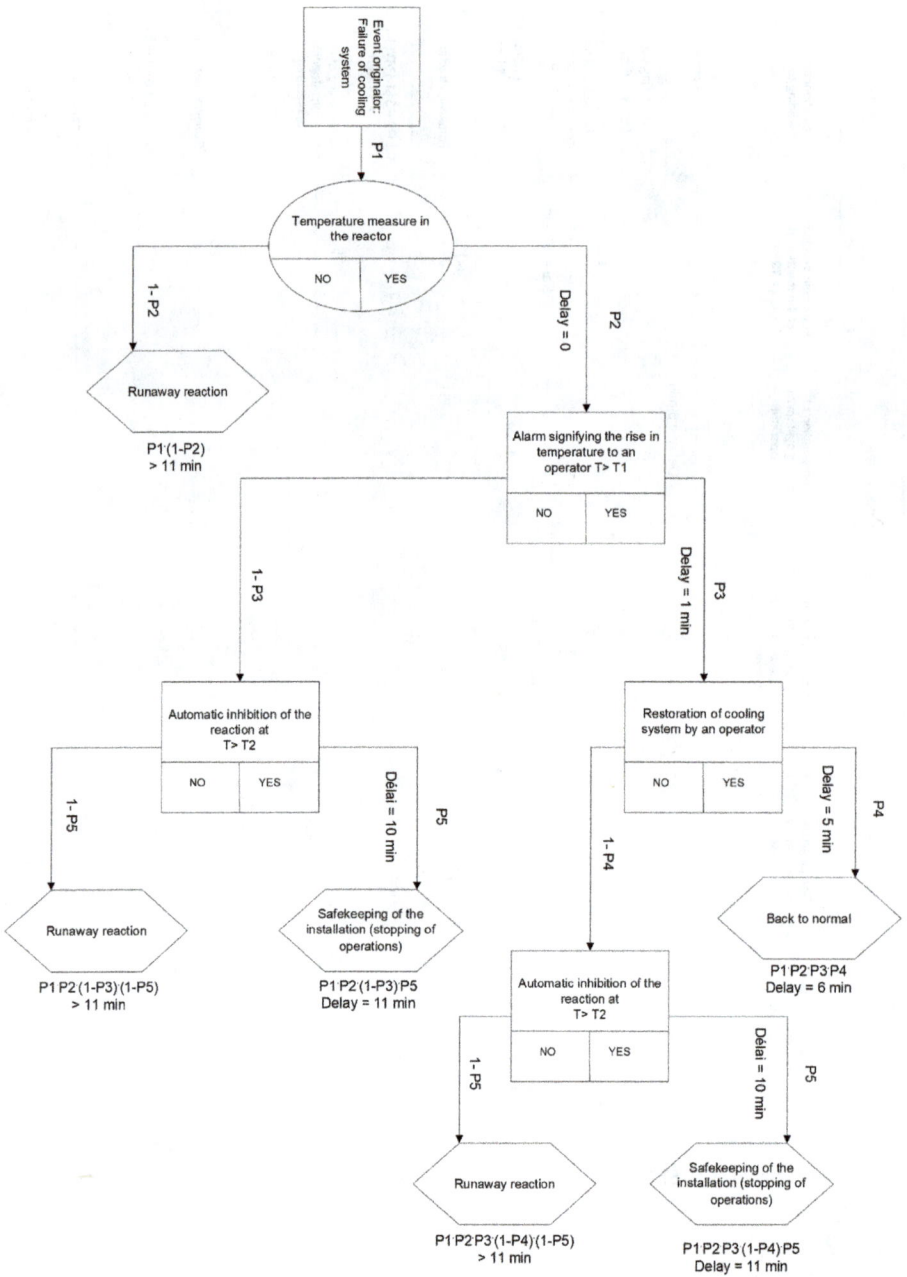

Fig. 4.24b: Example of a condensed ETA tree, the same example as presented in (a).

ETA is a method to examine, from an initiating event, the sequence of events that may or may not lead to a potential accident. It is thus particularly useful for studying the architecture of safety resources (prevention, protection, response) that already exist or that may be considered on a site. It can be used for the analysis of accidents after the fact. This method can quickly become cumbersome to implement. Accordingly, one must carefully define the initiating event that will be the subject of the analysis.

> ETA is an inductive failure analysis performed to determine the consequences of a single failure for the overall system risk or reliability. It shows all possible outcomes resulting from an accidental (initiating) event, taking into account whether installed safety barriers are functioning or not, and additional events and factors. **!**

4.7.3 Cause–consequence analysis (CCA) and bowtie, a combination of FTA and ETA

CCA (cause–consequence analysis) is a tree-like approach for illustrating the possible outcomes arising from the logical combination of selected input events or states. CCA was developed at Risø National Laboratory, Denmark, in the 1970s, mainly to improve the reliability of risk analysis of nuclear power plants in Scandinavian countries.

CCA is a methodology for identifying and evaluating the sequence of events leading to and resulting from the occurrence of an initiating event. The objectives of a CCA are to determine the development of the possible outcomes and if they will be sufficiently controlled by the safety systems and the implemented procedures (system design, safety barriers, organization, etc.).

Figure 4.25a illustrates an example of a CCA tree, starting from the main event of a leakage when unloading a tank car of chlorine. In addition, corrective measures in terms of organizational, passive and technical in prevention or protection are also depicted. The tree representation allows not only for the calculation of probabilities when necessary (FTA) but also clearly indicates where the problems are and where corrective measures should apply.

As FTA is not inductive, it cannot give an overview of a system and is limited to studying specific aspects. Conversely, uninterested in the causes of failures, ETA is often difficult to quantify as such, and although its inductive nature, the depth of analysis is often not sufficient to ensure a sufficiently accurate assessment of the level of risk.

The complementarities of these two approaches are reflected in the combined methodology FTA/ETA = bowtie, which is one of the most rigorous and powerful methods, as presented in Fig 4.25b. The simultaneous inductive-deductive approach allows for combining the general and the particular and has extremely good coverage of the considered situations.

The beginning of the procedure is identical to the ETA method. One selects the desired initiating event to study and constructs the event tree qualitatively. All intermediate failure causes can be determined by constructing a fault tree that will be linked directly to a branch of the event tree.

Quantifying the structure is performed either directly at the event tree for the branches that are not bound to a fault tree, or via the quantification of fault trees for the other branches.

In conclusion, although powerful, the implementation of a CCA analysis or a bowtie can be very heavy and needs to be performed by experts. These methodologies are not systemic, as they suffer the same limitation as ETA analysis, namely that it can consider a single initiating event at a time. This tool clearly highlights the action of the safety barriers opposing these accident scenarios and can provide a demonstration of enhanced risk management. For a complex system comprising several tens, even hundreds, of specific failures (initiating events), it would be illusory to perform a risk analysis of the total system by the FTA/ETA or CCA approach, because it will represent years of work. It provides a concrete visualization of accident scenarios that could occur, starting from the initial causes of the accident to the consequences at the level of identified targets.

> **!** CCA or bowtie analysis are similar techniques that combine the ability of fault trees to show how various factors may combine to cause a hazardous event with the ability of event trees to show the various possible outcomes. Sequences and, therefore, time delays can be illustrated in the consequence part of the diagram. A symbolism similar to that in fault trees is used to show logical combinations. The technique has considerable potential for illustrating the relationships from initiating events through to end outcomes. It can be used fairly directly for quantification, but the diagrams can become extremely unwieldy.

4.8 The risk matrix

The "risk assessment decision matrix," often shortened to "the risk matrix," is a systematic approach for estimating and evaluating risks. This tool can be employed to measure and categorize risks based on informed judgment regarding both probability and consequence as well as relative importance. An example of the risk matrix is shown in Tab. 4.15.

Once the risks have been identified, the question of assigning consequence and likelihood ratings must be addressed. A common, basic example of assigned ratings by a team on a generalized basis can be found in Tab. 4.16.

The probability level F, "impossible," makes it possible to assess residual risks for cases in which the hazard is designed out of the system.

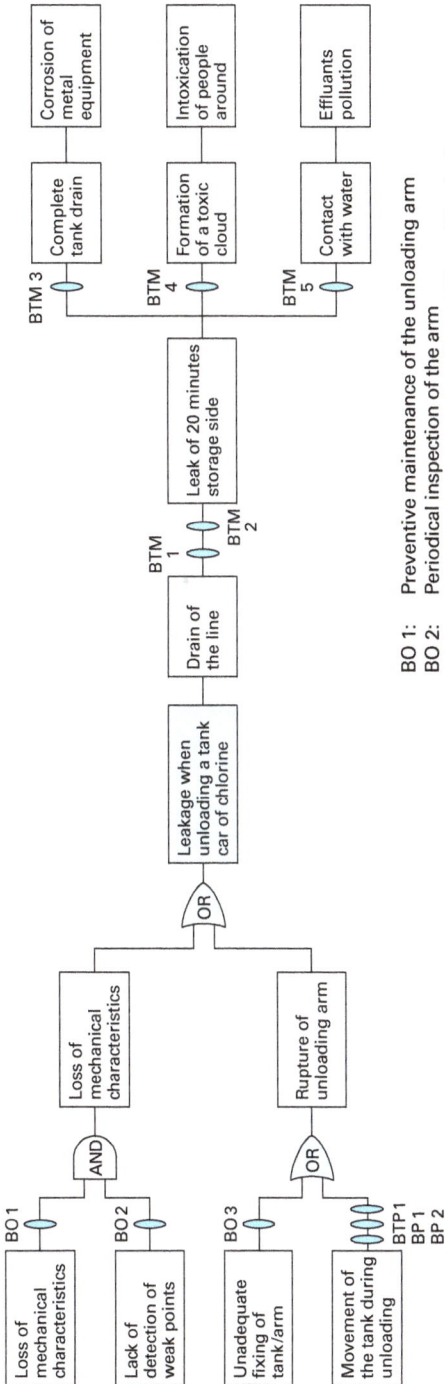

Fig. 4.25: (a) Example of a CCA tree.

BO 1: Preventive maintenance of the unloading arm
BO 2: Periodical inspection of the arm
BO 3: Arm mounting procedure (assembly verification)
BTP 1: Unloading signalization by a red light
BP 1: Guardrail along the unloading route
BP 2: Equipment avoiding the tank to slip off rails
BTM 1: Safety valve on the stock side
BTM 2: Safety valve on tank
BTM 3: Manual valve on tank
BTM 4: Watering by water spray
BTM 5: Retention basin isolated from the drainage

Acronmys:
BO: Organizational prevention barrier
BP: Passive barrier
BTP: Technical prevention barrier
BTM: Technical protection barrier

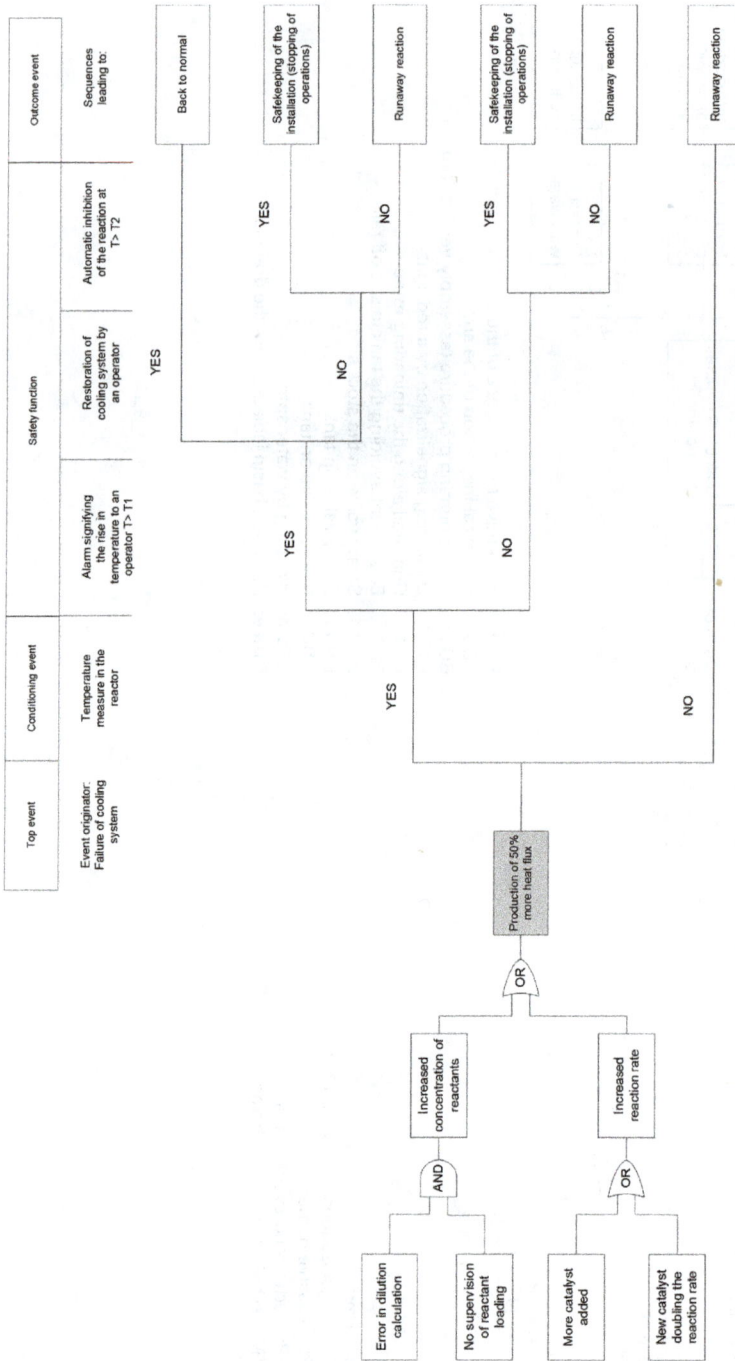

Fig. 4.25: (b) Example of a bowtie, a combination of FTA/ETA.

Tab. 4.15: The risk assessment decision matrix.

Severity of consequence	Likelihood of risk					
	F	E	D	C	B	A
	Impossible	Improbable	Remote	Occasional	Probable	Frequent
I Catastrophic						
II Critical						
III Marginal						
IV Negligible						

Source: Based on Department of Defense [27].

Tab. 4.16: Criticality and frequency rating for the risk assessment decision matrix.

Severity of consequences – ratings

Category	Descriptive word	Results in either
I	Catastrophe	– An on-site or off-site death – Damage and production loss greater than €750,000
II	Critical	– Multiple injuries – Damage and production loss range between €75,000 and €750,000
III	Marginal	– Single injury – Damage and production loss between €7,500 and €75,000
IV	Negligible	– No injuries – Damage and production loss less than €7,500

Hazard probability – ratings

Level	Descriptive word	Definition
A	Frequent	Occurs more than once per year
B	Probable	Occurs between 1 and 10 years
C	Occasional	Occurs between 10 and 100 years
D	Remote	Occurs between 100 and 10,000 years
E	Improbable	Occurs less often than once per 10,000 years
F	Impossible	Physically impossible to occur

Source: Based on Department of Defense [26].

The rankings provide a quick and simple priority sorting method. The rankings are then given definitions that include, e.g., definitions and recommended actions similar to those in Tab. 4.17.

Such a safety risk assessment methodology is especially implemented in cases of low-likelihood, high-consequence risks.

Tab. 4.17: Definitions and recommended actions for rankings.

Ranking	Description	Required action
1	Unacceptable	Should be mitigated with technical measures or management procedures to achieve a risk ranking of three or less within a specified time period such as 6 months.
2	Undesirable	Should be mitigated with technical measures or management procedures to a ranking of three or less within a specified time period such as 12 months.
3	Acceptable with controls	Should be verified that procedures or measures are in place.
4	Acceptable	No mitigation action required.

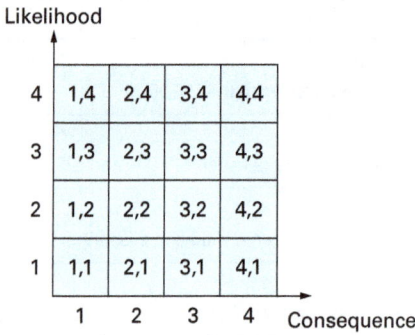

Likelihood

4	1,4	2,4	3,4	4,4
3	1,3	2,3	3,3	4,3
2	1,2	2,2	3,2	4,2
1	1,1	2,1	3,1	4,1
	1	2	3	4	Consequence

Fig. 4.26: A 4 × 4 ordinal risk matrix.

It is very important that likelihood estimates, as well as consequence estimates, are very well-considered and carried out by experienced risk managers. When this has been done, a risk is "binned" into one of the squares of the risk matrix as a function of the level of its consequence and the level of its likelihood (occurrence probability). How then can risks be ranked or prioritized across the squares of the risk matrix? To answer this question, consider, e.g., the 4 × 4 ordinal risk matrix of Fig. 4.26.

Denote an (i, j) risk event as one that has a level i consequence and a level j likelihood, where $i, j = 1, 2, 3, 4$. A common approach to prioritizing risks is to multiply the consequence and likelihood levels that define each square and use the resultant product to define the square's score. Figure 4.27 displays a matrix with risk-resultant scores.

Risks binned into a specific square then receive that square's score and are ranked accordingly. Risks with higher scores have higher priority than risks with lower scores. However, there are some disadvantages to this approach.

The first disadvantage that could possibly lead to problems is the multiplication of ordinal numbers, which is not a permissible arithmetic operation. The second disadvantage is that a risk with a level 4 impact and a level 1 likelihood receives the same score as a risk with a level 1 impact and a level 4 likelihood. Although these are two very different risks, they are equally valued and tie in their scores if the multiplication method is used. In industrial practice, risk managers should not lose visibility in the presence of high-consequence risks (types II) regardless of their likelihood. The third disadvantage is that there are numerous ties in the scores of Fig. 4.27, and, in view of the problem just discussed, the question should be posed whether risks that tie in their scores should be equally valued, especially when these ties occur in very different consequence-likelihood regions of the risk matrix. Hence, how to solve these problems?

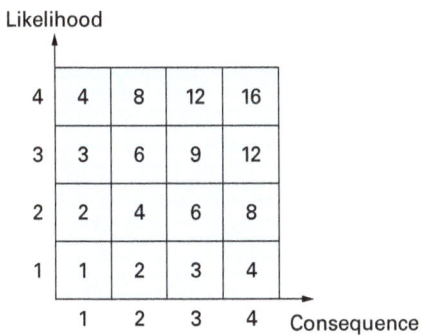

Likelihood

4	4	8	12	16
3	3	6	9	12
2	2	4	6	8
1	1	2	3	4
	1	2	3	4	Consequence

Fig. 4.27: A 4 × 4 ordinal risk matrix with risk resultant scores based on multiplication.

As Garvey [28] indicates, one way is to first examine the risk attitude of the risk management team. Is the team consequence-averse or likelihood-averse? A strictly impact-averse team is not willing to trade-off consequence for likelihood. For such a team, low-likelihood, high-consequence risks should be ranked high enough so that they remain visible to management. But how can risk prioritizations be assigned in this regard? The following presents an approach to address these considerations.

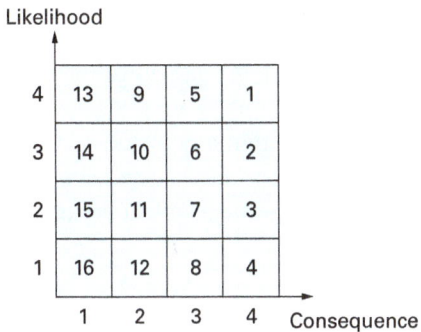

Likelihood

4	13	9	5	1
3	14	10	6	2
2	15	11	7	3
1	16	12	8	4
	1	2	3	4	Consequence

Fig. 4.28: A strictly consequence-averse risk matrix.

To start, the ordinal risk matrix should be redefined in a way that prioritizes risks along a strictly consequence-averse track. Figure 4.28 shows a matrix where this has been carried out.

Each square is scored in order of consequence and criticality by the risk management team. The lower the score, the higher the priority. Risks categorized in the upper-right corner of the matrix have the highest priority and have a score of 1. Note that in this risk matrix, risks with a level 4 consequence will always have higher priority than those with a level 3 and so forth. Thus, this matrix is one where risks with the highest consequence will always fall into one of the first four squares – and into a specific square as a function of their judged likelihood.

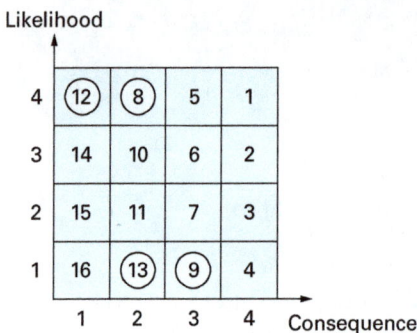

Likelihood

	1	2	3	4	
4	(12)	(8)	5	1	
3	14	10	6	2	
2	15	11	7	3	
1	16	(13)	(9)	4	
	1	2	3	4	Consequence

Fig. 4.29: First iteration of risk ranking.

Now, a risk management team might want to deviate from this strictly consequence-averse risk matrix (see also [28]). To illustrate this, imagine that the team decided that any risk with a level 4 consequence should remain ranked in one of the first four squares of the risk matrix (i.e., the right-most column). For all other columns, trade-offs could be made between the bottom square of a right-hand column and the top square of its adjacent left column (first iteration). This is shown by the circled squares in the risk matrix illustrated in Fig. 4.29.

Suppose that the team decides to further refine the rank-order of the squares in the matrix. This second iteration is illustrated by circles in Fig. 4.30.

This discussion illustrates one approach for directly ranking or prioritizing risks based on their consequences and likelihood.

Finally, it is common practice to assign color bands within a risk matrix. These bands are intended to reflect priority groups. Table 4.16 uses the matrix from Tab. 4.15 to illustrate how the matrix from Fig. 4.30 (after two iterations) might be colored with respect to priority groups. Thus, Tab. 4.18 is created.

Tab. 4.18: The risk assessment decision matrix.

Severity of consequences	Probability of hazard					
	F Impossible	E Improbable	D Remote	C Occasional	B Probable	A Frequent
I Catastrophic						1.
II Critical					2.	
III Marginal				3.		
IV Negligible			4.			
Risk code/ actions	1. Un-acceptable	2. Un-desirable	3. Acceptable with controls	4.		Acceptable

Likelihood

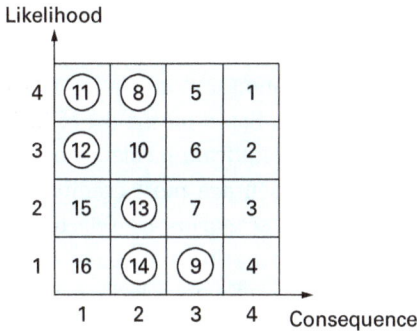

Fig. 4.30: Second iteration of risk ranking.

> Risk ranking uses a matrix that has ranges of consequence and likelihood as the axes. The combination of a consequence and likelihood range gives an estimate of risk or a risk ranking. Without adequate consideration of risk tolerability, a risk matrix can be developed that implies a level of risk tolerability much higher than the organization actually desires. Risk tolerability should thus be carefully considered by the risk management team.

4.9 Quantitative risk assessment (QRA)

Quantitative risk assessment (QRA) is a very systematic approach to calculate individual fatality risks and societal risks from industrial facilities or parts thereof. The generic stages for the QRA procedure for the risk assessment of hazards are

- System description – context and scope
- Hazard/threat identification
- Incident enumeration

- Selection: worst-case and/or credible accident scenario identification
- Consequence assessment
- Likelihood assessment
- Risk assessment
- Visualization and utilization of risk estimates

In stage 1, the scope of the study is established. Information regarding site locations, environs, weather data, PFDs, layout drawings, operating and maintenance procedures, technology documentation, etc., is gathered, as well as data regarding (for example) the use, storage, processing, etc., of hazardous materials. Note that not only information from within the study scope area but also from outside the scope area needs to be collected. This information is then further employed during the different consecutive steps of the QRA. Stage 2 identifies all possible hazards and threats that could possibly lead to incidents and accidents. There are many possible HAZID techniques available such as experience, engineering codes, checklists, detailed knowledge of the situation/plant/process/installation, equipment failure experience, hazard index techniques, what-if analysis, HAZOP studies, FMEA and PHA (see also previous sections). Stage 3 of a QRA procedure comprises the identification and tabulation of all incidents, without regard to importance or initiating event. Based on the list of incidents from the previous stage, stage 4 selects one or more significant incidents to represent all identified incidents. Mostly, the incidents chosen are most credible or worst-case. Hence, the credible and/or worst-case accident scenarios are selected in this stage. In industrial practice, often a large number of scenarios (sometimes up to 1,500, depending on the scope of the study) still remain, out of the very large number of possible incidents. Having established the scenarios that need further processing, stage 5 first determines the potential for damage or injury from specific events. CCPS [29] indicates that a single accident (e.g., rupture of a pressurized flammable liquid tank) can have many distinct accident outcomes (e.g., vapor cloud explosion, BLEVE and flash fire). These possible outcomes can then be analyzed using source and dispersion models, explosion and fire models, and other effect models, where the consequences to people and infrastructure are determined. In stage 6, a methodology is used to estimate the frequency or probability of occurrence of an accident. Estimates may, e.g., be obtained from historical incident and accident data on failure frequencies or from failure sequence models such as fault trees and event trees. Most systems require consideration of factors such as common-cause failures, human reliability and external events. Stage 7 combines the results obtained in the previous two stages, i.e., the consequences and the likelihood of all accident outcomes from all selected scenarios, and uses these to obtain a risk estimate for every scenario. The risks of all selected accident scenarios are consecutively summed up to provide an overall measure of risk. In the last stage the results are tabulated and visualized using, e.g., iso-risk contours (for individual fatality risks) or FN curves (for societal risks), to help decision-makers. Decisions are made either through relative ranking of risk reduction strategies or through comparison with specific risk targets. The last stage obviously requires some risk acceptability criteria, or risk guidelines, to be elaborated in advance by the user or the authorities.

Obviously, a QRA study is not a "simple" study, and the stepwise plan explained above indicates that the aim of the study is to narrow down an initial abundance of possible scenarios to an intelligently considered and justifiable number of accident scenarios, calculate them through and obtain a (relatively) limited number of results where one is able to make objective choices. There are some disadvantages attached to this approach. The main critiques of QRA are that it is unrealistic: it often deals with absurdly small numbers and statistics, which can often lead observers to question the validity of the approach. QRA tries to find out what circumstances have to happen simultaneously to lead to a serious problem, assesses which of the circumstances has the greatest importance in the hazard and suggests that this would be the primary focus of risk management. Another critique is that it is not reproducible. This is because a QRA relies on the application of generic data where no specific data is available and largely depends on the choices of the user. The many different choices that have to be made (regarding the consequence models, scenario selection, frequency estimates, etc.) may lead to different results, even for an identical scope/situation. This is an important drawback as the technique is developed to provide consistent results. Compared to other techniques, the results will be more consistent, but, as mentioned, they should still be interpreted with caution. Furthermore, there are arguments that the results of a QRA are best used to compare the relative safety of different systems and not look at the absolute magnitude of the risk in relation to risk criteria.

Despite all the shortcomings of a QRA, and despite the obvious fact that it is not an exact description of reality, it can be the best available tool to date to assess the risks of complex facilities and situations where a large number of accident outcome cases are at hand, and at the same time, sophisticated models are required.

Using the cube depicted in Fig. 4.31, the QRA risk analysis technique can be positioned in relation to other methodologies. The cube is based on three essential parameters:
- Focus
- Complexity
- Number of incident scenarios investigated

An example of a QRA study is difficult to provide, as the technique is characterized by a very elaborate study involving several steps. Incident enumeration, scenario selection, consequence assessment/calculation and likelihood assessment/determination can all be very complicated and time-consuming studies. Moreover, different software packages can be chosen and used to help the user carry out several required calculations of a QRA.

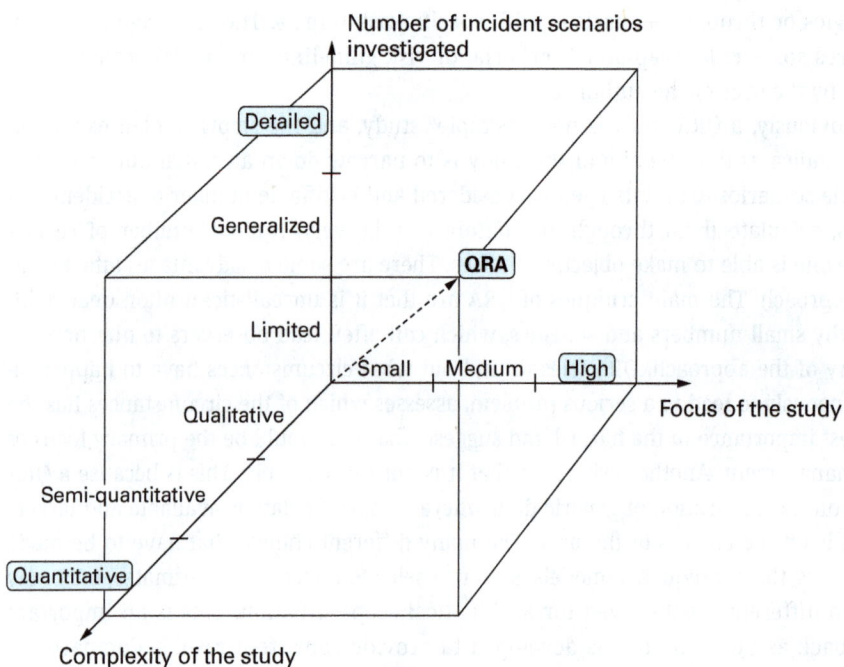

Fig. 4.31: Positioning of QRA relative to other risk analysis techniques.

> **!** QRA is a formalized specialist method used for calculating numerical individual and societal (employee and public) risk level values for comparing risks or for comparison with regulatory risk criteria (e.g., in the case of land-use planning). To conduct a QRA, the value of the potential losses needs to be determined. Then the probability of the occurrence of the risk failure needs to be estimated. Finally, the (annual) loss expectancy is calculated.

4.10 Layer of protection analysis (LOPA)

Layer of protection analysis (LOPA) is an analytical procedure that looks at the safeguards to see if the protection provided is adequate for every known risk. It ishows whether additional controls or shutdown system(s) are required by comparing the risks with and without these additional elements against a set of predetermined criteria [30]. LOPA basically answers three questions:

1. How safe is safe enough?
2. How many layers of protection are needed?
3. How much risk reduction does every layer of protection realize?

If a LOPA is carried out effectively, an accident scenario screened and assessed by the technique should not be able to lead to an accident.

The objective of LOPA is thus to determine if there are sufficient layers of protection against a certain accident scenario, and thus to answer the question of whether a risk can be accepted or tolerated. Despite the fact that often many layers of protection exist to prevent accident scenarios, no layer is perfectly effective (e.g., the Swiss cheese model by Reason and the holes in the cheese) and therefore accidents may still occur.

By using order of magnitude categories for initiating event frequency, consequence severity and the likelihood of failure of independent protection layers, the risk of a well-defined accident scenario is determined. Hence, LOPA is not a fully quantitative risk assessment approach but is rather a simplified method for assessing the value of protection layers for a certain accident scenario; by assessing the risks dependent on various protection layers, LOPA assists in deciding between alternative choices for risk mitigation [31]. The typical layers of protection that can be used in LOPA are depicted in Chapter 3.

An accident scenario is typically identified during a qualitative hazard evaluation, management of change evaluation, or design review. LOPA is limited to evaluating single cause–consequence pairs as a scenario and provides an order of magnitude approximation of the risk of any scenario. Once a scenario is determined, the risk analyst can use LOPA to establish which controls (often called "safeguards") meet the definition of IPLs, and then estimate the scenario's order of magnitude risk. In LOPA, the scenario that is chosen is usually the worst-case one.

In many LOPA applications, the risk manager has to identify all scenarios: cause–consequence pairs that exceed a certain predetermined tolerance of risk. In other applications, the risk manager chooses the scenario that likely represents the highest risk from many similar scenarios. As already mentioned, LOPA typically starts with a list of possible scenarios determined by qualitative risk analyses. The further stages of the technique are as follows:

– Identify the consequence to screen the scenario.
– Select an accident scenario.
– Identify the initiating event of the scenario and establish the initiating event's frequency (events per year).
– Identify the IPLs and estimate the probability of failure on demand for each IPL.
– Estimate the risk of a scenario by mathematically combining the consequence, initiating event and IPL data.

In stage 1, the outcome/impact/magnitude of the consequence of a scenario (provided by, e.g., a HAZOP study) is estimated and evaluated. Different possible outcomes can be taken into account from an initiating event (e.g., a release): e.g., the impact on people, the environment and production. Stage 2 is concerned with choosing one scenario that can be represented by a single cause–consequence pair. Each scenario consists of the following elements: (i) an initiating event that initiates the chain of events; (ii) a consequence that results if the chain of events continues without interruption; (iii) enabling events or conditions that have to occur or be present before the initiating event can result in a consequence; (iv) the failure of safeguards (note that not all safeguards are IPLs, but all IPLs are safeguards!). Scenarios may also include probabilities of a different kind (ignition probability of flammable materials, probability that a fatal injury will result from exposure to the of fire, explosion, etc.). In stage 3, a frequency has to be estimated where information and casuistic may be insufficient. Therefore, companies and sometimes authorities provide guidance on estimating the frequency to achieve consistency in LOPA results. Stage 4 forms the heart of the LOPA method. Here, the existing safeguards that meet the requirements of IPLs for a given scenario need to be recognized. Some accident scenarios will require only one IPL, while other accident scenarios may require many IPLs, or IPLs of very low or high PFDs, to achieve a tolerable risk for the scenario. In the final and fifth stage, the scenario risk is calculated. This may be carried out by different approaches. Regardless of the formulae and graphical methods available to do so, companies usually have a standard form for documenting the results.

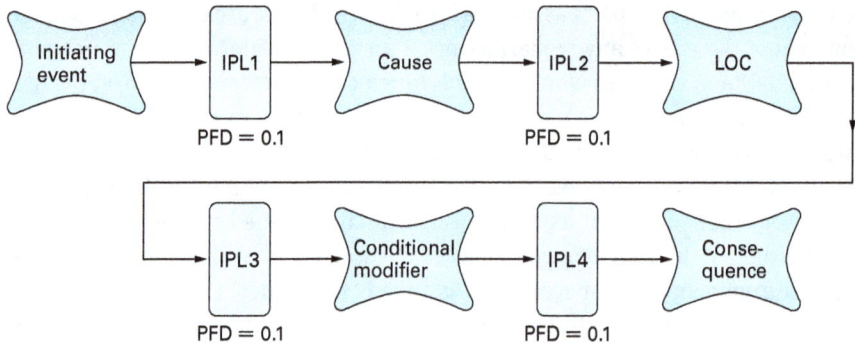

Fig. 4.32: Use of IPLs in a LOPA scenario.

Figure 4.32 illustrates the use of IPLs in a LOPA scenario. One of the main characteristics of any IPL is its reliability, which is expressed by its PFD. As indicated in Chapter 3, the PFD is dimensionless and may vary between 10^{-1} (weak protection) and 10^{-5} (very strong protection). Mostly, a value of 10^{-3} is considered strong protection. A value of PFD equal to 1 indicates that the layer of protection does not contribute to the LOPA scenario.

In the illustrative example of Fig. 4.32 (where every IPL is characterized by a PFD equal to 0.1), the probability that the initiating event eventually leads to the cause–consequence pair envisioned, that is, the probability that the scenario takes place, is estimated at 0.0001 or 10^{-4}.

In summary, a LOPA analysis is aimed at assessing the efficiency (or the lack thereof) of safety and protection measures to prevent accident scenarios. It allows for comparing a variety of technical, organizational and other protective measures. The measure leading to the highest risk reduction level can be determined for any single cause–consequence pair where sufficient information is available or retrievable.

LOPA is an analytical procedure that looks at the safeguards to see if the protection provided is adequate for every known risk. It is a powerful analytical tool for assessing the adequacy of protection layers used to mitigate process risk. LOPA builds upon well-known process hazards analysis techniques, applying semiquantitative measures to the evaluation of the frequency of potential incidents and the probability of failure of the protection layers.

4.11 Bayesian networks (BNs)

From the 1960s onward, when the first risk assessment techniques were conceptualized, the application and implementation of such methods have proven very important for continuously improving safety in all industrial sectors and all over the world. Indeed, despite the fact that they do not provide an exact description of reality, FMEA, FTA, QRA or LOPA, for instance, all well-known and state-of-the-art risk analysis techniques have proved to be effective predictive tools in the past decades to assess many risks related to industrial processes and make a difference with respect to safety.

Nonetheless, such methods remain static, and thus they only provide a snapshot or a proxy of the risk picture of any scope they are used for, every 3 years or so (whatever the frequency is with which the risk analysis technique is reused). Meanwhile, we all know that an organizational environment, a workplace, is very dynamic: things (work processes, storage activities, transport tasks, humans, critical assets, etc.) change all the time, and so do the risks. In these times, with social media, the Internet of things, real-time information based on sensor data, big data, etc., the proxy of the risk picture can, and should, be more timely, more updated all the time, more dynamic. Therefore, techniques such as Bayesian networks (BNs) can be used. Probabilities of scenarios can be updated relatively easily with new "evidence" that becomes available. This section explains how a BN works.

A BN is a network composed of nodes and arcs, where the nodes represent variables and the arcs represent causal or influential relationships between the variables. Hence, a BN can be viewed as an approach used to have an overview of relationships between causes and effects of a system. In addition, each node/variable has an associated conditional probability table (CPT). Thus, BNs can be defined as directed acyclic graphs, in which the nodes represent variables, arcs signify direct causal relationships between the linked nodes and the CPTs assigned to the nodes specify how strongly the linked nodes influence each other [32]. The key feature of BNs is that they enable us to model and reason about uncertainty [33].

The nodes without any arc directed into them are called "root nodes," and they are characterized by marginal (or unconditional) prior probabilities. All other nodes are intermediate nodes, and each one is assigned a CPT. Among intermediate nodes, the nodes with arcs directed into them are called "child nodes" and the nodes with arcs directed from them are called "parent nodes." Each child has an associated CPT, given all combinations of the states of its parent nodes. Nodes without any child are called "leaf nodes."

As an illustrative example, and to keep it as simple as possible, we assume in Fig. 4.33 that the variables are discrete; hence, they are either "true" or "false." Of course, in industrial practice, some variables, such as "No water available from source X," could be continuous on a scale from zero to a certain possible maximum flow of water.

Figure 4.33 illustrates that a technical failure of the fire hose F can cause Fireman A to be unable to extinguish the fire. Also, water not being available from source X may cause (or influence) both Fireman A and Fireman B to be unable to extinguish the fire. In this example, no water available from source X does not imply that Fireman B will definitely be unable to extinguish the fire, as he may tap water from a different source – e.g., source Y – but there is an increased probability that he will not be able to extinguish the fire. The CPT for the node/variable "Fireman B cannot extinguish fire" holds this information, providing the conditional probability of each possible outcome given each combination of outcomes for its parent nodes. Figure 4.34 shows a CPT for "Fireman B cannot extinguish fire."

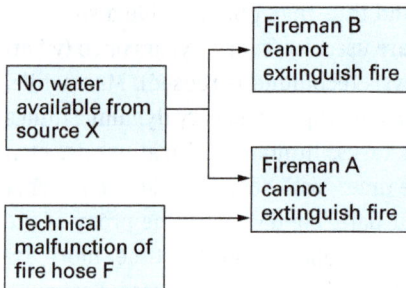

Fig. 4.33: An illustrative example of a Bayesian network.

Figure 4.34 illustrates that, e.g., the probability that Fireman B cannot extinguish the fire, given that there is water available from source X is 0.1. Obviously, the CPTs for the root nodes are very simple (they have no parents), and we only have to assign a probability to the two states "true" and "false" (see Fig. 4.35). On the contrary, the CPT for Fireman A is more complicated as this node has two parents, leading to four combinations of parent states (see Fig. 4.36).

The probabilities that have to be assigned to the different states of the variables can be determined in different ways. One way is to use historical data and statistical information (e.g., observed frequencies) to calculate the probabilities. Another way, especially when no or insufficient statistical information is available, is to use expert opinion. Hence, both probabilities based on objective data and subjective probabilities can be used in BN.

		No water available from source X	True	False
Fireman B cannot extinguish fire	True		0.8	0.1
	False		0.2	0.9

Fig. 4.34: CPT for "Fireman B cannot extinguish fire."

	True	0.4
Technical malfunction of fire hose F		
	False	0.6

	True	0.1
No water available from source X		
	False	0.9

Fig. 4.35: CPTs for "Technical malfunction of fire hose F" and "No water available from source X."

		Technical malfunction of fire hose F	True		False	
		No water available from source X	True	False	True	False
Fireman A cannot extinguish fire	True		0.8	0.6	0.6	0.3
	False		0.2	0.4	0.4	0.7

Fig. 4.36: CPT for "Fireman A cannot extinguish fire."

BN are based on Bayes' theorem and Bayes' theory to update initial beliefs or prior probabilities of events using data observed from the event studied. Bayes' theorem can be expressed as follows:

$$P(\text{Event}, \text{info}) = \frac{P(\text{Event}) \cdot P(\text{info} \mid \text{Event})}{P(\text{info})}$$

In this equation, P(Event) is the prior probability of an event, P(Info|Event) is the likelihood function of the event, P(info) is the probability of the info/data observed (commonly called evidence) and P(Event|Info) is the posterior probability of the event. Hence, we start with a prior probability of the event, but we are interested in knowing what the posterior probability of the event is, given the evidence "info." This can be done using Bayes' theorem.

Let us apply this theorem to our example. Assume that we find out that Fireman B is indeed not able to extinguish the fire. Then, intuitively, we feel that the probability of water not being available from source X must have increased from its prior value of 0.1 (see Fig. 4.34). But by how much? Bayes' theorem provides the answer. We know, e.g., from the CPT that P (Fireman B cannot extinguish fire | no water available from source X) = 0.8 and that P (no water available from source X) = 0.1. So the numerator in Bayes' theorem is 0.08. The denominator, P (Fireman B cannot extinguish fire), is the marginal (or "unconditional," as already mentioned) probability that Fireman B is not able to extinguish the fire: it is the probability that Fireman B cannot extinguish the fire when we do not know any specific information about all variables or events that influence it (in the case of our example, the availability of water from source X). Although this value cannot be directly determined, it can be calculated using probability theory and equals 0.17. Hence, substituting this value in Bayes' theorem, we obtain:

P(no water available for source X | Fireman B cannot extinguish fire)

$$= 0.08/0.17 = 0.471$$

Therefore, the observation of Fireman B not being able to extinguish the fire significantly increases the probability that there is no water available from source X: from 0.1 to 0.471. In a similar way, we are able to use the information to calculate the revised belief of Fireman A not being able to extinguish the fire. Once we know that Fireman B is not able to extinguish the fire, the prior probability of 0.446 (for Fireman A not being able to extinguish the fire) increases to 0.542. In fact, the calculations (which can be quite cumbersome) are done automatically in any BN tool.

Updating the probabilities with evidence and new information is called "propagation." Any number of observations can be input anywhere in a BN, and propagation can be employed to update the marginal probabilities of all the unobserved variables.

As Fenton and Neil [33] explain, BNs offer several important benefits when compared with other available techniques. In BNs, causal factors are explicitly modeled. In contrast, in regression models, historical data alone are used to produce equations relating dependent and independent variables. No expert judgment is used when insufficient information is available, and no causal explanation is carried out. Similarly, regression models cannot accommodate the impact of future changes. In short, classical statistics (e.g., regression models) are often good for describing the past but poor for predicting the future. A BN will update the probability distributions for every unknown variable whenever an observation or evidence is entered into any node. Such a technique of revised probability distributions for the cause nodes as well as for the effect nodes is not possible in any other approach. Moreover, predictions are made with incomplete data. If no observation is entered, then the model simply assumes the prior distribution. Another advantage is that all types of evidence can be used: objective data as well as subjective beliefs.

This range of benefits, together with the explicit quantification of uncertainty and the ability to communicate arguments easily and effectively, makes BNs a powerful solution for all types of risk assessment.

Whereas the essential BN is static, in many real-world problems, the changes in values of uncertain variables need to be modeled over successive time intervals. Dynamic Bayesian networks (DBNs) extend static BN algorithms to support such modeling and inference. Hence, temporal behavior can be captured by DBNs, and the necessary propagation is carried out on compact (rather than expanded, unmanageable and computationally inefficient static) models.

> The scenario in a risk analysis can be defined as the propagating feature of a specific initiating event which can go to a wide range of undesirable consequences. If we take various scenarios into consideration, the risk analysis becomes more complex than it does without them. BNs allow for considering the effects of predictions by incorporating knowledge with data to develop and use causal models of risk that provide powerful insights and better decision-making.

4.12 Functional resonance analysis method (FRAM)

Since two decades, safety scientists have been looking for ways to understand how performance shaping factors or performance conditions could "force" people to fail. Insights into "unsafety," or how things go wrong, have shifted from inherent human "error mechanisms" and behavioral safety more toward workers being a product of working conditions and work pressures. Although this change enabled people to understand accidents of a more complex nature for a while, it still fell short in a number of situations. This led to the recognition, strongly supported by resilience engineering, that failures and successes have the same source and that metaphorically speaking, they are two sides of the same coin.

The FRAM [34] provides a way to describe outcomes using the idea of resonance arising from the variability of everyday performance. To arrive at a description of functional variability and resonance and to lead to recommendations for damping unwanted variability, a FRAM analysis consists of five steps:

i.   Identify and describe essential system functions and characterize each function using the six basic characteristics (aspects). The following aspects are used by FRAM: input, output, precondition, resource, control and time. In the first version, only use and describe the aspects that are necessary or relevant. The description can always be modified later. Functions are represented by hexagons.

ii. Check the completeness/consistency of the model.

iii. Characterize the potential variability of the functions in the FRAM model as well as the possible actual variability of the functions in one or more instances of the model.

iv. Define the functional resonance based on dependencies/couplings among functions and the potential for functional variability, thereby modeling it by using hexagons.

v.   Identify ways to monitor the development of resonance either to dampen variability that may lead to unwanted outcomes or to amplify variability that may lead to desired outcomes.

Remark that FRAM is a method to analyze how work activities take place either retrospectively or prospectively. This is done by analyzing work activities in order to produce a model or representation of how work is done. This model can then be used for specific types of analysis, whether to determine how something went wrong, to look for possible bottlenecks or hazards, to check the feasibility of proposed solutions or interventions or simply to understand how an activity (or a service) takes place. The FRAM is a method for modeling nontrivial sociotechnical systems. In this sense, the technique should not be regarded as a typical risk assessment method or an accident analysis method. Nonetheless, the model produced by a FRAM analysis can serve as the basis for a further risk analysis or an event investigation.

4.13 Tips for using risk diagnostic tools

Managing projects, situations, processes, equipment or things you are doing without addressing the fundamental risks that threaten them can be disastrous. There are many risk analysis methodologies available. None of them are universal, and a careful selection should be made depending on the purpose, the data, the maturity of the process, the internal expertise and the object of the study. These techniques are a process for identifying and analyzing scenarios, which can be seen as undesirable events or results of a process, and determining whether the risks (calculated by using a formula that the risk analysis technique user agrees with) are acceptable. If risks are unacceptable, the process may include recommendations and assessments of risk control measures. The methodology may include the following steps:

–   Description of activity or process
–   HAZID (including categorization into type I/type II risk)

– Accident and incident scenario generation (sometimes the worst-case scenario is a conservative option; other options are, for instance, the most probable case scenario or the worst credible case scenario)
– Likelihood of occurrence or frequency estimation
– Consequence or impact estimation
– Risk evaluation

Further steps may include the implementation, the generation of risk control measures and a repeat of the steps to evaluate the new risk resulting from the implementation of the measures suggested by the analysis.

4.14 Conclusion

The field of risk analysis has assumed increasing importance in recent years, given the concern by both the public and private sectors regarding safety, health, security and environmental problems. The approaches to risk assessment normally center on identifying a risk and then assessing the extent to which the risk would be a problem (the severity of the risk) and calculating how likely it is that the risk would occur (the probability). Following this, attempts are made to mitigate the risk (ideally eliminate the source or take measures to reduce the likelihood of it happening). Where the possibility of the risk occurring can be reduced no further, steps should be taken to monitor the risk (detection), preferably as an early warning, so that action can be taken.

There is no universal risk analysis techniques were presented in this chapter. Qualitative methodologies, though lacking the ability to account for the dependencies between events, are effective in identifying potential hazards and failures within the system. The tree-based techniques address this deficiency by taking into consideration the dependencies between each event. The probabilities of occurrence of the undesired event can also be quantified with the availability of operational data. Methods such as QRAs and BNs use quantitative data to make as accurate predictions as possible.

There is no universal risk analysis method. Each situation analyzed will require an adequate method, whether it is descriptive, qualitative, semiqualitative, quantitative or predictive as well as static and dynamic or analytic and systemic. Successful risk analyses require scientists and engineers to undertake assessments to characterize the nature and uncertainties surrounding a particular risk. It is up to the team performing the risk analysis to use the correct tool in a similar way that a carpenter would not use a screwdriver to hammer a nail even if he would succeed eventually.

Regardless of the prevention techniques employed, possible threats that could arise inside or outside the organization need to be assessed. Although the exact nature of potential disasters or their resulting consequences are difficult to determine, it is beneficial to perform a comprehensive risk assessment of all threats that can realistically occur at – and to – the organization.

We underlined that in undertaking risk assessments, it is important to attempt risk mitigation and to attempt to lower the risk until the risk can be lowered no further. This involves identifying actions to reduce the probability of an event and to reduce its severity. When this can be taken no further, the focus should move toward providing a more reliable detection method designed to initiate a reliable response to a risk event. A further important consideration is that risk-assuming actions should be periodically reassessed.

References

[1] Reniers, G.L.L., Dullaert, W., Ale, B.J.M., Soudan, K. (2005). Developing an external domino accident prevention framework: Hazwim. J. Loss Prev. Proc. 18: 127–138.
[2] Groso, A., Ouedraogo, A., Meyer, T. (2011). Risk analysis in research environment. J. Risk Res. 15: 187–208.
[3] Ericson, C.A. (2005). Hazard Analysis Techniques for System Safety. Fredericksburg, VA: John Wiley & Sons, Inc.
[4] Ahmadi, A., Soderholm, P. (2008). Assessment of operational consequences of aircraft failures: Using event tree analysis. IEEE Aerosp. Conf. 1(9): 3824–3837.
[5] CCPS. (2008). Guidelines for Hazard Evaluation Procedures. New York: John Wiley & Sons, Inc.
[6] Swann, C.D., Preston, M.L. (1995). Twenty-five years of HAZOPs. J. Loss Prevent. Process Ind. 8: 349–353.
[7] Warner, F. (1975). Flixborough disaster. Chem. Eng. Prog. 71: 77–84.
[8] Cagno, E., Caron, F., Mancini, M. (2002). Risk analysis in plant commissioning: The multilevel HAZOP. Reliab. Eng. Syst. Safety. 77: 309–323.
[9] Cocchiara, M., Bartolozzi, V., Picciotto, A., Galluzzo, M. (2001). Integration of interlock system analysis with automated HAZOP analysis. Reliab. Eng. Syst. Safety. 74: 99–105.
[10] Labovsky, J., Svandova, Z., Markos, J., Jelemensky, L. (2008). HAZOP study of a fixed bed reactor for MTBE synthesis using a dynamic approach. Chem. Pap. 62: 51–57.
[11] Mushtaq, F., Chung, P.W.H. (2000). A systematic HAZOP procedure for batch processes and its application to pipeless plants. J. Loss Prev. Proc. 13: 41–48.
[12] Ruiz, D., Benqlilou, C., Nougues, J.M., Puigjaner, L., Ruiz, C. (2002). Proposal to speed up the implementation of an abnormal situation management in the chemical process industry. Ind. Eng. Chem. Res. 41: 817–824.
[13] McCoy, S.A., Zhou, D.F., Chung, P.W.H. (2006). State-based modeling in hazard identification. Appl. Intell. 24: 263–279.
[14] Zhao, C., Bhushan, M., Venkatasubramanian, V. (2005). Phasuite: An automated HAZOP analysis tool for chemical processes. Part I: knowledge engineering framework. Process Saf. Environ. Prot. 83: 509–532.
[15] Chin, K.S., Chan, A., Yang, J.B. (2008). Development of a Fuzzy FMEA based product design system. Int. J. Adv. Manuf. Tech. 36: 633–649.
[16] Franceschini, F., Galetto, M. (2001). A new approach for evaluation of risk priorities of failure modes in FMEA. Int. J. Prod. Res. 39: 2991–3002.
[17] Scipioni, A., Saccarola, G., Centazzo, A., Arena, F. (2002). FMEA methodology design, implementation and integration with HACCP system in a food company. Food Control. 13: 495–501.
[18] Thivel, P.X., Bultel, Y., Delpech, F. (2008). Risk analysis of a biomass combustion process using MOSAR and FMEA methods. J. Haz. Mat. 151: 221–231.

[19]   Su, C.T., Chou, C.J. (2008). A systematic methodology for the creation of six sigma projects: A case study of semiconductor foundry. Expert. Syst. Appl. 34: 2693–2703.

[20]   Price, C.J., Taylor, N.S. (2002). Automated multiple failure FMEA. Reliab. Eng. Syst. Safety. 76: 1–10.

[21]   D'Urso, G., Stancheris, D., Valsecchi, N., Maccarini, G., Bugini, A. (2005). A New FMEA Approach Based on Availability and Costs. In Advanced Manufacturing Systems and Technology. Kuljanic, E., (ed.). New York: Springer-Verlag, p. 703–712.

[22]   Ericson, C.A. (1999). Fault Tree Analysis – A History. A History from the Proceedings of the 17th International System Safety Conference. Orlando, Florida, USA, p.1–9.

[23]   Hauptmanns, U., Marx, M., Knetsch, T. (2005). GAP – A fault-tree based methodology for analyzing occupational hazards. J. Loss Prev. Proc. 18: 107–113.

[24]   Khan, F.I., Abbasi, S.A. (1998). Techniques and methodologies for risk analysis in chemical process industries. Process Safety Prog. 11: 261–277.

[25]   Ekaette, E., Lee, R.C., Cooke, D.L., Iftody, S., Craighead, P. (2007). Probabilistic fault tree analysis of a radiation treatment system. Risk Analysis. 27: 1395–1410.

[26]   Ale, B.J.M. (2009). Risk: An Introduction. The Concepts of Risk, Danger, and Chance. London, Routledge: Taylor & Francis Group UK.

[27]   Department of Defense. (2000). Standard Practice for System Safety. MIL-STD-882D. Wright-Patterson Air Force Base. OH: HQ Air Force Material Command.

[28]   Garvey, P.R. (2009). Analytical Methods for Risk Management. A Systems Engineering Perspective. Boca Raton, FL: CRC Press.

[29]   Center for Chemical Process Safety. (2000). Guidelines for Chemical Process Quantitative Risk Analysis. 2nd edn. New York: American Institute of Chemical Engineers.

[30]   Crawley, F., Tyler, B. (2003). Hazard Identification Methods, European Process Safety Centre. Rugby, UK: Institution of Chemical Engineers.

[31]   Center for Chemical Process Safety. (2001). Layer of Protection Analysis. Simplified Process Risk Management. New York: American Institute of Chemical Engineers.

[32]   Torres-Toledano, J.G., Sucar, L.E. (1998). Bayesian networks for reliability analysis of complex systems. Lect. Notes Comput. Sci. 1484: 195–206.

[33]   Fenton, N., Neil, M. (2007). Managing Risk in the Modern World. Applications of Bayesian Networks. London: London Mathematical Society.

[34]   Hollnagel, E. (2012). FRAM, the functional resonance analysis method. Modelling complex socio-technical systems. Ashgate.

5 Risk treatment/reduction

5.1 Introduction

Risk treatment encompasses all strategies, whose goal is to limit and/or decrease risks and damages related to a specific domain. Risk reduction is the promotion of health and assets. It is empowerment and is quoted in a Canadian concept as follows: "Its role is to give the community the means to assume its destiny" [1].

Risk treatment, control, and reduction techniques are designed to decrease the likelihood of an unwanted event (considering exposure to risks leading to the event) and to reduce the severity of potential losses resulting from the event or to make the likelihood or impact more predictable. Examples include the following:
- Substitution: replacing substances and procedures with less hazardous ones, by improving construction work, etc.
- Elimination of risk exposure: consisting of not creating or of completely eliminating the condition that could give rise to the exposure.
- Prevention: combining techniques to reduce the likelihood/frequency of potential losses. Observation and analysis of past accidental events enable the improvement and intensification of prevention measures.
- Reduction/mitigation: assessing techniques to reduce the severity of accidental losses when an accident occurs.
- Measures applied before the occurrence of the event (often also have an effect on the likelihood/frequency).
- Measures applied after the occurrence of the event (often aim to accelerate and enhance the effectiveness of the rescue).
- Segregation: summarizing the techniques that are to minimize the overlapping of losses from a single event. It may imply very high costs.
    - Segregation by separation of high-risk units.
    - Segregation by duplication of high-risk units.

Another option is so-called transfer, i.e., risk transfer by
- Contractual transfer of the risk financing, essentially insurance.
- Risk financing by retention (auto financing) and financial planning of potential losses by your own resources.
- Alternative risk: the alternative risk transfer solutions comprise both elements of auto financing and contractual transfer and so cannot be classified in any of the above categories.

In summary, to discuss a systematic categorization of the risk treatment options, there are several ways to deal with risks:

https://doi.org/10.1515/9783111493633-005

– Risk acceptance comprising (i) risk transfer and (ii) risk retention
– Risk reduction comprising (i) risk control and (ii) risk avoidance

Risk ownership increases from "acceptance" toward "reduction," and more specifically from low to high ownership of the risk, the risk treatment options can be ordered as follows: transfer → retention → control → avoidance.

Risk avoidance and risk control can be achieved in an optimal way by inherent safety or design-based safety. There are five principles of design-based safety. The possible implementation of these five principles is explained for the case of the chemical industry as illustrated in Fig. 5.1.

Principle 1: Intensification

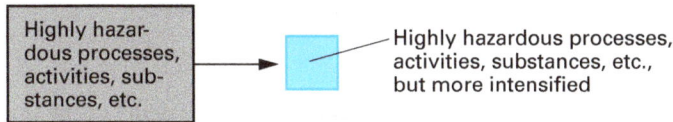

Principle 2: Substitution

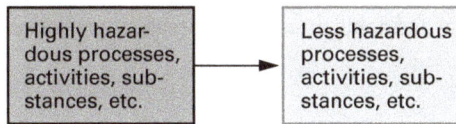

Principle 3: Attenuation by moderation

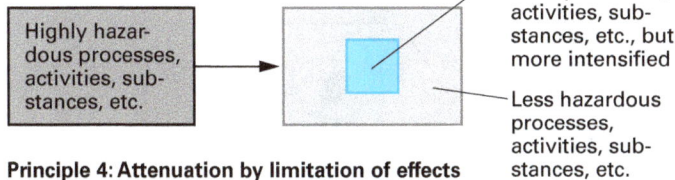

Principle 4: Attenuation by limitation of effects

Principle 5: Simplification

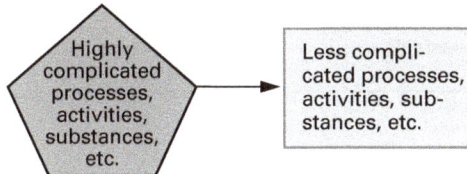

Fig. 5.1: Five principles of design-based safety implemented in the chemical industry (source: based on [2, 3]).

The simplified model of events describes the nondesired event in a concise way. It is composed of distinct parts, of which the succession leads to the event as illustrated in Fig. 5.2.

To physically model an accidental pathway, we must have a sequence starting from a potential energy and a process using this potential energy. Then an incident may happen as an escape of this energy in an uncontrolled flux and direction. If this escape of energy reaches a vulnerable resource (hazardous phenomenon), an accident happens. In the end, the vulnerable resource suffers from damage or injury.

Potential energy can be
- Physical: mechanical, electrical, thermal, radiations (ionizing or not), pressure, etc.
- Chemical: reactivity, activity, corrosivity, flammability, etc.
- Physicochemical: phase change (volatility), division state (granulometry), etc.
- Physiological: sensitization, irritation, skin penetration, toxicity, etc.
- Biological: potential pathogen, mutagen, etc.

Risks/scenarios are estimated according to the two criteria: the probability of the unwanted event/scenario and the severity of the consequences of the unwanted event/scenario. Employers have a responsibility to take adequate prevention and/or protection and mitigation measures to avoid these accidents and/or to decrease their consequences.

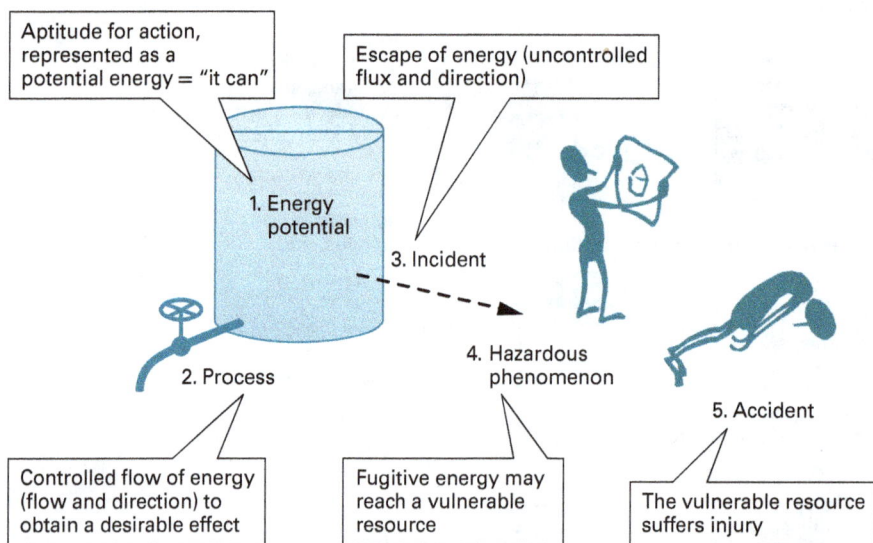

Fig. 5.2: Physical modeling of unwanted events.

Different safety measures are used to mitigate risks [4]:
– Disincentives for avoiding the implementation of potential threats.
– Preventive measures inhibiting a threat realization (they act on causes and reduce the probability of occurrence).
– Protective (and mitigation) measures limiting the impact of a threat by reducing
  the direct consequences (they act on consequences without taking into account
  the occurrence).
– Remedial actions limiting the consequences of the threat implementation and indirect consequences.
– Recovery measures aiming to recover damages by transferring the risk (e.g., insurance) to limit final losses.

Risk reduction aims to influence the occurrence of a hazard and/or its consequences.
The two dimensions of vulnerability (occurrence and severity) are used to character-
ize the instruments and actions:
– Preventive instruments
  – They aim to reduce the likelihood of occurrence of a disaster.
  – They act on the causes and reduce the probability of occurrence.
– Protective/mitigation instruments
  – They aim to limit the consequences of a disaster.
  – They act on the consequences without taking into account the occurrence.

The goal of all reduction methods is summarized in Fig. 5.3, where we can observe
that risks that are considered unacceptable (with the visual help of the risk matrix,
see Section 4.8) have to be brought down to an acceptable or at least tolerable level.
Four areas can be focused upon
– When severity and occurrence are low, then we should concentrate on periodi-
  cally reviewing the situation in order to keep this low level of risk.
– When severity is low and occurrence is high, good housekeeping is necessary.
– When severity is high and occurrence is low, contingency plans should be adopted.
– When both severity and occurrence are high, then we should manage actively the
  risk in order to bring it to a safer level.

We note in the middle zone of the matrix a diagonal intermediate region between the
unacceptable zone (needing immediate action) and the acceptable zone. Below the acceptable zone, there is the negligible zone. The middle zone – the "tolerable zone" – is
the one where the ALARA ("as low as reasonably achievable") principle occurs (meaning that either we have to prepare for mitigation measures or we have to tolerate the
risk knowingly). ALARA means making every reasonable effort to decrease vulnerabilities in order that the residual risk shall be as low as reasonably achievable or
practicable (in the latter case, the principle is called "ALARP," see Section 3.5.2 on soci-

etal risk). This concept was published in the UK in 1974, in the Health and Safety at Work Act [5]. Type II risks can be situated in the unacceptable risk zone as well as in a part of the ALARA zone (the upper part). Another part of the ALARA zone (the lower part) together with the acceptable zone is the region where type I risks are located.

In principle, the ALARA risk problems are decision problems and require a choice among alternative options. Investing in one safety improvement option implicitly means that another safety investment option cannot be chosen since budgets are not infinite. This is where microeconomics come into play and can be used to provide economic-based information as an input for the decision-making process. The "most attractive" or "most acceptable" option has to be determined. Strictly speaking, one does not simply "accept risks." One seeks for accepting investment options that entail some level of risk, related to a variety of still possible accident scenarios (after implementation of the option) as well as option-related avoided accident scenarios. One of the options can be "do nothing," obviously. The attractiveness of a safety/risk reduction investment option depends upon its full set of relevant positive and negative consequences.

The decision-making process for the options is inherently situation-specific: there are no universally acceptable options (or risks, costs or benefits). The choice of an option (and it associated risks, costs and benefits) depends on the set of options, consequences, likelihoods, values and facts examined in the decision-making process, and in different circumstances, different options, values and information may be relevant. Moreover, over time, any of a number of changes could lead to a change in the relative attractiveness of any given option: errors in the analysis may be discovered, new safety devices may be invented, values may change, additional information may come to light and so forth. Even in the same situation and at a single time, different people with different values, beliefs, objectives or decision methods might disagree on which option is best. In brief, "absolute acceptability," however tempting the concept is, does not exist [6].

The risk reduction process can be described by 10 successive steps:

1. Build a multidisciplinary team that includes specialists in different subjects and sectors, users and a moderator. Sometimes a contribution from someone "innocent" (with no or limited experience) is desired. He/she asks questions no one else would think of.
2. Identify and define the risks to be reduced (hazard source, situation, requirements, etc.)
3. Analyze the process and operating procedures (description, interaction, objectives, need, constraints, etc.)
4. Generate ideas by brainstorming (a priori, all ideas are good). The multidisciplinary nature of the team reveals its importance. It is crucial not to discriminate against ideas at this stage, and it is the moderator's role to guarantee this. From this step will come an accepted solution, and discriminating against it too early would be prejudice, especially without selection criteria.

5. Establish criteria to support the evaluation of the suggestions and ideas, decision help and a referential for the different group members.
6. Evaluate all alternatives [costs (direct and indirect), risk reduction, feasibility, collateral effects, acceptation, opportunity, new hazard apparition, etc.]
7. Retain risk reduction alternatives based on criteria.
8. Implement the decided modification (definition of the person in charge, schedule and deadlines and which necessary means and resources).
9. Information and training – this step is often forgotten but is crucial. Communication and training is an inevitable need in order for the users to accept and use corrective or improvement measures during a process or activity change. Without communication and training, the will or acceptance of the people who are concerned by the result cannot be guaranteed.
10. Control – it is indispensable to evaluate the new situation, evaluate the new hazards and risks that have appeared during the implementation of the measures, check the acceptance of the measure by the users and validate the project.

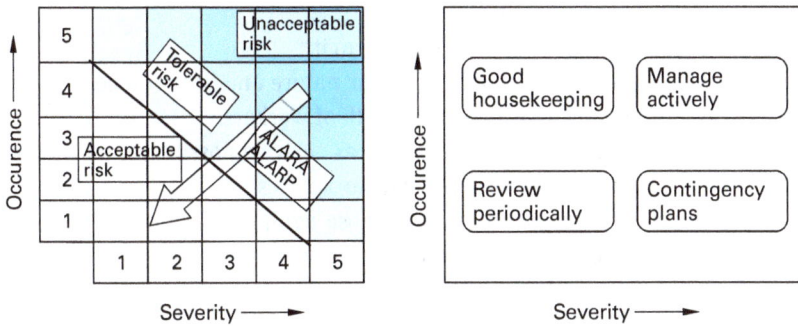

Fig. 5.3: The effect of risk reduction is represented in the risk matrix.

Risk reduction encompasses all strategies whose goal is to limit the risks and damages related to a specific domain. Risk reduction is the promotion of health and assets. It is an integral part of the risk management process of identifying, assessing and controlling risks arising from operational factors and making decisions that balance risk cost with mission benefits.

5.2 Prevention

Prevention is an attitude and/or a series of measures to be taken to avoid the degradation of a certain situation (social, environmental, economical, technological, etc.) or to prevent accidents, epidemics or illnesses. It has the following characteristics:

- Prevention limits risk with measures aiming to prevent it by eliminating or decreasing the occurrence probability of a hazardous phenomenon. It can be without effect on the severity of potential accidents and disasters.
- Preventive measures act upon the causality chain that leads to a loss (a break at a given chain location), e.g., the prohibition of smoking in a room containing flammable materials.
- The effective prevention measures must be based on careful analysis of the occurring causes.

Prevention measures imply
- Organizing actions for preventing professional risks, information and training
- Implementing an organization with adapted means
- Taking into account changes that are likely to happen (new products, new work rhythms, etc.)
- The improvement of existing situations

The nine principles of prevention include:
1. *Avoid* risks: remove the hazard or the exposure to it.
2. *Assess* risks that cannot be avoided: assess their nature and importance, identify actions to ensure safety and guarantee the health of workers.
3. *Fight* risks at the source: integrate prevention as early as possible, from the design of processes, equipment, procedures and workplaces.
4. *Adapt* work to man: design positions and choose equipment, methods of work and production to reduce the effects of work on health.
5. *Consider the state* of technological developments: implement preventive measures in line with technical and organizational developments.
6. *Replace* the hazardous by what is less hazardous: avoid the use of harmful processes or products when the same result can be obtained by a method with fewer hazards.
7. *Plan* prevention integrated in a coherent package: (i) technique, (ii) work organization, (iii) working conditions, (iv) social relations and (v) environment.
8. Take *collective protection* measures and give them priority over individual protective measures: use of personal protective equipment (PPE) only to supplement collective protection or their defaults.
9. Give *appropriate instructions* to employees: provide them the necessary elements for understanding the risks and thus involve them in the preventive approach.

Preventive measures are developed, taking into account the additional costs generated over the risk. They constitute the *prevention plan*.

5.2.1 Seveso Directive as prevention means for chemical plants

After the emotion induced by the release of dioxin in Seveso, Italy, in 1976, European member states introduced a directive on the control of major accident hazards involving dangerous substances. On June 24, 1982, the Seveso Directive asked the member states and enterprises to identify the risks associated with industrial activities and the necessary measures to face them. The Seveso Directive was modified several times, and its field of action was progressively enlarged, specifically after the Schweizerhalle (Basel, Switzerland) accident in 1986.

The framework of this action is from then on the 96/82/CE Directive, also called the Seveso II Directive, regarding the control of hazards linked to major accidents involving hazardous substances. Amendments (2003/105/CE), which modified the 96/82/CE (Seveso II) Directive, were published in the official EU journal on December 31, 2003 [7]. The new directive strengthened the concept of major disaster prevention by requiring a company to include the implementation of a management system and organization (or safety system management) proportionate to the risks inherent to the plant. In 2010, the European Commission adopted a proposal for a new directive on the control of major accident hazards involving dangerous substances that replaced Seveso II including the global harmonized system for chemicals (GHS). Meanwhile, the Seveso III European Directive (2012/18/EU) [8] has been published, introducing the GHS for the classification of hazardous substances. The Directive applies to more than 12,000 industrial establishments in the European Union (EU), where dangerous substances are used or stored in large quantities, mainly in the chemical and petrochemical industry as well as in fuel wholesale and storage (including LPG and LNG) sectors.

But what is the procedure to change legislation concerning major accidents within Europe? Vierendeels et al. [9] provide the answer, as illustrated in Fig. 5.4.

Some background information is given in Fig. 5.4. Research, debates and analyses on the prevention and protection of human and environment take place on a regular basis, as a "standard procedure" to continuously improve existing regulations. This long-term knowledge (upper left corner of Fig. 5.4), based on reliable scientific evidence such as expert knowledge, scientific literature, new insights in managing safety and managing chemical products and processes, new best available practices or techniques, international conferences and seminars, lead to changing regulations. This can be considered the "standard procedure" within the EU to change and improve existing legislation. But there are other, "nonstandard," factors that lead to new legislation within Europe. The full process is therefore described here.

The European Commission is responsible for writing a proposal and sending it to the European Parliament. Once written by the Commission, a proposal is made public, and a debate arises among all stakeholders such as industry federations and nongovernmental organizations. Advice from different parties (see "advisory organs" in Fig. 5.4) is provided to the European Commission, officially as well as in more indirect ways (i.e., through influencing and lobbying activities even before the first official text is written).

The European Commission subsequently sends an "influenced proposal" to the European Parliament and the Council, where it is discussed by the politicians. A lot of contacts exist between the Parliament and the Council via the so-called Coreper, which is a Committee of Permanent Representatives. The proposal can either be accepted or amended by the Parliament. If amendments are made to the proposal, it is sent back to the European Commission. The Commission then takes all remarks, comments, etc., into account and writes a second proposal. This process of amendment by the Parliament can be repeated once. Then the Coreper is responsible for making a draft of a compromise document. If this fails, the proposal is definitely rejected. If the Coreper succeeds in making a compromise text, the Commission may send this compromise proposal to the Parliament and the Council for a third reading. A European Directive or Amendment then finally comes into force after an agreement (by votes) is settled between politicians of the European Parliament. If in this final stage there is no majority for the proposal, it is definitely rejected.

As the Seveso legislation concerns and includes highly specific and technical topics, the European politicians are advised by experts in various working groups. In and between these working groups, as well as between the working groups/experts and the European institutions, a continuous reciprocal exchange of information and data takes place. This can be found in the middle of Fig. 5.4. Experts are categorized into three working groups: (a) technical working groups (TWGs), (b) the Committee of Competent Authorities (CCA) and (c) working groups of the Council. In TWGs, technical topics that should be defined more accurately by the legislator or that should be technically elaborated more in depth in the regulations are discussed. Findings and recommendations of TWGs are reported to the CCA. The CCA, composed of member states representatives, discusses the Seveso legislation implementation and prepares the texts to be used by the European Commission to write a proposal (i.e., for changing the legislation). The working groups of the Council are composed of government officials of the EU and of the member states' ministries. These government representatives seek advice from technical experts of their own country.

Although the Seveso legislation is extremely technical (concerning chemical products, their characteristics and tiers, etc.) and the technical experts from the different working groups contribute in an essential way with their expertise and their knowledge, political influences are present throughout the entire legislation process. For example, the experts are influenced by a variety of pressure groups, local politicians and private companies. Local politicians influence European politicians who will meet with the CCA. During these negotiations, lobbying activities are ever present. Such lobbying can be seen as some kind of cost–benefit analysis, which ultimately leads to a societal optimum between ecology, safety, health and economy. The media and the press also play an important role in this regard and may/will influence politicians in particular [10].

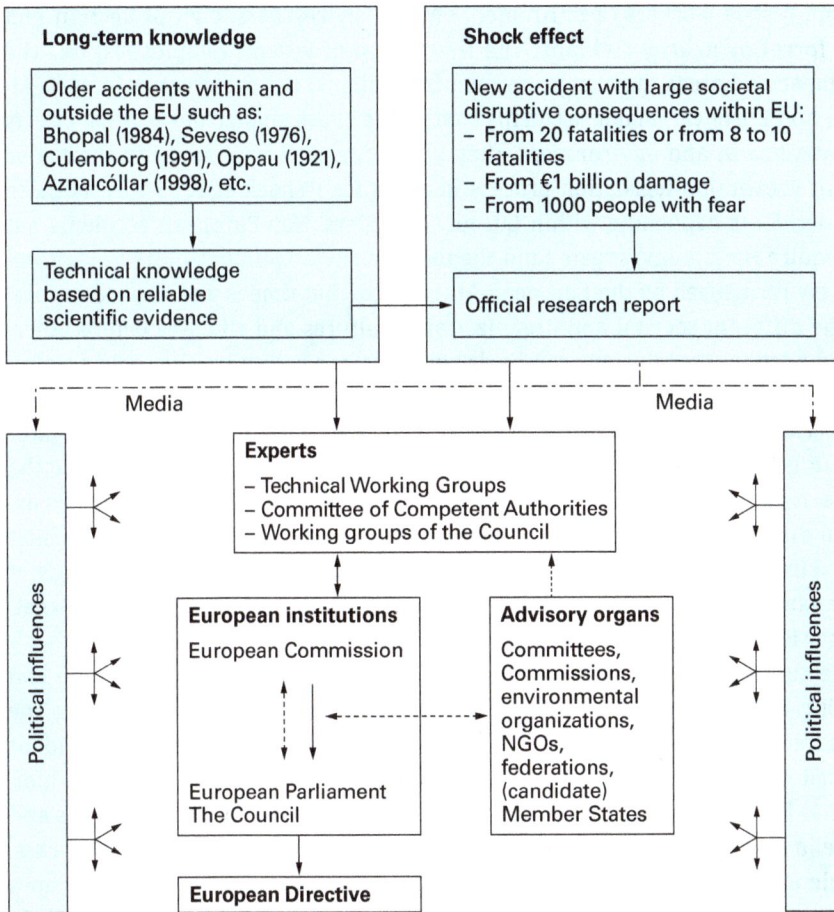

Fig. 5.4: Major accident prevention legislation model within Europe [7].

Then, besides the continuously updating and improving regulations via the experts/tech-
nicians and the working groups, a "shock effect" (via a major accident) may also lead to
changing Seveso regulations. The left upper part of Fig. 5.4 illustrates the long-term continuous improvement of the Seveso legislation, whereas the right upper part denotes the
shock effect that could possibly lead to changes. But when can an accident be labeled as a
shock effect? An important element is the number of fatalities accompanying the accident. Mac Sheoin [11] indicates that the number of people killed has an impact on the
speed of the legislation change process. Research revealed that 20 casualties will cause a
shock effect for certain. However, societal disruption can already take place from eight to
ten fatalities, in coherence with other factors. A parameter that has a certain influence on
the perceived impact of an accident (and thus indirectly on the level of pressure to
change the regulations) is the type of victims: employees of the company suffering the

accident, rescue workers (e.g., firemen, emergency rescuers, civil protection and police force) or nearby civilians. The first group of victims (employees) has the least impact, whereas the last group (nearby civilians) has the greatest impact. Although non-European major accidents sometimes cause similar devastating effects to human health and environment, they are not taken into account in the major accident prevention legislation process in a similar ("shock effect") way as those major accidents happening within European borders. Non-European accidents not only require time to investigate (and the incident investigation reports are not immediately recognized by the European Authorities), but time is also needed to compare the different societal and organizational cultures and climates before objective and adequate conclusions can be drawn. They are certainly taken into account in the major accident prevention legislation change process in the long run. The older accidents, which did not play a role at first but became influential at later stages in time, are denoted in Fig. 5.4 in the upper left corner. They belong to the "long-term knowledge" influential factor and are taken into account by various experts in direct and indirect ways. Research reports from various "older accidents" are used in official accident research reports of "new accidents," usually to pinpoint similar causes and the lack of lessons learned from previous accidents (e.g., Aznalcóllar for Baia Mare, Culemborg for Enschede and Oppau for Toulouse).

The number of fatalities is clearly not the only cause of societal disruption and hence it is not the sole "shock effect" inducement factor for legislative changes. The Baia Mare accident, e.g., did not include any fatalities; nonetheless, this accident was cited by the 2003 Seveso Amendment. Research by Vierendeels et al. [9] indicates that fear for large-scale diseases in the long-term caused by heavy metals and cyanide in the aquatic environment led to societal disruption in this particular case. A visible and traceable disposal of chemical substances in air, water or soil does make people afraid of major long-term health effects, especially if carcinogenic substances are involved.

The cost of an accident may also induce legislative changes. One billion euros seem to be a psychological minimum tier that leads to more or different regulations. Accident costs include property damage and production losses, losses because of image problems and external effects. The latter external effects – the financial consequences of an accident imposed on society – are especially pivotal in this regard.

Media attention because of the shock effect parameters of a major accident will always lead to societal pressure on the political decision-makers. In fact, societal disruption (caused by the shock effect parameters) will influence the legislative process as pictured in Fig. 5.4. The result is ad hoc legislative action, based on the research reports' and research commission's conclusions and recommendations.

5.2.2 Seveso company tiers

Company sites are classified "Seveso" in terms of quantities and types of hazardous materials they host. There are two different thresholds, see annex 1 of Seveso III Directive [5]: the "Seveso low" ("tier 1") and the "Seveso high" ("tier 2") threshold.

The overflow threshold is calculated based on the type of products and their risk phrases. The thresholds are present in the directives' annexes. For example, the thresholds for combustive substances (section Physical Hazards P8; Hazard statements H270, H271 and H272) are 50 and 200 tons. A company that can store on its site 40 tons of oxygen (combustive H270) and 50 tons of peroxides (combustive H271) is classified "Seveso low tier" or "Seveso tier 1" as the mass of combustive substances is superior to 50 tons but inferior to 200 tons. Without peroxides, this organization would not be concerned by the directive.

As well as the thresholds of combustive substances (H270, H271 and H272), the directive also suggests different thresholds for explosive substances (section Physical Hazards P1a: Hazard Statements H200–H205), liquid flammables (section Physical Hazards P5a Categories 2 and 3: Hazard Statements H226–227), easily flammable substances (H225), highly flammable substances (H224) and substances acute toxic for humans (section Health Hazards H1 Categories 1 and 2: Hazard statements H300–H301, H310–H311, H330–H331) and toxic for the environment (section Environmental Hazards H1 Categories 1 and 2: Hazard Statements H400, H410–H411).

As indicated in the major accident prevention legislation model, the directive took into account the different accidents that happened throughout history in different countries following the awareness of the hazards of major accidents. After this, some accidents became the basis of the addition of amendments to the initial directive. As a reminder, some major industrial accidents are listed below:

- The Flixborough catastrophe (UK, 1974): explosion following the rupture of a cyclohexane conduit, 28 dead [12].
- The Seveso disaster (Italy, 1976): dioxin release following a runaway reaction induced by malfunctioning heating. No deaths and 200,000 affected by dioxins [13, 14].
- The Bhopal catastrophe (India, 1984): one of the largest industrial disasters, releasing methyl isocyanate following an unwanted reaction with water and tens of thousands of deaths (3,787 immediate and 15,000 thereafter) [15–17].
- Romeoville (Illinois, USA, 1984): refinery explosion caused by an amine absorption unit, 17 dead [18].
- Mexico City (Mexico, 1984): boiling liquid expanding vapor explosion or BLEVE of LPG, followed by a domino effect in a petrol terminal, 650 dead and more than 6,400 wounded [19].
- Schweizerhalle disaster (Switzerland, 1986): a chemical warehouse catches fire. The Rhine gets polluted as a side effect. The pollution is perceptible in several countries [20].

- Phillips catastrophe (Pasadena, TX, USA, 1989): polyethylene plant explosion, 23 dead, 314 injured [21].
- Enschede disaster (the Netherlands, 2000): fire and explosion of a fireworks warehouse, 22 dead and 950 wounded [22].
- AZF factory explosion in Toulouse (France, 2001): this is caused by ammonium nitrate, 31 dead and 2,442 injured [23]. The 2003/105/CE Directive modified in 2003 shows the threshold limits of several substances including ammonium nitrate.
- Buncefield disaster (UK, 2005): because of the domino effect it affected several petrol stocking warehouses [24], 43 injured and billions of euros of financial losses.
- Texas City disaster (Texas City, TX, USA, 2005): refinery explosion and fire, 15 dead and 180 injured [25].
- Georgia disaster (Port Wentworth, GA, USA, 2008): dust explosion in a sugar refinery, 13 dead and 42 wounded [26].
- Foxconn disaster (Chengdu, China, 2011): dust explosion in iPad2 polishing workshop, four dead and 15 wounded [27].
- Tianjin disaster (Tianjin, China, 2015): chemical warehouse explosion, over 170 fatalities and 800 wounded (latest available translated reports, Hong Kong free press).
- Yancheng disaster (Yancheng, China, 2019): an explosion at a pesticide plant, 78 people killed and more than 610 injured ("Death toll rises to 78 in Chinese chemical plant explosion." UPI. Retrieved 26 March 2019).
- Waukegan explosion (Chicago, USA, 2019): explosion and fire at AB Specialty Silicones facility in Waukegan, 3 dead and 1 injured (CSB News release, 1/24/2020).

Many more major accidents can be mentioned. In fact, looking at major accidents with consequences of at least 25 fatalities simultaneously, estimates indicate that some 25,000 people died due to such accidents since 1917 [28]. Some elaborated and explained examples of disasters are provided in Chapter 7.

> Prevention is an attitude and/or a series of measures to be taken to avoid the degradation of a certain situation (social, environmental, economical, technological, etc.) or to prevent accidents, epidemics or illnesses. It acts mainly on the likelihood of occurrence and the causality chain, trying to lower the probability that an event happens. Prevention actions are also intended to keep a risk problem from getting worse. They ensure that future development does not increase potential losses.

5.3 Protection and mitigation

Protection and mitigation consist of all measures reducing consequences, severity or development of a disaster.

There are two types of protection measures (safeguards):
- *Before the event ("protection")*: reducing the size of the object of risk exposure when an event occurs.

– *After the event ("protection by means of mitigation")*: there are for instance emergency measures to stop the damage accumulation or counteract the effects of the disaster.

There is also a difference between active and passive protection. Let us take the example of fire protection:
– Build firewalls = passive protection.
– Establish a system for detection and/or sprinkler = active protection.
– Establish an evacuation plan + drills = active protection.

Remedial actions reducing the impact of a proven risk are developed, taking into account extra cost versus occurring risk. Together they form the *emergency plan*.

5.4 Safety barriers

Safety barriers (see Tab. 5.1) are all the adopted available measures regarding conception, construction and exploitation modalities including internal and external emergency measures, in order to prevent the occurrence and/or limit the effects of a hazardous phenomenon and the consequences of a potential associated accident [29].

Tab. 5.1: An overview of types of safety barriers, can be either preventive, protective/mitigative or both.

Safety barrier	Definition	Example	
Technical	Passive safety devices	Unitary elements whose objective is to perform safety function, without an external energy contribution coming from outside the system of which they are part and without the involvement of any mechanical system	– Retention basin – Rupture disk
	Active safety devices	Nonpassive unitary elements whose objective is to fulfill a safety function, without contribution of energy from outside the system of which they are part	– Discharge valve – Excess flow lid
	Safety instrumented systems	Combination of sensors, treatment units and terminal elements, whose objective is to fulfill a safety function or subfunction	– Pressure measure chain to which a valve or a power contractor is linked

Tab. 5.1 (continued)

Safety barrier	Definition	Example
Organizational	Human activities (operations) that do not include technical safety barriers to oppose the progress of an accident	– Emergency plan – Confinement
Manual action systems	Interface between a technical barrier and a human activity to ensure the success of a safety function	– Pressing an emergency button – Low flow alarm, followed by the manual closing of a safety valve

There are two main types of safety measures regrouping:

1. *Preventive* measures (Tab. 5.2) are the first barriers to put into place because they aim to prevent a risk by reducing the occurrence likelihood of a hazardous phenomenon.

2. *Protection* measures (Tab. 5.3) are the ones that limit the spread and/or severity of consequences of an accident on vulnerable elements without modifying the likelihood of occurrence of the corresponding hazardous phenomenon. Protection measures can be implemented "as a precaution" before the accident, e.g., a confinement. We can distinguish limitation barriers within the protection measures (enfeeblement or mitigation), which aim to limit the effects of a hazardous event after its occurrence (e.g., a bund wall).

Tab. 5.2: Examples of preventive measures.

Preventive measures	Example
Fire risk elimination	Using nonflammable materials
Limiting hazardous functioning parameters	Continuous following and automatic control of the functioning parameters classified as critical
Installing isolation, blocking and restriction devices	Rendering flammable liquids inert and locking down electrical equipment
Installing a fail-safe after a failure	Use of electrical fuses and circuit breakers
Reducing the likelihood of breakdowns and errors	Oversizing important elements, using redundancy
Substance leak recuperation	Careful floor cleaning

Tab. 5.3: Examples of safety barriers.

Protection measures	Human	Technical	Organizational
Passive, which act by their presence alone	– Mastering the urbanization – Seclusion rooms – Escape ladders	– Firewalls – Storage underground – Retention basin	– Emergency plan existence
Active, which act only with a specific human or material action	– Following orders – Wearing personal protective equipment – Using portable extinguishers	– Sensor activating safety systems: cutoff valves, water curtain – Valves	– Operator training – Activating the emergency plan – Implementing crisis cell

Instrumental fail-safes are made up of the three following elements:
– Detection elements (sensors and detectors), whose role is to measure the drift of a parameter by signal emission ("Sensors").
– Elements ensuring the gathering and treatment of the signal coming from the implementation of the safety logic (programmable electronic system, relay, etc.) in view of giving orders to the associated actuator ("Logic solvers").
– Action elements (actuators and motors), whose role is to put the system into a positive safe state and to maintain it that way ("Actuators").

For each of these three functions, instrumental protection may fail. Availability is measured by the probability of failure on demand (see also Section 3.4). In fact, every active safery barrier, also a human, will be subject to these three elements and their reliability will depend on them. In case of human barrier, "sensors" can be translated into the question whether the human has noticed the situation deviating from normal operations, "logic solvers" may be viewed as humans assessing the situation and coming to a conclusion about what is happening and "actuators" may be seen as humans taking action upon what they have noticed and concluded. In each step something might go wrong.

Examples of available barriers are given, respectively, in Tabs. 5.4 and 5.5 for protection and prevention of technical barriers.

Protection and mitigation consist of all measures that reduce consequences, severity or development of a disaster. Mitigation means taking action to reduce or eliminate long-term risk from hazards and their effects.

Tab. 5.4: Probability of failure on demand (PFD) of some technical protection barriers.

Protection barrier	PFD	Commentary
"All or nothing" valve	5×10^{-3} to 10^{-1}	Valid with complete and planned maintenance
Prevention or overpressure/ subpressure protection valve	10^{-3} to 10^{-1}	Safety valve function (no opening when asked)
Rupture disk	10^{-3}	Typical values

Tab. 5.5: Probability of failure on demand (PFD) of some technical prevention barriers.

Prevention barrier	PFD	Commentary
Fix fire-fighting equipment	10^{-2}	Typical value if tested regularly
Catalytic gas detection with associated alarm	5×10^{-3}	Valid with tests and calibration once every 2–3 months for a redundant mechanism
Alarm triggering	10^{-3}	Typical value (probability of failure on siren request)
Confinement against explosion/ toxic risks	0	Value close to 0

5.5 Risk treatment methodology

The treatment of risks is the central phase of engineering risk management. Thanks to action taken at this stage, the organization can concretely reduce the (negative) risks it faces. These actions should act on the hazard, the vulnerability of the environment or both when possible. It requires an organization to identify, select and implement measures to reduce risks to an acceptable level.

Three specific steps are involved in the treatment of risk:
1. Identification of potential measures under the prevention, preparedness, response and recovery domains
2. Evaluation and selection of measures
3. Planning and implementation of chosen measures

Risk treatment is hence described as a selection and implementation process of measures that are destined to reduce (negative) risks.

The ISO 27001 [30] norm imposes the implementation of an analysis method and risk treatment capable of producing reproducible results. Most of the available methods rely on tools that enable an answer to these constraints and aim to treat the steps described in Fig. 5.5.

As observed, the risk treatment answers to the following five consecutive questions:

1. What are the issues? Define the list of sensitive process.
2. Why and what to protect? Give a list of sensitive assets.
3. From what to protect? Enumerate the list of threats.
4. What are the risks? Propose a list of impacts and potential.
5. How to protect from them? Resume the list of safety measures to be implemented.

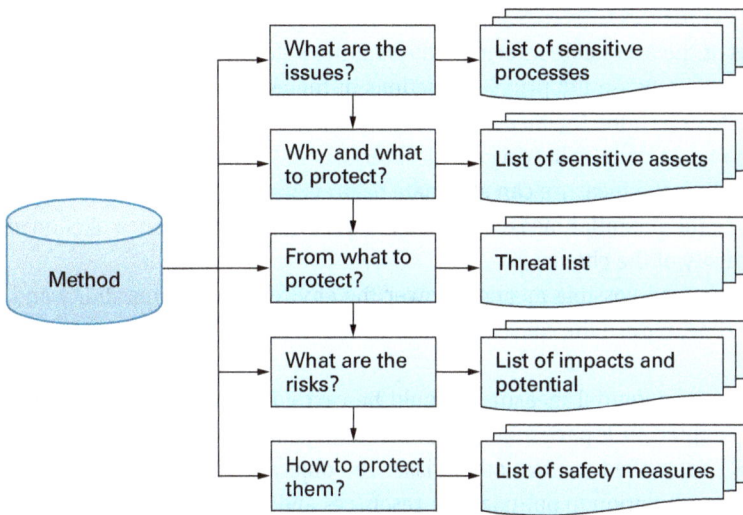

Fig. 5.5: Risk treatment method.

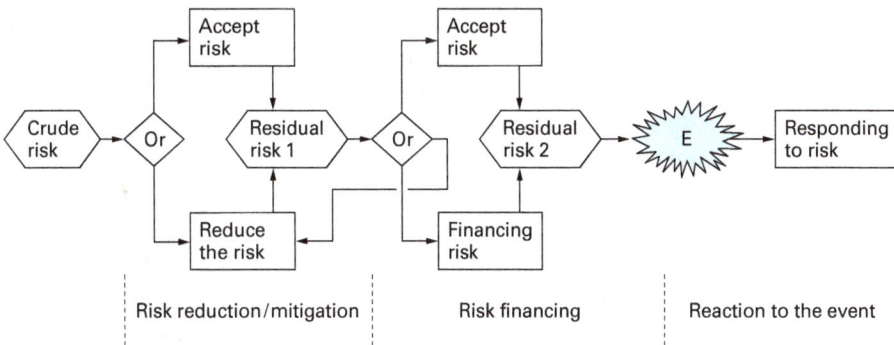

Fig. 5.6: Risk treatment process (E = event).

Putting this methodology in a flowchart, at every step, following questions can be asked, as depicted in Fig. 5.6:
– Is the risk acceptable?
– Should the risk be reduced?

Once an event has occurred, it is clear that asking these questions is no longer relevant. Instead, the focus must shift to responding to the risk, its occurrence and its consequences. It is convenient to emphasize that during risk treatment, a more detailed risk analysis can be necessary in order to dispose of the necessary information for an appropriate identification, evaluation and selection of the measure to be taken. We must count on an adequate analysis level to assure that the taken measure really treats the causes of the risks.

On the basis of the accessible information about the risks and the existing control measures, and depending on the priority functions of the established treatments, the first step will be to determine the measure to be implemented to reduce risks.

Thinking should be done, following a logical sequence. As an example:
1. *Evaluate* whether the measure can eliminate or prevent the risk.
2. Then, *analyze* the potential measures to aim to reduce the occurrence likelihood and the intensity of the chance.
3. Finally, *examine* the possible means to lower the environment vulnerability and, at the same time, reduce the consequences.

This identification of potential measures should be carried out while looking at the (any) measure adapted to a particular risk, and taking into consideration that the measure can also be applied to most other risks. The objective must be to dispose of the best possible combination to optimize the resources and assure adequate management of all the risks to which the community or the organization is exposed. It must also target the implementation of situations where several risks are interacting in one and the same environment.

Finally, this treatment option evaluation should also consider the different legal, social, political and economic factors. It should also be part of a perspective that contemplates the establishment of measures in the short and long term.

Tables 5.6–5.8 expose selected categories of safety measures or barriers appropriate for evaluation and implementation in the risk treatment step.

Risk reduction is primarily a matter of removing or containing the source of the risk (the hazard or "threat" (the latter in case of security)). There are a number of alternatives for preventing exposure ("vulnerability" in case of security), and these could be applied where appropriate. An analysis of the actual risks posed in particular situations would be required on a site-specific basis to determine the appropriate risk reduction method. Once the treatment measures are defined, they still need to be challenged and assessed for different criteria as depicted in Tab. 5.9.

Tab. 5.6: Examples of safety measures for different categories.

Measure categories	Examples
The legal and normative requirements	Adoption of laws, regulations, policies, orders, codes, standards, proceeded to certification, etc., intended to govern or to supervise the management of a chance, a risk or a domain of wider activity
The consideration of the risks in the town and country planning and development	Rules governing land-use in the exposed zones, the standards of immunization, maximal densities of land-use, rules of compatibility, prescriptions toward materials and specific techniques of construction to increase the resistance of infrastructures and buildings, the realization of technical studies inside an exposed zone, development evacuation paths, etc.
The elimination or the reduction of the risk at the source	Modification of industrial processes, use of a less hazardous product, modification of transportation routes, installation of equipment or realization of works to decrease risks, to limit their probability of occurrence or to decrease their potential intensity, establishment and application of directives and procedures of reduction of the risk, etc.
The rehousing of the people and the displacement of the exposed goods	Population, residences, infrastructures, etc. (it is a means of last recource that intervenes generally when a risk is considered unacceptable by a community and when the other measures of prevention and preparation do not represent a valid option)

Tab. 5.7: Examples of safety measures for structural actions.

Measure categories	Examples
Infrastructures, developments and equipment intended to avoid the appearance of risks	Protection wall for a reservoir of hazardous materials, works to prevent the release of an avalanche, rocks to avoid a landslide, etc.
Mechanical or physical means to reduce the probability of an occurrence or reduce intensity	Retention basin for rainwater, intensification of tanks to avoid leaks of hazardous materials, fast severing mechanisms for leaks and emergency stop in industrial installations, etc.
Inspection and maintenance programs	Measures to prevent the development of conditions convenient to the appearance of risks and to assure the preservation of safe conditions in the execution of risky activities: order in the equipment, state of buildings and infrastructures, hygiene and public health, etc.
Financial and fiscal capacities	Dissuasive measures to prevent or to limit the risk by the imposition of a penalty (tax or higher price rate), incentive measures to encourage the realization of actions allowing risk reduction, subscription to insurance to obtain compensation in the eventuality of losses consecutive to a disaster, etc.

Tab. 5.7 (continued)

Measure categories	Examples
Research and development programs and activities	Research and development on varied subjects such as fast alert systems, surveillance and forecast mechanisms, methods and tools of risk appreciation, materials and techniques of construction allowing an increase in the resistance of buildings and infrastructures, etc.
Public awareness programs, risk communication and population preparation	General campaigns to raise awareness, communication on the nature and the characteristics of risks, exposed territory, predictable consequences, measures taken to avoid the disaster, the means the citizens have to protect themselves, records to follow in case of disaster, etc.

Risk treatment consists not only of trying to reduce or mitigate risk but also on how to finance the risk or potential losses and finally how to react when an event occurs. There is a need to integrate the principles and practices of sustainability with the principles and practices of risk treatment. Only by adopting a sustainable approach can effective, equitable and long-term approaches to treating risks and building resilience be developed.

Tab. 5.8: Examples of modalities to assure intervention and restoration.

Measure categories	Examples
Modes and procedures of alert and mobilization	Measures intended to warn the population, the responsible authorities and the participants in an emergency situation or a disaster and to put into reserve the resources necessary for the management of the situation, etc.
Help measures for the population, goods protection and the natural environment	Search and rescue, evacuation, put under cover, health care, control, protection of the goods and the natural environment, etc.
Measures aiming to preserve the essential services and operations (continuity of the operations) and the protection of economic activities	Drinking water, energy, transport, telecommunications, financial services, emergency services, health system, food supply and essential governmental services, support for companies, resumption of activities and return to good conditions of public health levying of protective measures, cleaning and reassurance of places, restoration of services, etc.
Help measures for the population	Services to the disaster victims, psycho-social care, management of the needs of the whole community, support following a disaster: financial, psychological, technical support, etc.

Tab. 5.8 (continued)

Measure categories	Examples
Modes and mechanisms of public information	Communications at the time of and following a disaster: instructions to the citizens, relations with the media, etc.
Procedures for the experience of feedback	Production and broadcasting of reports on the causes and the circumstances of the event, holding of evaluation sessions of the operations following a disaster or following an exercise, an analysis of the answer to the disaster and the measures of reduction of the risks of setting up, etc.
Exercise programs	Exercises of alert, traffic of information, mobilization, activation of a center of coordination, coordinated management, operational and general evaluation with or without deployment, etc.
The administrative and logistic modalities	Agreements, procedures and administrative directives for the mobilization of human, material and informative resources, the acquisition of material resources, the preparation of installations: centers of coordination, accommodation, etc.
The follow-up and revision capacities of the level of preparation	Check on the functioning and maintenance of the installations, the equipment and the intervention equipment, programs of information for participants, the procedures of update and periodic revision measures, and the periodic check of the level of preparation: reports, questionnaire of self-assessment, audit, etc.

Tab. 5.9: Some criteria for assessing risk treatment options (adapted from [31]).

Criteria	Question to be answered
Cost	Is this option affordable? Is it the most cost-effective?
Timing	How long does it take to be able to implement the treatment option?
Equity	Do those responsible for creating the risk pay for its reduction?
Fairness	Where there is no man-made cause, is the cost fairly distributed?
Leverage	Will the application of this option lead to further risk-reducing action by others?
Administrative efficiency	Can this option be easily administered or will its application be neglected because of difficulty of administration or lack of expertise?
Continuity of effect	Will the effects of the application of this option be continuous or merely short term?
Compatibility	How compatible is this option with others that may be adopted?

Tab. 5.9 (continued)

Criteria	Question to be answered
Jurisdictional authority	Does this level of government have the legislative authority to apply this option? If not, can higher levels be encouraged to do so?
Effects on the economy	What will be the economic impacts of this option?
Effects on the environment	What will be the environmental impacts of this option?
Risk creation	Will this option itself introduce new risks?
Risk reduction potential	What proportion of the losses due to this risk will this option prevent?
Political acceptability	Is this option likely to be endorsed by the relevant authorities and governments?
Public and pressure group reaction	Are there likely to be adverse reactions to implementation of this option?
Individual freedom	Does this option deny basic rights?

5.6 Risk reduction approach

If during a risk assessment it is observed that the work system is not safe enough, or that the risk is too high for the group of people examined, then appropriate measures must be sought to eliminate or reduce risks. In assessing once again the risk with the selected measure, we check whether the chosen measure effectively reduces the risk. It must also be checked at that time whether the implementation of new measures of protection involves additional or new hazards.

If that is the case, these hazardous phenomena have to be added to the list of already noticed hazardous phenomena, and a new risk appreciation must be performed. Figure 5.7 presents an iterative process of risk reduction.

As mentioned before in this chapter, risk control measures, or safety barriers, are based on

- *Prevention* measures, whose role is to reduce the probability of the occurrence of feared events that are the source of hazard for damaging targets.
- *Protection* measures, whose role is to protect targets against the effects of such things as heat flow, pressure or projectiles, which are associated with the release of dangerous phenomena.
- *Mitigation* measures, whose role is to limit the effects of the appearance of feared events.

Prior to analysis, the limits of the system that forms the business or process to consider have to be defined. We should also define precisely what is in the system, what is therefore taken into account in the identification of hazards and what lies outside that system and the various modes of operation.

Large areas or processes should be divided into smaller components. If a sector of activity or a process includes a whole line of production composed of several facilities, then the different sectors or partial processes should, when possible, correspond to a process phase. The interfaces between the whole system and the environment, as well as the interface between some parts of the sector or parts of the process, have to be well-defined and highlighted.

It is necessary to specify the type of hazardous phenomena under consideration and to indicate to whom and to what they apply (employees, facilities, environment, etc.). It is also important to clarify if there are eventual interactions with the neighboring facilities that must be taken into account and which aspect does not need inspection (construction static, process chemistry, etc.).

We must make a particular effort not to forget the operating modes of a system as follows:

– Normal operation: The facility fulfills the function for which it was designed.
– Special operations: Prepare, convert, install and adjust, errors, clean.
– Maintenance:
  – Control (measure, control and record): determine the real state and compare with the predicted state.
  – Maintenance (cleaning and maintenance): measures to conserve the desired state.
  – Replacement (replacement and improvement): restore to the previous state.

Let us take chemistry as an example. The choice of preventive action follows a parallel hierarchy to the exposure process (see Tab. 5.10):

1. Priority should always focus on the search for measures that limit the use or emission of harmful agents.
2. Then the measures that prevent the spread of these agents to workers.
3. Finally, personal protective measures and community protective measures.

Risk reduction includes methods by which firms evaluate potential losses and take action to reduce or eliminate such threats. It is a technique that utilizes findings from risk assessments (identifying potential risk factors in a firm's operations, such as technical and nontechnical aspects of the business, financial policies and other policies that may impact the well-being of the firm) and implementing changes to reduce risk in these areas.

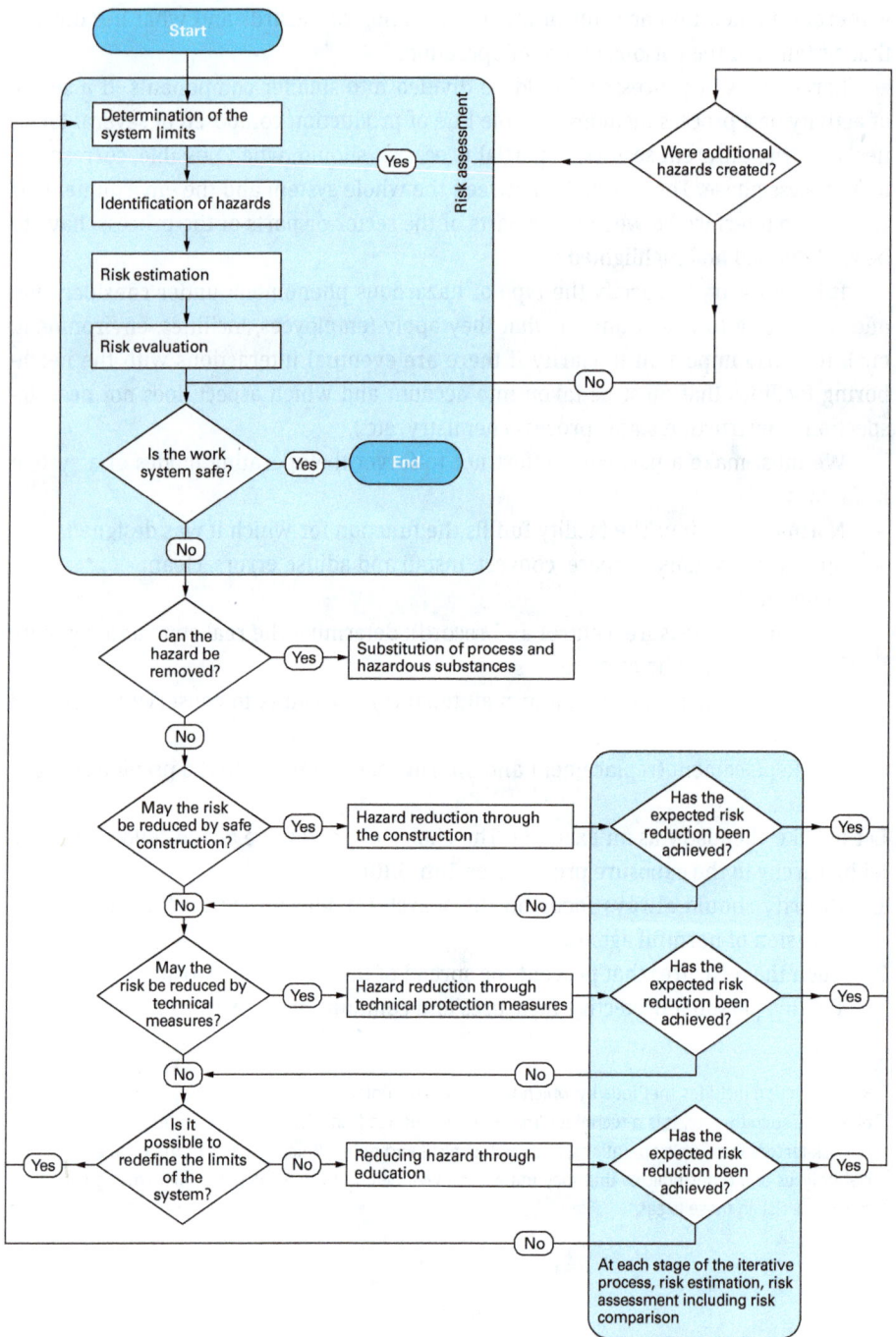

Fig. 5.7: The iterative process of risk reduction inspired from SUVA [32].

Tab. 5.10: Exposure and priority of preventive actions.

Exposure process		Priority of preventive actions	
1.	Emission	1.	At the source of hazard
2.	Transmission in the environment of the target	2.	At the interface between the target and the source
3.	Exposition	3.	At the target

5.7 STOP principle

When looking for appropriate solutions to safety problems, we have to, first of all, clarify whether the hazardous phenomenon can be deleted by replacing certain substances and some dangerous processes. If it is not possible to delete the hazard by improving the construction work or by using less hazardous substances, we must then proceed with technical and organizational measures and as a last resort, measures relative to people.

The STOP (strategic, technical, organizational and personal) measures[1] principle underlines this approach by giving priority to
- The measures in the following order:
    i. **S** measures: *Strategic*, substitution of processes or substances giving a less hazardous result (e.g., substituting, eliminating, lowering, modifying and abandoning); abandon the process or product and modify the final product.
    ii. **T** measures: *Technical* protection against hazardous phenomena that cannot be eliminated, lowering the likelihood of occurrence of an event and reducing the spread of the damage (e.g., replacing, confining, isolating/separating, automating, firewall, EX zones and bodyguards).
    iii. **O** measures: *Organizational* modifications of the work, training, work instructions, information concerning residual risk and how to deal with it (e.g., training, communicating, planning, supervising and warning signs)
    iv. **P** measures: *Personal*, relative to people (e.g., PPE, masks, gloves, training, communication, coordinating and planning)
- The hierarchy of the priorities is in the following order:
    i. Acting at the *source*: Deleting the risk (substituting product or process, in situ neutralization), limiting leak risks (re-enforcing the system and lower-

---

1 We also refer to the P2T model mentioned in Chapter 3. The analogy between "TOP" and "P2T" is obvious: P2T stands for procedures (or organizational), people (or personal), and technology (or technical).

ing the energy levels), predictive measures (rupture disk, valves) and surveillance (integrity and functionality of the system, energy levels).

ii. Acting at the *interface* (on the trajectory between the source and the target): Limiting the propagation (active barriers/passive barriers), catching/neutralizing (local or general ventilation, air purification and substance neutralization), people control (raising barriers, access restrictions and evacuation signs) and surveillance [energy levels in the zone, excursions or deviations (alarms)].

iii. Acting at the *target*: Lowering the vulnerability (PPE selection and special training), reducing exposure (e.g., automation), reducing the time (job rotation) and supervising (individual exposure, biological monitoring, medical survey, correct PPE use and following rules).

In general, we must combine measures to obtain the required safety. It is important that the choice of safety measures enables the reduction of the likelihood and severity of the hazardous events. To make this choice, we must not only take into account the short-term costs but also the long-term profitability calculations. Once the priorities have been established, it is possible to determine the correct method to master each of the identified risks. These methods are often regrouped in the following categories:
– Elimination (including substitution)
– Engineering measures
– Administrative measures
– Individual protective equipment

Table 5.11 presents a recap of the ordering of measures and the considered environment, illustrated by few examples for each category. Directions of approach are from top to down and then from left to right.

Eliminating the hazard is the most favorable approach when reducing risks; substitution is interesting as long as it does not generate new hazards. No hazard, no risk. In the STOP principle, the elimination and substitution phases are included in the strategic measure S. They are, however, rarely possible in practice, thus eliminating and substituting may sometimes not be applicable.

Let us take the example of substituting a solvent (benzene) in a chemical reaction with a solvent with little toxicity (1,2-dichlorobenzene). Thus, the toxic effect of benzene is eliminated and replaced by a lower toxicological effect. From this point of view, the problem is over. However, during the synthesis with the new solvent, side products have appeared whose toxicity could cause other dramatic effects. So what must we do? One problem has been replaced by another. If another hazard of equal importance had not been introduced, the suggested solution would have won the vote. This example shows that the strategic measure is not always applicable in the field. An easy measure, from the risk management point of view, would be to say that the synthesized product is no longer interesting, so no more synthesis, no more solvent and the problem is solved. But can we live without this product? This is the strategic question!

The principle of the STOP concept is a continuous and consecutive approach and is summarized in Fig. 5.8. Note that the S part (strategical) is missing as it consists of substituting, eliminating or changing a process. Strategic measures are not dependent on the process. Often, this principle is shortened to TOP because strategic or substitution measures have already been taken whenever possible.

The definition of "hazard" (which is to be eliminated or substituted) can be established with the help of two questions, that is, "Why?" and "How?," in relation to the product and the product necessity.

The principle of the method in three steps for each TOP measure is depicted in Fig. 5.8:
1. Researching facts
2. Deducing the relative problems
3. Searching for the appropriate measures

Tab. 5.11: The STOP table.

	At the source	At the interface	At the target
Measures **S** (strategic)	– Substitution – Change process	– Automation, telemanipulation – Room subdivision	– Criteria for selection of licensed operators
Measures **T** (technical)	– Reactant production or use in continuous mode – Safety relief valves	– Fumes extraction process enslaved – Physical access restrictions	– Lab coats – Safety goggles – Eye shower
Measures **O** (organizational)	– Work instructions for process – Emergency instruction available	– Extraction of fumes manually controlled – Information about hazards and threats – Pictogram for restricted access	– Prescription for PPE and CPE – Organization of first aid
Measures **P** (personal)	– Education/training of the process operation	– Personal access denial to certain rooms/process areas	– Use of PPE

PPE, personal protective equipment; CPE, community protective equipment.

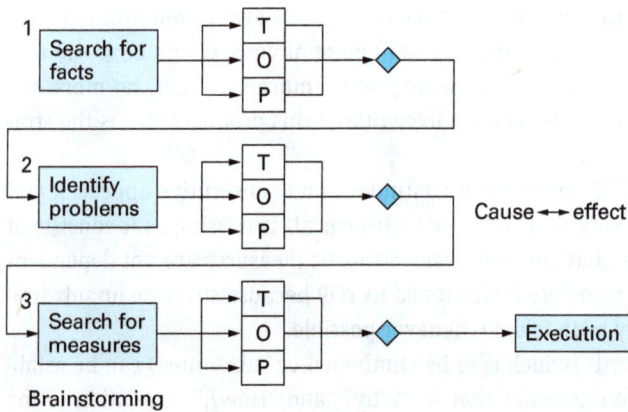

Fig. 5.8: The TOP iterative process.

As for every principle, there are not only pros or only cons. However, we will not discuss the method itself, although quite logical, but rather the different types of measures it recommends. The advantages and inconveniencies of the different types of measures (STOP) are summarized in Tab. 5.12.

Note that in practice, personal protection measures are put into place before the technical and organizational measures. This happens for many different reasons including costs, delays, implementation simplicity, loss of responsibility, having or taking no time to analyze the situation and the simplicity.

Tab. 5.12: Pros and cons of the STOP principle.

	Pros	Cons
Measures **S** (strategical)	– Cancel or reduce the considered hazard – Intervene at the beginning of the process	– In case of a substitution, it is possible to create other hazards or risks – Deletion needing a strategic decision
Measures **T** (technical)	– Fixed – Difficult to bypass	– Costs – Deadlines
Measures **O** (organizational)	– Quick – Moderate costs	– Controllability – Easy to bypass
Measures **P** (personal)	– Quick – Moderate costs – Simple implementation	– Controllability – Acceptability – Convenience – Omission

Many organizations have invested heavily in personnel, processes and technology to better manage their risk. But these investments often do not address the strategy and processes that should be implemented. To successfully turn risk into results, we need to become more effective at managing scarce resources, making better decisions and reducing the organization's exposure to negative events by implementing the four-level steps comprising strategic, technical, organizational and personal aspects.

5.8 Resilience

The first definition describing resilience was established by Holling in 1973 as "a measure of the persistence of systems and of their ability to absorb change and disturbance and still maintain the same relationships between populations or state variables" [35].

Resilience is often defined in engineering as the ability to return to the "steady state" following a perturbation. Ecological resilience emphasizes conditions far from any stable steady state, where instabilities can flip a system from one regime of behavior into another. Resilience is here the system's ability to absorb disturbances before it changes the variables and processes that control behavior. Resilience adds a dynamical and proactive perspective into risk governance by focusing (i) on the evolution of system performance during undesired system conditions and (ii) on surprises ("known unknowns" or "unknown unknowns"), i.e., disruptive events and operating regimes which were not considered likely design conditions. Resilience encompasses the concept of vulnerability as a strategy to strengthen the system response and foster graceful degradation against a wide spectrum of known and unknown hazards. Moreover, it expands vulnerability in the direction of system reaction/adaptation and capability of recovering an adequate level of performance following the performance transient [36].

Erik Hollnagel defined, in Resilience Engineering in Practice [33], resilience as "the intrinsic ability of a system to adjust its functioning prior to, during, or following changes and disturbances, so that it can sustain required operations under both expected and unexpected conditions." Since resilience is about being able to function, rather than being impervious to failure, there is no conflict between productivity and safety. Resilience is a protective strategy to build in defenses to the whole system against the impact of the realization of an unknown or highly uncertain risk. Resilience strategies will primarily aim to reduce exposure and vulnerability.

The continued development of resilience engineering has focused on four abilities that are essential for resilience. These are the abilities:
a) to respond to what happens,
b) to monitor critical developments,
c) to anticipate future threats and opportunities and
d) to learn from past experience – successes as well as failures.

The purpose of resilience engineering is to achieve full or partial recovery of a system following an encounter with a threat that disrupts the functionality of that system. The time evolution of resilience is presented in Fig. 5.9 where the different phases are represented. One should start to understand the risks in order to anticipate and prepare for the worst. When the event or the "shock" happens, it is the time to absorb and withstand the effects and then to respond in order to recover. The learning and adaptive phase could help to make the resilience better. As depicted in Fig. 5.10, the loss and recovery of the functionality of a system evolve with time, and the state of recovery is often never the same as the one before the event. Only "smart" systems having the capability to evolve with the time could be better than the starting situation. For example, living species, i.e., insects, are so resilient that they could evolve and be specifically resistant to insecticides when they are exposed to. We could say that they are "stronger" and have more functionalities than what they had before. Their resilience brings them to a new state using the active learning capacities.

So how can smartness support future-proof resilience? Smartness is the road to future-proof resilience since it enables novel paradigms like proactive dependability and self-healing. Those paradigms are completely different from the current implementations of safety-critical and dependable systems where threats, vulnerabilities and consequences are supposed to be known in advance during risk assessment or somehow updated and uploaded later (e.g., threat signatures/patterns, vulnerability information, troubleshooting instructions and repair workflow). Resilience in future smart systems will increasingly leverage on embedded intelligence in order to anticipate and detect unknown threats and automatically compute the most appropriate and safe solutions by using approaches based on machine learning, heuristics, fuzzy logic, Bayesian inference and artificial neural networks driving real-time cosimulation and online model checking. Smart machines represent a new field which is still to be thoroughly researched. Recent research shows that the creation of smart machines is bringing to a higher level of automation in many fields and that trend will continue in the future.

A resilient system possesses four attributes [34]:

1) **Capacity** is the attribute of a system that allows it to withstand a threat. Resilience allows that the capacity of a system may be exceeded, forcing the system to rely on the remaining attributes to achieve recovery. The following design principles apply to the capacity attribute: absorption (withstand a design-level threat), redundancy (physically critical components are physically redundant), functional (critical functions to be duplicated by different means) and layer defense (single points of failure should be avoided).

2) **Flexibility** is the attribute of a system that allows it to restructure itself in the face of a threat. The following design principles apply to the flexibility attribute: reorganization design (system is able to change its own architecture, applicable particularly to human systems), human backup (backup automated systems), complexity avoidance (minimization of complex systems) and drift correction (make corrections before the encounter with the threat, early detection).

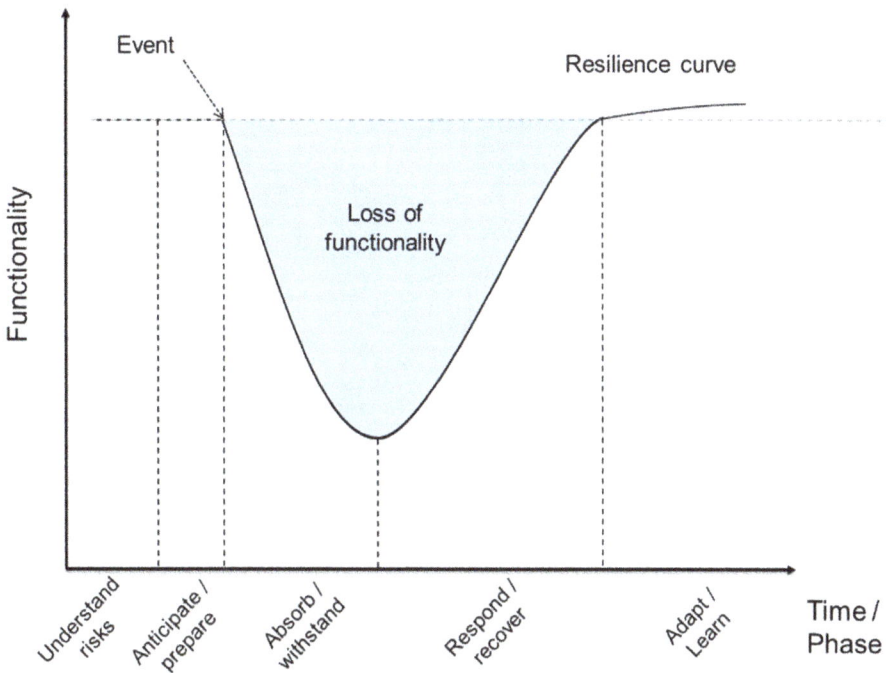

Fig. 5.9: Time evolution of a resilient system.

3) **Tolerance** is the attribute of a system that allows it to degrade gracefully following an encounter with a threat. The following design principles apply to the tolerance attribute: localized capacity (concentrate on individual nodes of the system),
loose coupling (check cascading failures), neutral state (bring system into neutral
state before taking actions) and reparability design (system should be reparable).
4) **Cohesion** is the attribute of a system that allows it to operate before, during and
after an encounter with a threat. Cohesion is a basic characteristic of a system.
The following global design principle applies to the cohesion attribute: internode
interaction (nodes of a system should be capable of communicating, cooperating
and collaborating with each other).

Since humans are indispensable in all situations involving change, resilience engineering naturally has strong links with human factors and safety management. It is
based on the following premises:

a. Performance conditions are always underspecified. Individuals and organizations
must therefore adjust what they do to match current demands and resources. Because resources and time are finite, such adjustments will inevitably be approximate.

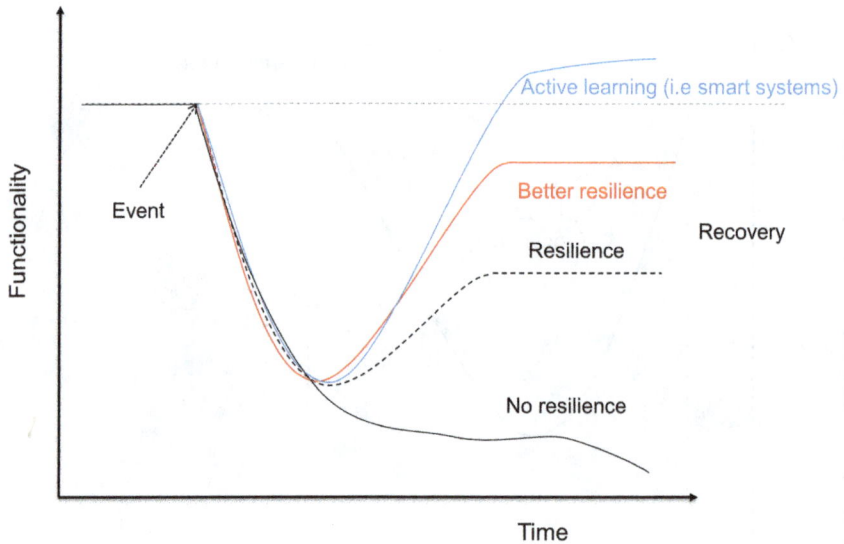

Fig. 5.10: Disruption evolution diagram.

b. Some adverse events can be attributed to a breakdown or malfunctioning of components and normal system functions, but others cannot. The latter can best be understood as a result of unexpected combinations of performance variability.
c. Safety management cannot be based exclusively on hindsight nor rely on error tabulation and the calculation of failure probabilities. Safety management must be proactive as well as reactive.
d. Safety cannot be isolated from the core (business) process or vice versa. Safety is the prerequisite for productivity, and productivity is the prerequisite for safety. Safety must therefore be achieved by improvements rather than by constraints.

Adopting this view creates a need for an approach that can represent the variability of normal system performance, and for methods that can use this to provide more comprehensive explanations of accidents as well as identify potential risks.

Hollnagel presents the management of resilience, according to his resilience analysis grid, with four basic abilities as natural starting points for understanding how an organization functions: how it responds, how it monitors, how it learns and how it anticipates [33]:

1. Systems of the first kind: A system that passively reacts whenever something happens – whenever a situation passes a certain threshold – will by definition always be surprised and therefore always reactive. While systems of the first kind may survive, at least for a time, they are not really resilient.
2. Systems of the second kind: Systems that can manage something not only when it happens but also after it has happened. This means that the system can learn

from what has happened and can use this learning to adjust both how it monitors – what it looks for – and how it responds – what it does.

3. Systems of the third kind: Those that can manage something before it happens, by analyzing the developments in the world around and preparing itself as well as possible. Systems of the third kind are able to respond, monitor, learn and anticipate and may therefore seem to meet all criteria to being called resilient – and be able to manage their resilience.

4. Systems of the fourth kind: The anticipation includes the system itself – not only in the sense of monitoring itself or learning about itself but also considering how the world responds or changes when the system makes changes, how these responses may affect the changes and so on. This is the recursive type of anticipation and represents the pinnacle of resilience management.

According to Hollnagel, safety efforts under Safety-I are often initiated by what is defined as an adverse event or unexpected outcome. The more important and serious the event is, the more urgent and extensive the response becomes. The primary objective is to prevent the adverse event from happening again, either by trying to identify and eliminate the causes or by finding ways to reduce the consequences of the accident. This involves comparing what actually happened with what was prescribed and classifying the actual actions as one form of noncompliance.

These same efforts under Safety-II lead to a different practice. It is first and foremost a question of examining the work done, i.e., paying attention to what happens when "nothing" happens, and recognizing the usual working methods. One will work upstream of the accident to identify the small adjustments that are made that make the results acceptable and unacceptable.

In Safety-I, learning is accident-based, and learning efforts are proportional to the severity of the consequences of the accident. In Safety-II, learning must be based on the frequency of events rather than their severity. It is simpler and less costly to make small changes to everyday work/process than to make large changes to infrequent ones. Safety management is often based on accidents and incidents, which represent snapshots of unacceptable performance. Instead, modern safety management should be concerned with what happens all the time and the continuous flow of activities that make up the daily activity. We almost know how many times something has failed or gone wrong, but we rarely know how many times something simply works.

Resilience engineering does not advocate a complete replacement of Safety-I with Safety-II, but rather proposes a combination of the two ways of thinking. Safety-II is foremost a different understanding of what safety is, and therefore a different way of applying many familiar methods and techniques.

Resilience can enhance safety by strengthening a system's capacity to anticipate, absorb, adapt to, and recover from unforeseen events or disruptions. Here's how resilience supports safety:

1. **Proactive risk management**: Resilience encourages systems and organizations to identify potential risks and vulnerabilities in advance, promoting preventive measures to avoid incidents.
2. **Adaptability and flexibility**: Resilient systems are designed to adapt quickly to changing conditions, which is crucial during emergencies or when conditions deviate from expected norms. This adaptability reduces the likelihood of incidents escalating.
3. **Enhanced recovery and continuity**: Resilience enables systems to recover more quickly from disruptions, minimizing downtime and reducing the potential for further safety hazards that can arise during recovery.
4. **Robust design and redundancy**: Building resilience often includes adding redundancies and strengthening key components, which helps prevent single points of failure from causing a total system collapse.
5. **Effective communication and training**: Resilience promotes a culture where continuous learning, training and communication are prioritized, ensuring that people are better prepared to respond to and manage incidents safely.
6. **Learning from past incidents**: Resilience emphasizes learning from previous failures and near-misses to improve safety practices continuously, creating a cycle of improvement that enhances safety over time.

By integrating resilience into safety planning, organizations create systems that are better equipped to manage unexpected challenges, ultimately resulting in safer environments for people, processes and infrastructure.

Important knowledge fields strongly related to resilience in industrial practice are crisis management, emergency management, business continuity planning and the alike. Chapter 8 discusses these distinct engineering management domains in a separate chapter.

5.9 Tips for implementing risk treatment

Risk treatment strategies are action plans you conceptualize after making a thorough evaluation of the possible threats, hazards or detriments that can affect a project, a business operation or any form of venture. The purpose of such strategies is to lessen or reduce the adverse impacts of the known or perceived risks inherent in a particular undertaking, even before any damage or disaster takes place.

Best practices require that known and perceived risks be analyzed in terms of the degree and likelihood of expected negative outcomes. Then, all of these analyzed risks should be documented according to their level of priority in a risk mitigation plan. Thereafter, the development and integration of the corresponding risk mitigation strategies follows and is referenced against the previously prepared risk management plan.

The main important point is to "Know Your Risk." Better is the level of knowledge, better will be the response in terms of mitigation and therefore in risk reduction. Then implement a risk mitigation strategy in place. All this could not be done without the help of (specific/local) experts, use the (specific/local) resources and get external support when needed.

Last but not least, risk reduction without implementation and control is worth nothing. It is not because it is planned that it will be realized the way it was intended.

> Resilient individuals or organizations are able at any time to adjust their performance to the current conditions and, due to time and resources being finite, understand that adjustments are approximate. Successful resilience in terms of engineering risk is thus the ability to anticipate the changing shape of the risk before failures and harm occur as well as the ability to recuperate if an encounter with a threat occurs.

5.10 Conclusion

Risk treatment is the selection and implementation of appropriate options for dealing with risk. It includes risk avoidance, reduction, transference and/or acceptance. Risk reduction is used as a preferred term to risk termination. Often, there will be residual risk, which cannot be removed totally as it is not cost-effective to do so. Risk acceptance is sometimes referred to as risk tolerance.

Risk treatment involves identifying the range of options for treating risks, assessing these options and preparing and implementing treatment plans. The risk management treatment measures are summarized as follows:
- Avoid the risk and decide not to proceed with the activity likely to generate risk.
- Reduce the likelihood of harmful consequences occurring by modifying the source of risk.
- Reduce the consequences occurring by modifying susceptibility and/or increasing resilience.
- Transfer the risk by causing another party to share or bear the risk.
- Retain the risk by accepting the risk and planning to manage its consequences.

Risk reduction should be integrated into an economic analysis (see also Chapter 9 on economics). It can play a pivotal role in advocacy and decision-making on risk reduction by demonstrating the financial and economic value of incorporating risk reduction initiatives into aid planning.

Risk reduction is suitable for risks with significant costs, human or environmental impacts, and a relatively high likelihood of occurrence, where the business may be unable to bear the uncertainty alone. In such cases, the organization is compelled to strategically minimize both the probability of occurrence and the severity of the risk's impact. We cannot ensure success merely by preventing failures. Success comes from understanding the nature of everyday performance and learning to recognize aspects that might otherwise go unnoticed.

References

[1]   Operations Directorate Public Safety Canada (2011). Public Safety of Canada, National Emergency Response System. Ottawa, Canada: Operations Directorate Public Safety Canada.

[2]   Kletz, T. (1998). Process Plants. A Handbook for Inherently Safer Design. Ann Arbor, USA: Braun-Brumfield.

[3]   Kletz, T., Amyotte, P. (2010). Process Plants. A Handbook for Inherently Safer Design. 2nd Ed. Boca Raton, USA: CRC Press.

[4]   Frost, C., Allen, D., Porter, J., Bloodworth, P. (eds.). (2001). Risk Management Overview. In Operational Risk and Resilience. Oxford, UK: Butterworth-Heinemann.

[5]   Health and Safety at Work etc. Act (1974) (c37). www.legislation.gov.uk/ukpga/1974/37/contents, latest revision 2012.

[6]   Fischhoff, B., Lichtenstein, S., Slovic, P., Derby, S.L., Keeney, R.L. (1981). Acceptable Risk. Cambridge, UK: Cambridge University Press, 521241642.

[7]   Council of the European Communities (2003). Directive 2003/105/EC. Official Journal of the European Union, L 345. 31/12/2003. 97–105.

[8]   Directive 2012/18/EU of the European Parliament and of the Council of 4 July 2012 on the control of major accident hazards involving dangerous substances, amending and subsequently repealing Council Directive 96/82/EC.

[9]   Vierendeels, G., Reniers, G., Ale, B. (2011). Modeling the major accident legislation change process within Europe. Safety Sci. 49: 513–521.

[10] Zoeteman, B.J.C., Kersten, W., Vos, W.F., Voort, L.V.D., Ale, B.J.M. (2010). Communication management during risk events and crises in a globalised world: Predictability of domestic media attention for calamities. J. Risk Res. 13: 279–302.

[11] Mac Sheoin, T. (2009). Waiting for another Bhopal. Glob. Social Pol. 9: 408–433.

[12] Venart, J.E.S. (2007). Flixborough: A final footnote. J. Loss Prevent. Proc. 20: 621–643.

[13] Cardillo, P., Girelli, A., Ferraiolo, G. (1984). The Seveso case and the safety problem in the production of 2,4,5-trichlorophenol. J. Haz. Mat. 9: 221–234.

[14] Bertazzi, P.A. (1991). Long-term effects of chemical disasters. Lessons and results from Seveso. Sci. Total. Environ. 106: 5–20.

[15] Jasanoff, S. (1988). The Bhopal disaster and the right to know. Social Sci. Med. 27: 1113–1123.

[16] Chouhan, T.R. (2005). The unfolding of the Bhopal disaster. J. Loss Prevent. Proc. 18: 205–208.

[17] Eckerman, I. (2011). Bhopal Gas Catastrophe 1984: Causes and Consequences. In Encyclopedia of Environmental Health. Nriagu J.O. (ed.), Amsterdam, The Netherlands: Elsevier, 302–316.

[18] Nolan, D.P. (ed.). (1996). Historical Survey of Fire and Explosions in the Hydrocarbon Industries. In Handbook of Fire & Explosion Protection Engineering Principles for Oil, Gas, Chemical, and Related Facilities. Burlington, MA: William Andrew Publishing.

[19] Pietersen, C.M. (1988). Analysis of the LPG-disaster in Mexico City. J. Haz. Mat. 20: 85–107.

[20] Giger, W. (2009). The Rhine red, the fish dead – The 1986 Schweizerhalle disaster, a retrospect and long-term impact assessment. Environ. Sci. Pollut. Res. 16: 98–111.

[21] U.S. Chemical Safety and Hazard Investigation Board (2007). Refinery Explosion and Fire. Investigation Report No. 2005-04-I-Tx. 1–341.

[22] World Health Organization (2009). Public Health Management of Chemical Incidents. Geneva, Switzerland: WHO.

[23] Dechy, N., Bourdeaux, T., Ayrault, N., Kordek, M.A., Le Coze, J.C. (2004). First lessons of the Toulouse ammonium nitrate disaster, 21st September 2001, AZF plant, France. J. Haz. Mat. 111: 131–138.

[24] Buncefield Major Incident Investigation Board (2008). The Buncefield Incident 11 December 2005, The final report. J. Loss Prevent. Proc. 1. 2a and 2b.

[25] U.S. Chemical Safety and Hazard Investigation Board (2007). Refinery Explosion and Fire. Investigation Report No. 2005-04-I-TX. 1–341.

[26] U.S. Chemical Safety and Hazard Investigation Board (2009). Sugar Dust Explosion and Fire. Investigation Report No. 2008-05 I-GA. 1–90.

[27] Apple (2012) Apple Supplier Responsibility 2012 Progress Report, 1–27. Cupertino, CA:Apple Inc.

[28] Mihailidou, E.K., Antoniadis, K.D., Assael, M.J. (2012). The 319 major industrial accidents since 1917. Int. Rev. Chem. Eng. 4(6): 529–540.

[29] Sklet, S. (2006). Safety barriers: Definition, classification, and performance. J. Loss Prevent. Proc. 19: 494–506.

[30] International Organization for Standardization (2022) Norm n° ISO/IEC 27001.

[31] Foster, H.D. (1980). Disaster Planning: The Preservation of Life and Property. New York: Springer-Verlag.

[32] SUVA, (2016). Méthode Suva D'appréciation Des Risques À Des Postes de Travail Et Lors de Processus de Travail. Document no. 66099 1–47. SUVA, Luzern, Switzerland.

[33] Hollnagel, E. (2011). Resilience Engineering in Practice. Burlington, USA: Ashgate.

[34] Jackson, S. (2010). Architecting Resilient Systems: Accident Avoidance and Survival and Recovery from Disruptions. Hoboken, NJ: John Wiley & Sons, Inc.

[35] Holling, C.S. (1973). Resilience and stability of ecological systems. Ann. Rev. Ecol. Systemat. 1–23.

[36] IRGC (2016). Resource Guide on Resilience. Lausanne: EPFL International Risk Governance Center, v29-07-2016, 1–290.

6 Event analysis

In this chapter, we will use the term "event analysis" to mean accident, incident or near-miss analyses. Human and organizational factors are an important cause of accidents. As the design of electromechanical equipment becomes more and more safe, the causes of accidents are more likely to be attributed to human and organizational factors. Event analysis is conducted in order to discover the reasons why an accident or incident occurred, and to prevent future incidents and accidents. Investigators have long known that the human and organizational aspects of systems are key contributors to accidents, indicating the need for a rigorous approach to analyzing their impacts. Safety experts strive for blame-free reports that will foster reflection and learning from the accident, but struggle with methods that require direct technical causality; they also do not often consider systemic factors, and they seem to leave individuals looking culpable.

It is hard to keep your cool when trying to analyze the objective causes of an incident or accident and to produce a common report. This is where the difficulty arises, as it is hard to find a common language that is limited to the objectivity of the facts. Indeed, when an accident happens, a sensitivity aggravating climate is installed, and thinking gives place to arguing and everyone tries to find who or what is responsible without first trying to understand anything. The event is often considered as a result of a combination of unfortunate circumstances: it is such and such at fault... it could have been worse, how unlucky... inhibiting any ulterior analysis. This phenomenon is illustrated in Fig. 6.1 – there is a huge divergence between facts, their interpretation and the suppositions.

An event analysis method is therefore needed to guide the work, aid in the analysis of the role of human and organizations in accidents and promote blame-free accounting of accidents that will support learning from the events and thus becoming a proactive process.

We may still raise the question: "Why analyze accidents?" The answers are multiple and can be listed as follows:
– To understand by analyzing the objective causes related to the accident
– To prevent a recurrence
– To determine which measures will improve safety
– To act by implementing adequate solutions
– To show employees that safety and health protection must be taken seriously
– To communicate and therefore temper debates
– To create value for companies
– To meet regulations

https://doi.org/10.1515/9783111493633-006

The common denominators among the major analytical techniques are that
- We do not seek for responsibilities but for solutions.
- An accident is always a combination of several factors.
- We should retain only facts, no judgments nor interpretations.
- The analysis is interesting only if it leads to the implementation of solutions.
- The analysis of accidents is a collective work.

We should never forget that near-misses, even if they have not concretized in an incident or accident, are crucial to investigate in order to prevent the realization of the risk, leading to damages.

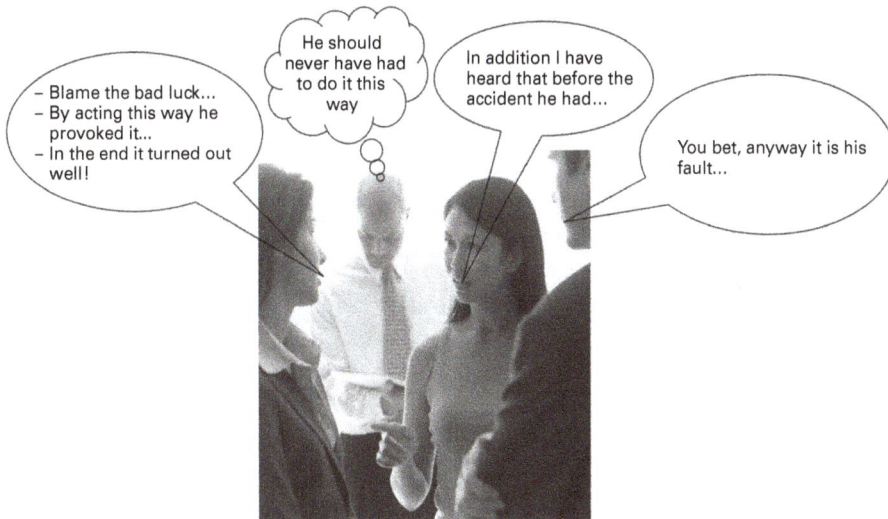

Fig. 6.1: Rumors and facts.

Different types of accident models exist. Each model is based on some underlying premises and paradigms. The various models can be summarized as follows [1]:
- the accident as a probability phenomenon,
- the accident as a human error/mistake/failure,
- the accident as a falling domino tile,
- the accident as a contagious disease (epidemiological approach),
- the accident as a system deviation/disruption and
- the accident as an organizational shortcoming.

The latest insight, viewing an accident as an organizational shortcoming, has the premise that safety is mainly an organizational task, and incidents and accidents are actually an indication of a failing organization. The way that an organization is managed, its culture, the working conditions, the context, the social factors, the policies, etc. all are intrinsically linked to the performance of any organization, both for the positive (production and innovation) and the negative (incidents and accidents). As a consequence, accidents can be avoided, or their repercussions can be decreased by adequate organizational safety management.

6.1 Traditional analytical techniques

Traditional analytical techniques deal mainly with the identification of an accident sequence and seek unsafe acts or conditions that led to the accident. Such techniques include the sequence of events, multilinear event sequencing and root cause analysis (RCA), among many other available techniques. The methods presented below represent only a portion of the available accident analysis techniques. The primary objectives are to
- learn to identify and stop destructive behaviors,
- understand the benefits of the observation process,
- understand how to positively communicate observations,
- build on employee empowerment,
- learn the monitoring and tracking system and
- develop a prioritized observation and strategy plan.

6.1.1 Sequence of events

This technique was originated by Herbert William Heinrich in 1929 [2], based on the premise that accidents result from a chain of sequential events, metaphorically like a line of dominoes falling over. When one of the dominoes falls, it triggers the next one, and the next . . . but removing a key factor (such as an unsafe condition or an unsafe act) prevents the start or the propagation of the chain reaction (see also Section 3.3).

In the newest version of the domino theory model, five labeled dominoes – (i) lack of control by management, (ii) basic causes, (iii) symptoms, (iv) incident and (v) loss of property and people – form the basis of the chain of dominoes. Each domino represents one event. A row of dominoes representing a sequence of events leading to the mishap is lined up. When one domino falls (when an event in the sequence occurs), the other dominoes will follow [2, 3]. However, should a domino in the sequence be removed, no injury or loss will be incurred. Note that such an analysis is usually confined to accidents happening in the exact sequencing.

James Reason [4] (see also Chapter 3) proposed a similar model to that of the domino theory model and the accident analysis framework through the incorporation of the pathogen view on the causation of accidents (the likelihood of an accident is a function of the number of pathogens within the system). Together with triggering factors, the pathogens will result in an accident when defenses in the system are breached. The model is based on a productive system, which comprises five elements: (i) high-level decision-makers, (ii) line management, (iii) preconditions, (iv) productive activities and (v) defenses. The pathogens in a typical productive system originate from either human nature or an organization's strategic apex (high-level decision-makers). The associated types of pathology to each respective productive element are fallible high-level decisions, line management deficiencies, the psychological precursors of unsafe acts and inadequate defenses [4].

6.1.2 Multilinear event sequencing

Multilinear event sequencing was conceived by Benner [5]. Basically, this technique charts the accident process. The model is based on the fact that the first event to create unbalance in a system constitutes the start of a chain of events that ends in damage or injury. The accident sequence is thus described as an interaction between various actors in the system. A description of the accident sequence constitutes the starting point for identification of a situation that can explain why the accident occurred. Every event is a single action by a single actor. The actor is something that brings about events, while actions are acts performed by the actor. A timeline is displayed at the bottom of the chart to show the timing sequence of the events, while conditions that influence the events are inserted in the time flow in logical order to show the flow relationship as depicted in Fig. 6.2 [2]. With this chart, countermeasures can be formulated by examination of each individual event to see where changes can be introduced to alter the process.

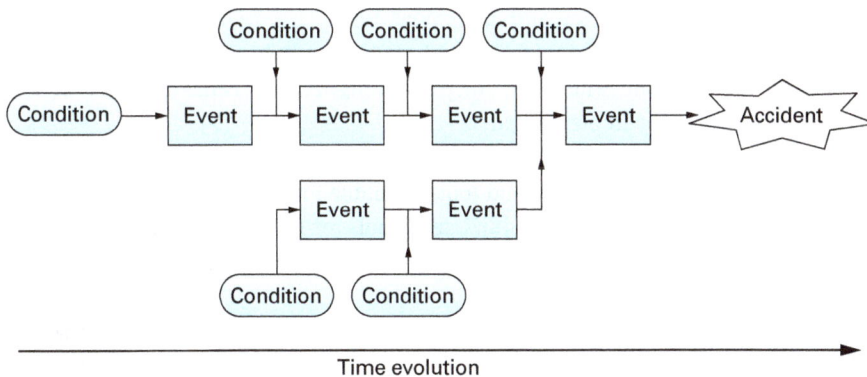

Fig. 6.2: General event sequencing chart.

6.1.3 Root cause analysis

RCA is a process designed for use in investigating and categorizing the root causes of events with safety, health, environmental, quality, reliability and production impacts. The term "event" is used to generically identify occurrences that produce or have the potential to produce these types of consequences. Simply stated, RCA is a tool designed to help identify not only what and how an event occurred but also why it happened. Only when investigators are able to determine why an event or failure occurred can they specify workable corrective measures that prevent similar future events. Understanding why an event occurred is the key to developing effective recommendations.

The RCA process involves four steps [6]:

1. Data collection and gathering
   - Without complete information and an understanding of an event, the causal factors and root causes associated with the event cannot be identified.
2. Causal factor charting
   - The causal factor chart is simply a sequence diagram with logic tests that describes the events leading up to an occurrence, in addition to the conditions surrounding these events.
3. Root cause identification
   - After all the causal factors have been identified, we can begin the root cause identification. This step involves the use of a decision diagram called a "root cause map" to identify the underlying reason or reasons for each causal factor. The map structures the reasoning process to help with answering questions about why particular causal factors exist or occur.
4. Recommendation generation and implementation
   - Following identification of the root causes for a particular causal factor, achievable recommendations for preventing its recurrence are then generated and must be implemented.

There are many analytical methods and tools available for determining root causes to unwanted occurrences and problems [6]. Useful tools for RCA are, e.g., the "five whys" [7], the Ishikawa diagrams, also called fishbone diagrams [8] or the FMEA.

> ! All incident/accident or near-miss analyses tend to discover the main causes that led to the event. Causes may be unsafe acts, unsafe conditions or technical failures. They are generally not self-standing but they link together either in sequential or bijective mode. The sequential interconnection is important in order to correctly analyze the event.

6.2 Causal tree analysis

The causal tree accident analysis method elaborated by the INRS (Institut national de recherche et de sécurité, France) [9] is based on an original work initiated by the European Coal and Steel Community and was attempted for the first time in a practical way in 1970 in the iron mines of Lorraine. This method aims at being situated beyond the debates and the opinions. It offers a way to analyze fine circumstances that have led to an incident/accident, and to transform the causes of these incidents/accidents into predictable facts, hence leading to prevention. It is an investigation and analysis technique used to record and display in a logical, tree-structured hierarchy, all the actions and conditions that were necessary and sufficient for a given consequence to have occurred.

Causal tree analysis provides a means of analyzing the critical human errors and technical failures that have contributed to an incident or accident in order to determine the root causes. It is a graphical technique that is simple to perform and very flexible, allowing for mapping out exactly what we think happened rather than being constrained to an accident causation model. The diagrams developed provide useful summaries for inclusion in incident and accident reports that give a good overview of the key issues.

It has to be noted that this method has many similarities with fault tree analysis (FTA). However, FTA focuses on failure determination (technical and process), whereas the causal tree is mainly looking for accident causes.

6.2.1 Method description

Tree structures are often used to display information in an organized and hierarchical fashion. Their ability to incorporate large amounts of data, while clearly displaying parent–child or other dependency relationships, also makes the tree a very good vehicle for incident investigation and analysis. The combination of the tree structure with cause–effect linking rules and appropriate stopping criteria yields the *causal tree*, which is one of the more popular investigation and analysis tools in use today.

Typically, it is used to investigate a single adverse event or consequence, which is usually shown as the top or right item in the tree. Factors that were immediate causes of this effect are then displayed below it or on the left, linked to the effect using branches. Note that the set of immediate causes must meet certain criteria for necessity, sufficiency and existence. Proof of existence requires evidence.

Often, an item in the tree will require explanation, but the immediate causes are not yet known. The causal factor tree process will only expose this knowledge gap; it does not provide any means to resolve it. This is when other methods such as change analysis or barrier analysis can be used to provide answers for the unknowns. Once the unknowns become known, they can then be added to the tree as immediate causes for the item in question.

The method has four main steps:

1. Search for facts
   - Without any judgment
   - Without interpreting
   - By treating each fact one by one
   - One fact must be measurable and/or photographable
2. Build the tree
   - Begin with the last fact
   - Develop branches by asking the three following questions:
     - What is the direct cause that provoked this?
     - Was this cause really necessary for the occurrence of that fact?
     - Was this cause sufficient to provoke the event?
3. Search for measures
   - Begin with the first fact
   - Search measures for each fact
   - Accept all ideas
4. Define measures
   - Efficiency in use
   - Measures do not displace a risk
   - Simple, sustainable measures
   - Measures comply with laws and regulations

The overall process could be schematized as depicted in Fig. 6.3, which shows that building a comprehensive tree is not sufficient if it does not lead to solutions and implementation of corrective measures. In order to avoid the repetition of a similar accident, it is also crucial to install a follow-up procedure.

6.2.2 Collecting facts

A fact is an action or event that happened. The main aspects are
- **Form:** Collection of hot circumstances and additional elements
- **Retain a fact list:** Directly linked to the accident, validated by the team members
- **Objectives:** Provide an opportunity to identify and establish the circumstances that led to an event
- **Prerequisites:** A report should be made straight after the accident, and all team members should have been at the event location
- **Qualifications:** Be curious, careful; choose words with precision and accuracy.

1. Collect circumstances on the spot	2. Build the team analysis	3. Collect facts	4. Seek for causes, building the tree	5. Definition of preventive and correctives solutions	6. Action plan	7. Implementing solutions	8. Follow-up

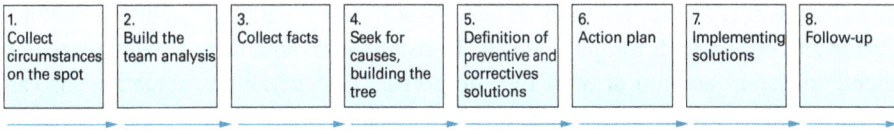

Fig. 6.3: Overall process of the causal tree.

They are several rules to follow that are vital when collecting facts so that no judgments or interpretations are brought to the analysis:
– No judgments.
– Do not make any assessment.
– Identify only the facts.
– Note one independent fact at a time.
– Facts are measurable and/or photographable, in simple terms.
– Describe the facts.
– Facts must be unambiguous.
– Do not write down feelings, except from an injured person when they are expressed spontaneously.
– Do not use adjectives (beautiful, dangerous, many, little, etc.).

Let us take an illustrative example to discover that facts are not always easy to retain. Read the description of the situation and then try to answer the following questions:

John Bely, a car body repairer, arrives at the garage from 5th Avenue. In the courtyard, a worker animatedly shouts in Spanish, gesturing dramatically and looking upward. Once inside, John begins replacing the tires on a car. Near the porch, a young man with a beard speaks gently to someone with long hair and dressed in jeans

Proposals	Answers		
	True	False	?
John Bely is garage mechanic.			
There are only four persons in the court.			
John Bely is on 5th Avenue.			
The contractor has heard a worker shouting.			
One of John Bely's workers was changing tires.			
A Spanish worker was in the court.			
The worker shouting in the court was speaking with a colleague located on an upper floor.			
He shouted to alert his colleagues of the arrival of John.			
A bearded man and a woman were talking.			

At times, interviewing witnesses will be essential to collect relevant facts. A witness is defined as anyone with information about the incident or accident. It is important to recognize that evidence obtained through witness testimonies and statements is often quite fragile. It is uncommon for a witness to recall all the details of an event – before, during and after – with complete clarity. In reality, each witness will likely provide a slightly different account of what they observed. These variations in statements can be attributed to several factors, such as:

–   No two individuals see details or objects and remember them the same way.
    –   Points of observation vary from witness to witness.
    –   Witnesses differ in both technical and personal backgrounds.
    –   Witnesses are interested in self-preservation and the protection of friends.
    –   Witnesses differ in their abilities to rationalize what they have seen and artic-
        ulate an explanation or description.

Some basic tips for the investigator to start an interview [3] are:

–   Explain who he/she is.
–   Explain why the accident is being investigated.
–   Discuss the purpose of the investigation (e.g., to identify problems and not to de-
    termine fault or blame).
–   Provide assurances that the witness is not in any danger of being compromised
    for testifying about the accident.
–   Inform the witness of who will receive a report of the investigation results.
–   Ensure witnesses that they will be given the opportunity to review their state-
    ments before they are finalized.
–   Absolutely guarantee the privacy of the witness during the interview.
–   Remain objective – do not judge, argue or refute the testimony of the witness.

To gather as much relevant information as possible, the questions should focus on *who*, *what*, *when*, *where* and *how* aspects of the incident or accident. Such focus will preclude the possibility that the witness will provide opinion rather than fact. While the *why* questions can be asked at the end of the interview, it is a straightforward rule that the majority of the interview should be based on facts and observations and not on opinions.

At least the following areas should be covered during an interview of a witness:

1. *What* was the exact or approximate time of the event?
2. *What* was the condition of the working environment at the time of the event, be-
    fore the event and after the event (e.g., temperature, noise, distractions and
    weather conditions)?
3. *Where* were people, equipment and materials located when the event occurred,
    and *what* was their position before and after the event?
4. *Who* are the other witnesses of the event (if applicable) and *what* is their job
    function?
5. *What* and/or *who* was moved from the scene, repositioned or changed after the
    event occurred?

6. *When* did the witness first become aware of the event and *how* did he or she become aware?
7. *How* did the response or emergency personnel perform, including supervisors and outside emergency teams?
8. *How* could the event have been avoided?

The first five questions establish the facts of the event. This information is required in order to establish an understanding of the facts as they happened or as they were perceived, but also to validate the credibility of the witness testimony. Question 6 is an indication of the point in time during the event when the witness became a witness. It is obvious that such a question may lead to vital information about the cause of the accident; e.g., if the witness claims that he or she became aware that something was wrong because of a strange sound or visual happening prior to the actual event. The purpose of question 7 is to obtain feedback on the effectiveness of the organization's existing loss control procedures. The last question serves to fix a point of reference at the end of the investigation. An opinion on the prevention of the event may actually provide a direction from which the investigator can proceed when providing recommended solutions, prevention and protection measures, mitigation measures, corrective actions, etc.

6.2.3 Event investigation good practice

The setting of the event needs to be unambiguously described: "what," "where," "when" and "how" questions need to be asked. Always try to avoid asking "why." If people are interviewed, they should be able to tell their story without interruptions. The interviewer should mainly listen.

Questions that need answering in case of an occupational accident involving one or more persons carrying out specific activities/tasks are:
– Are there instructions and/or procedures existing and available with respect to the activities/tasks?
– Were these instructions and/or procedures looked at and followed?
– Are there contradictions in the instructions or procedures?
– What about the specific competences of the person carrying out the tasks/activities?
– Was there adequate training of the person regarding the tasks/activities to be carried out?
– Were all means available to carry out the tasks/activities in a correct and safe way? If not, what means were available and what means were used?
– Was there any time pressure involved? If so, please elaborate/explain.
– Were there sudden hindrances or unforeseen circumstances? Explain.

Collect all the incident/accident facts (i) as soon as possible after the accident, (ii) on location, (iii) in a team (e.g., supervisor, victim if possible, witnesses) with sufficient and adequate competences and (iv) with pictures and/or images/movies, if possible. Investigators should be sufficiently independent from the incident/accident, and the investigation should be carried out within a reasonable period after the incident/accident. It is always a good idea to look for similar incidents/accidents within the company, as well as external to the organization. Make a timeline: unsafe situation, activities/tasks providing the occasion for the incident/accident, description of the incident/accident itself, intervention activities and recovery.

> **!** The criteria for the facts to be collected are, in order of importance/reliability (1 = most important/reliable):
> 1. Fact observed by (member of) the investigation team.
> 2. Material evidence at the location (physical evidence, work material, broken/defect/misused equipment, measurements, samples, spilled material, sewer contents, waste bin contents, used safety equipment, infrastructure of surroundings, hindrances, no safety barrier, safety material out of order).
> 3. Noninvolved and independent witnesses.
> 4. Try to recheck stories/interviews and have supporting evidence, if possible.

6.2.4 Building the tree

Once the immediate causes for the top item in the tree are known, then the immediate causes for each of these factors (see Fig. 6.4) can be added and so on. Every cause added to the tree must meet the same requirements for necessity, sufficiency and existence. Eventually, the structure begins to resemble a tree's root system. Chains of cause and effect flow upward or from left to right of the tree, ultimately reaching the top level. In this way, a complete description can be built on the factors that led to the adverse consequence. Remember that the three questions to be answered when identifying facts are (Fig. 6.4):
1. What direct cause provoked this fact?
2. Was this cause really necessary for this fact to occur?
3. Was this cause sufficient to provoke the event?

Once the facts are determined, there are three possibilities to connect them together as described in the following recap (Fig. 6.5).

The remaining task is to build the complete tree, either from the top down or from right to left depending on the tree size. Before illustrating the method with an example, let us try to find the right connection of facts through a small illustration.

In this brief case, four facts have been identified, one of them being the top event or the final fact:

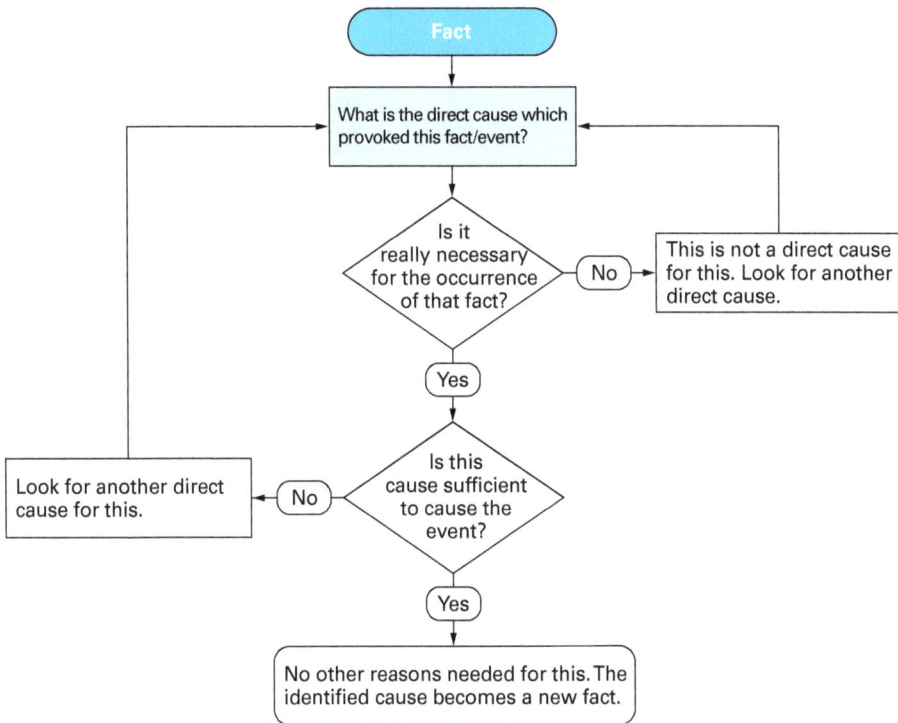

Fig. 6.4: How to define independent facts?

1. The faucet is left open.
2. The tub overflows.
3. The drain is blocked.
4. The bathroom carpet is completely wet.

The first step is to determine the final event: in this case, fact 4 is the top event.

There are several possibilities to interconnect the remaining facts with the top event, as illustrated in Fig. 6.6; however, only one is the reflection of what happened. Of four facts and three possible combinations, which one is the correct solution: A, B or C?

Let us find the correct solution. We should begin by asking what was the direct cause that led to "The bathroom carpet is completely wet," according to the procedure described in Fig. 6.4. The answer is that only fact 2 is sufficient for fact 4 to happen. So fact 2 is sequentially linked to fact 4. Thus, solution C cannot be the correct one.

Next we start with fact 2 and ask the same question as before: what was the direct cause that led to the tub overflowing? The answer is that both facts 1 and 3 are necessary and sufficient for fact 2 to happen. So facts 1 and 3 are linked by conjunction to fact 2. This excludes solution A, revealing that the correct causal tree is solution B.

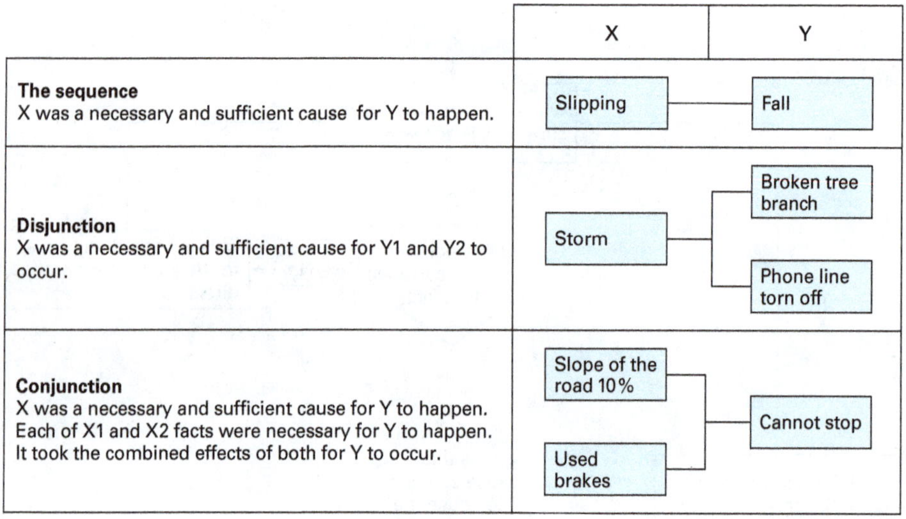

	X	Y
The sequence X was a necessary and sufficient cause for Y to happen.	Slipping ———	Fall
Disjunction X was a necessary and sufficient cause for Y1 and Y2 to occur.	Storm	Broken tree branch Phone line torn off
Conjunction X was a necessary and sufficient cause for Y to happen. Each of X1 and X2 facts were necessary for Y to happen. It took the combined effects of both for Y to occur.	Slope of the road 10% Used brakes	Cannot stop

Fig. 6.5: The types of connectors.

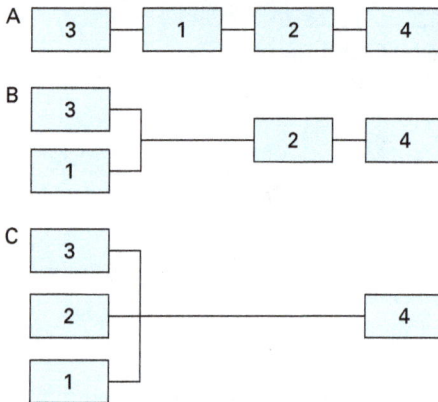

Fig. 6.6: Different possible causal trees.

6.2.5 Example

Now, we are ready to illustrate the causal tree analysis by evaluating the following accident description:

An employee was ordered to sort the bottles by glass color. Customers can return the empty bottles in crates. The delivery truck will take empty bottles from the customer, and the driver visually checks if the crates are filled. The customer does not receive any reimbursement for the missing bottles. The employee takes out every bottle from the crate and puts it on the conveyor belt installation. The bottles that do not correspond to the specified glass are set aside in a special crate for this purpose. By pulling out a defective bottle with a broken neck, the employee injured his right hand, causing a severe tendon cut of the right thumb. The collaborator will be out of work for 2 months.

When working on the facility, wearing gloves is mandatory as specified in the policy of the company. Consequently, employees receive personal protective gloves.

First, we have to collect the facts:
1. Workers must take the bottles out of crate to be sorted by type.
2. Mr. X does not wear protective gloves.
3. The customer is not refunded for broken and/or missing bottles.
4. Protective gloves are not designed for the specific task – hazard.
5. The customer puts the number per supplier in crates without separating by content.
6. Mr. X holds the broken bottle.
7. There is a broken bottle in the crate.
8. Mr. X has a reflex movement.
9. The driver does not control the bottles when taking them from the customer.
10. Mr. X does not meet the guidelines of wearing gloves.
11. Mr. X does not see that the bottle is broken.
12. Mr. X has a significant injury to his right hand.
13. The customer puts broken bottles in the crate.
14. There is an absence of regulation, "Control of crates by the driver."
15. There is a lack of control by the supervisor.

Second, we must build the tree according to the aforementioned rules. It seems obvious that the final event is number 12. Building the tree leads to the final causal tree presented in Fig. 6.7.

6.2.6 Building an action plan

Defining measures is easier when the tree is built. Safety features (safety barriers and corrective measures for both prevention and protection) must be provided as barriers in response to the initiating event. They generally aim to prevent, as far as possible, the initiating event as the cause of a major accident. It is usually sufficient to stop one branch of the tree in order to avoid the occurrence of the final event. They are summarized in the *action plan*, whose objective is to determine the measures by answering the following questions:
– What are the chosen measures to stop a repetition of an event?
– What is the measure implementation deadline?
– Who is responsible for their implementation and meeting the deadline?

The number of solution propositions has to be large enough to build a coherent action plan. It is essential to acquire the skills required to implement measures. The measures are chosen according to the following criteria:

- Efficient solution
- No risk displacement
- Extent consistent with law
- Effective solution
- Simple solution (strong acceptance by individuals)
- Eliminating the root causes of the event
- Possible use of the solution in other areas

When choosing measures, one must ensure that all the influences included in the causal tree are counteracted by the selected measures. The gathered proposed solutions should be classified depending on their efficiency level. Combinations of measures should be preferred and eligible; for example: technical modification of an apparatus + organization adjustment + personnel and superior formation. *The farther the neutralized influences are from the fact, the more effective they are.*

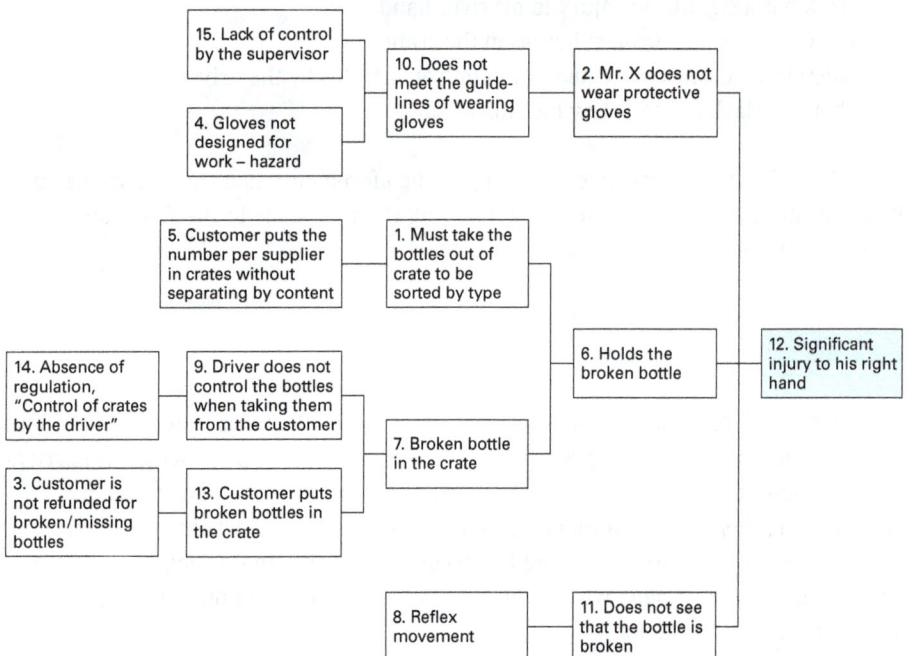

Fig. 6.7: Causal tree from the mentioned example.

In the example from Fig. 6.7, we can use our imagination to choose between facts 4, 15 and 10 to ensure that adequate protective gloves are worn. It is also possible to imagine that the driver looks carefully to identify broken bottles – facts 9, 13 and 7. There are no limits in defining measures; they should only be adequate, economically sustainable, effective and implemented.

6.2.7 Implementing solutions and follow-up

Once the action plan has been established, solutions must be implemented, and a follow-up and control must be performed. Communication with personnel is essential as it enables us:
– to value the study made by the team,
– to inform employees about what actually happened,
– to inform employees about the anomalies and
– to inform employees about the retained and adopted solutions.

> The causal tree method allows us, using a graphical representation, to investigate the root causes and subsequent facts that led to an event. It deserves not only the incident/accident analysis but also a teaching purpose to avoid the repetition of such an event by taking appropriate measures. Corrective measures and actions must be based on a careful analysis in order to implement adequate remedies. !

6.3 Tripod technique

The so-called tripod technique aims at identifying and treating latent factors that can be dealt with by the management of an organization. The accident investigation method provides a structured overview of how the accident outcome(s) came about and what latent factors can be attributed to the causal factors. The latent factors or so-called basic risk factors (BRFs) are very useful for management to decide on risk reduction measures. There are 11 BRFs defined and to be used while analyzing an accident:

1. Organization (structural difficulties with respect to responsibilities and authorities)
2. Incompatible goals (conflicting goals, e.g., with respect to production, safety, planning, economic interests, or between and within individuals and groups)
3. Communication (insufficient, wrong, inadequate, late messages, or, for instance, wrong interpretation of message)
4. Procedures (work instruction problems: correct, full, understandable, relevant, etc.)
5. Training (right training and up-to-date training)

6. Design (design of materials, processes, components, etc.)
7. Hardware (quality, condition, availability, actual versus expected lifetime of materials, tools and components of installations)
8. Maintenance management (effectiveness of maintenance strategies)
9. Housekeeping (order and cleanliness of working environment)
10. Error enforcing conditions (context, working environment: physical (e.g., heat/noise/cold/visibility), medical, psychological, social, etc.)
11. Defenses (failures with respect to detection, warning procedures and methods, repair, evacuation, use of PPM and CPM and emergency preparedness)

Tripod considers an accident as the top event. Starting from this event (the consequences of an accident), the different events/facts that have led to this event are put into a scheme. The idea of the tripod is to have an open triangle, a tripod, based on energy/energies/hazards, event(s) and object(s) (can be human or nonhuman). The tripod researcher first looks at all the objects (damaged materials or victims), and then tries to understand how energies/hazards and events are linked to the object(s). The reasoning is: energy damages object via event. The linking of the three elements is graphically represented as a tripod. By considering the timing into the picture (what happened at what time before the incident), the existing trios (energy, object and event) can be determined.

After constructing the various trios, barriers for every trio can be identified that could have avoided the trio. The barriers could have been present and failed, or they simply could have been absent. Once the barriers are known, the direct causes related to the failing or missing of these barriers can be determined. Finally, the tripod researcher can look at the list of 11 BRFs and define the context of the accident (the preconditions) and link the BRFs that have led to that context.

6.4 AcciMap technique

The AcciMap approach is a systems-based technique for accident analysis, specifically for analyzing the causes of accidents and incidents that occur in complex sociotechnical systems. AcciMap graphically maps the multiple contributing factors to an accident and their inter-relationships onto the following six levels:
1. Government policy and budgeting
2. Regulatory bodies and associations
3. Local health economy planning and budgeting (including hospital management)
4. Technical and operational management
5. Events, processes and conditions
6. Outcomes

The model describes the interactions between the different hierarchical levels being involved in the (un)safety related to an accident. The precise number of levels may vary depending on the nature and size of an organization, a municipality or beyond. Levels within a company are, for instance, the workers at the work floor, their supervisors, the middle management and the top management. Outside a company, the following levels can, for instance, be involved: local authorities, regional authorities and overarching organizations, country-wide legislation and the government, European Directives and guidance.

The AcciMap approach was developed by Rasmussen (1997) as a means of modeling the sociotechnical context to identify the combination of events and decisions that produce an accident. Each level is involved in safety management through laws, rules and instructions. For systems to function safely, decisions made at high levels should trickle down and be reflected in the decisions and actions occurring at lower levels of the system. Conversely, information at the lower levels (e.g., staff, work and equipment) regarding the system's status has to travel up the hierarchy to inform the decisions and actions occurring at the higher levels. Without this so-called vertical integration, systems can lose control of the processes and fail. AcciMap does not use predefined taxonomies of failures across the different levels. It is relatively simple to learn and use, but the analysis could be time-consuming and the output could become large and unwieldy.

6.5 Organizational learning

Argyris [10] emphasizes that learning should be embedded in the entire organization as a part of its normal operation. It is not an add-on extra! Translated into the field of safety, this means that there must be an intimate link between the risk assessment process, which specifies what are the hazard and threat scenarios, the management process (and the safety management system), which establishes control strategies and practices for them, the operational process which carries them out, and the learning process, which evaluates, improves and fine-tunes these controls.

Two types of learning exist: single- and double-loop learning. Single-loop learning affects the way operational goals are achieved without changing goals or values themselves. Organizational single-loop learning products are visible in the organization's theories of action, e.g., as minor modifications in a task protocol. Conversely, double-loop organizational learning affects norms, values and organizational targets that govern the organizational unit and its theory of action. Therefore, double-loop changes mean modifications to the constraints for operations run by the unit.

If the organizational unit can make the necessary changes to its working practices within its own resources and authority, single-loop learning is sufficient. A change in objectives and values is an example of double-loop learning. A decision to change the process or to impose new norms, e.g., on performance and sustainability, calls for double-loop learning. The need to invoke a more senior person to authorize expenditure on equipment may still be single loop.

Argyris and Schon [11] have identified a characteristic style adopted by people dealing with threatening problems in organizations. This style seems to be an almost automatic, defensive reflex of which the person seems to be unaware. The style is characterized by four rules of thumb and two strategies:

Rules of thumb:
- Strive to be in unilateral control
- Minimize losing and maximize winning
- Minimize the expression of negative feelings
- Be rational

Strategies:
- Advocate views without encouraging inquiry
- Unilaterally save face (own and others)

Obviously, these characteristics of behavior in case of potential problems strongly hinder individual learning as well as the organizational learning process. Looking at the obstacles and the ways to avoid learning, it is possible to understand what are the best practices and what should be the behavior, attitude and style to learn from mistakes and errors. Kingston [12] summarizes 20 obstacles and 10 ways to avoid organizational learning:

The 20 obstacles to learning:
- Perceived lack of time
- Blame culture
- Resistance to change
- Lack of accountability
- Wrong sort of accountability
- Few opportunities for lateral communication
- Too much top-down management
- "Not my problem"
- Sometimes: "shoot the messenger"
- Passive communication (i.e., do not require change in behavior or beliefs)
- Tendency to "dig the detail"
- Impersonal styles of communication
- Alienation
- Poor quality of relationships
- Not enough "time-out" to talk
- Poor commitment to lifelong learning
- Specialists "own the message"
- Specialists "own the problem"

- Tightly comfort zone (e.g., only engineering issues are thought legitimate to consider)
- Lack of trust

The 10 ways to avoid learning:
- Do not collect or preserve information.
- Believe that "error" is a sufficient explanation.
- Do not use appropriate methods of analysis.
- Ensure that only specialists do investigations.
- Do not debrief at each level of line, especially operational personnel.
- Rely on formal communication alone.
- Ensure that solutions are the sole output of investigations.
- Do not own or track remedial processes.
- Only debate the technical details.
- Do not question your methods and motivations.

Besides the above lists of topics, some other important hints on how not to learn and how to ensure to make the same mistakes over and over again in an organization are to not use performance indicators properly (with all information needed to build company memory), to have no systematic and harmonized incident investigation form and process, and – very important – certainly not to keep any lists of near-misses and investigate them.

6.6 Conclusions

Why should an accident analysis be conducted? The ultimate reason for conducting an accident analysis is to avoid future injuries through the identification of facts. The accident analysis process should not be used to place blame. No person wants to get hurt. We need to conduct a thorough accident and incident analysis to identify the multiple causes of the accident. Then, based on the facts, develop solutions. This activity is designed to stop damaging infrastructure and stop human suffering.

A thorough accident and incident analysis will determine multiple causation and potential trends by department or area of responsibility. With the facts and data gathered in the process of the analysis, corrective measures can be developed and similar injuries will be reduced or eliminated.

A side effect is that accident analysis could be largely used as an efficient educational tool for training and empowered awareness of safety. It is the main tool in enterprise safety partnerships for the development and the political prolongation of planned preventions, conceived as an element of enterprise management.

References

[1] Wienen H., Bukhsh F., Vriezekolk F. Wieringa R. (2018). Learning from Accidents: A Systematic Review of Accident Analysis Methods and Models. International Journal of Information Systems for Crisis Response and Management. 10(3). 42–62. DOI: 10.4018/IJISCRAM.2018070103

[2] Ferry, T.S. (1977). Modern Accident Investigation and Analysis. 2nd edn. Hoboken, NJ: John Wiley & Sons, Inc.

[3] Vincoli, J.W. (1994). Basic Guide to Accident Investigation and Loss Control. New York: John Wiley & Sons, Inc.

[4] Reason, J. (1997). Basic Guide to Accident Investigation and Loss Control. Aldershot, UK: Ashgate Publishing Company.

[5] Benner, L. (1975). Accident investigations. Multilinear events sequencing methods. J. Saf. Res. 7: 67–73.

[6] Vanden Heuvel, L.N. (2007). Root Cause Analysis Handbook: A Guide to Effective Incident Investigation. 3rd edn. ABS Consulting. Brookfield, CT: Rothstein Associates Inc.

[7] Ohno, T. (1977). Toyota Production System: Beyond Large-Scale Production. Productivity Press. New York: Taylor and Francis Group.

[8] Ishikawa, K. (1990). Introduction to Quality Control. Productivity Press. New York: Taylor and Francis Group.

[9] Monteau, M. (1974) Méthode pratique de recherche des facteurs d'accidents. Principes et application expérimentale, Rapport no. 140/RE INRS.

[10] Argyris, C. (1992). On Organizational Learning. Massachusetts: Blackwell Publishers.

[11] Argyris, C., Schön, D. (1996). Organizational Learning II: Theory, Method and Practice. Reading, MA: Addison Wesley.

[12] Kingston, P. W. (2001). The unfulfilled promise of cultural capital theory. Sociology of education, Vol. 74. Extra issue Currents of Thought: Sociology of Education at the Dawn of the 21st Century 88–99.

7 Major industrial accidents and learning from accidents

7.1 Link between major accidents and legislation

Taleb [1] indicates that there are catastrophes with outsized consequences, which are so rare and unthinkable that people were not prepared to face them because they could not even conceive of their existence. He called them "black swans" since eighteenth-century Europeans could not picture a swan as being black. All European swans were white, so why should there exist black swans, or indeed blue or green swans? It turns out that black swans are not rare at all in some other parts of the world (but blue or green swans have not yet been discovered). Taleb could have called the phenomenon "platypus" without any problem as well, since this animal has even more strange characteristics than just a different color, unknown to contemporary science. Black swans (an extremum of type II events) (see Section 2.9) lurk outside of our ability to predict. Many catastrophes are, however, not as rare as one might think. They are what are called "gray rhinos." Gray rhinos are threats that people ought to see but often do not see, or that they see but willfully ignore [2]. Most gray rhinos are not the case of signals that are too weak but of listeners determined to ignore them and systems that encourage and accept as normal the failure to respond to such obvious threats. There will always be people that are stubborn enough to ignore even the most obvious threat. But, as a rule, if a threat is obvious enough that a reasonable person can see it coming, it is a gray rhino, and not a black swan.

Despite Taleb's railing against people's perception of their ability to accurately see into the future, most of the crises in the world are very likely occurrences. The biggest threats facing leaders are not highly improbable Black Swans, but highly probable gray rhinos. We may not be able to foresee the details or the timing, but the outlines of the biggest threats facing us are hard to ignore. As Wucker [2] puts it, "Why worry about an odd bird when you're facing a two-ton beast that is snorting, pawing the ground, and looking straight at you as it prepares to run you down?"

Gray rhino events are obvious and easy to picture. A good example of a gray rhino is the Covid-19 pandemic. Many pandemics have preceded this 2019/2020/2021 one and many lessons were learned a long time ago. Regretfully, many societies worldwide struggled with implementing adequate measures to deal with the pandemic. This should not be the case if adequate learning had happened in a systemic way. Gray rhinos happened before, and thus information is available, making them obvious type II risks and events. You cannot argue that a rhino does not exist because it would be the wrong color: all rhinos are actually gray. Their potential impact is massive, whether social/human, political, economic, environmental, etc. Most of the major accidents described hereafter in

https://doi.org/10.1515/9783111493633-007

this chapter are gray rhinos: we all know of them since they happened before. Actually, before charging, the gray rhino event has provided certain warning signals and precursors. People were just too blind to see them. Afterward, lessons are drawn. But too often, they are forgotten quickly.

The tendency of major accidents to force politicians to take political actions ad hoc is illustrated in Fig. 7.1, presenting a chronological overview of a non-exhaustive number of significant major accidents worldwide and the chronological developments in a non-exhaustive number of European and US safety regulations.

All political attention comes to focus on the accident that has just happened, while a proactive broad political view of the accident prevention issue as a whole that is needed is overlooked. The reason for this reactive political behavior is that politicians and company policymakers have limited imaginative power to fully understand the probabilities of accident estimates. Moreover, people are risk-averse for gains, but risk-taking for losses (see also Chapter 9). This phenomenon causing ad hoc prevention legislation is one of a number of biases in information processing that occurs when we make choices.

Originally described by two psychologists, Daniel Kahneman and Amos Tversky [3], the work has had a considerable impact on economic models of choice. Tversky and Kahneman discovered there is a strong tendency for individuals to be what they called risk-averse for gains, but risk-taking for losses. The choice experiment has been carried out many times, in many different situations, and the results are very robust. Whether the decision is presented as a loss or a gain will influence what decision is taken. People prefer to hang onto their gains, but gamble with their losses. In other words, losses and gains are not equally balanced in decision-making. Certain gains are weighted far more heavily than probable losses. As Sutherland et al. [4] point out, this can be easily seen in the balancing of "production" and "safety." Ignoring some safety aspects will lead to almost certain increases in production in the short term. Applying safety measures for highly improbable accidents will lead to additional costs in the short term. Hence, it takes a much larger, and more probable, loss to tip the prevention management decisions in favor of safety with respect to low probability, high-consequence risks.

Nonetheless, the potential for major industrial accidents, which have become more significant with the increasing production, storage and use of hazardous substances, has emphasized the need for a clearly defined and systematic approach to the control of such substances in order to protect workers, the public and the environment.

Major hazard installations possess the potential, by virtue of the nature and quantity of hazardous substances present, to cause a major accident in one of the following general categories:
- The release of toxic substances in tonnage quantities that are lethal or harmful even at considerable distances from the point of release
- The release of extremely toxic substances in kilogram quantities that are lethal or harmful even at considerable distances from the point of release

US Legislation

- Clean Air Act
- Occupational Safety and Health Act

Clean Air Act Amendments: emissions of particulate matter and sulfur dioxide are restricted

Amendments Clean Air Act (new accident prevention program)

OSHA Process Safety Management (PSM) Regulation

EPA Risk Management Program Rule

Wide range of post 9/11 legislation, including ISPS, CFATS, Patriot Act, etc.

Regulations on offshore drilling are in preparation

European Legislation

British Health and Safety at Work Act

Swiss Law on Protection of the Environment

Seveso I Directive (82/501/EEC)

Emergency Planning and Community Right-to-Know Act

Seveso I Directive Amendments (87/216/EEC) and (88/610/EEC)

Swiss Ordinance on Protection against Major Accidents

EU Framework Directive 89/39

Seveso II Directive (96/82/EC)

COMAH regulations (UK) (offshore oil platform safety)

Seveso II Directive Amendments (2003/105/EC)

Last revision of Swiss Ordinance on Protection against Major Accidents

Seveso III Directive (2012/18/EU)

Major Accidents

Kentucky, USA

Feyzin, France (18 fatalities, 81 injuries)

Texas City, USA (13 fatalities, 5 injuries)

Flixborough, UK (28 fatalities, 400 injuries)

Seveso, Italy (200.000 affected, long term)

Los Alfaques, Spain (216 fatalities, 100 injuries)

- Mexico City, Mexico (650 fatalities, 6400 injuries)
- Bhopal, India (3787 fatalities, 200.000 injuries)

Chernobyl, Ukraine (2 million affected, long term)

Piper Alpha, UK (167 fatalities)

Pasadena, Texas, USA (23 fatalities, 314 injuries)

Nagothane, India (31 fatalities, 63 injuries)

Vishakhapatnam, India (60 fatalities)

Enschede, The Netherlands (22 fatalities, 950 injuries)

- 9/11 WTC terrorist attacks (2996 fatalities)
- Toulouse, France (31 fatalities, 2442 injuries)

- Houston, USA (15 fatalities, 170 injuries)
- Buncefield, UK (43 injuries)

Ath, Belgium (24 fatalities, 120 injuries)

Deepwater Horizon, Gulf of Mexico (11 fatalities, 16 injuries)

Fukushima, Japan (tens of thousands affected, long term)

Wetteren, Belgium (1 fatalities)

Tianjin, China (173 fatalities, >800 injuries)

Sichuan, China (19 fatalities, 12 injuries)

Beirut, Lebanon (218 fatalities, 7000 injuries)

Andhra Pradesh, India (17 fatalities, 50 injuries)

1960 1962 1965 1966 1969 1970 1974 1975 1976 1977 1978 1980 1982 1983 1984 1985 1986 1987 1988 1989 1990 1991 1992 1995 1996 1997 1999 2000 2001 2003 2004 2005 2008 2010 2011 2012 2013 2015 2018 2020 2024

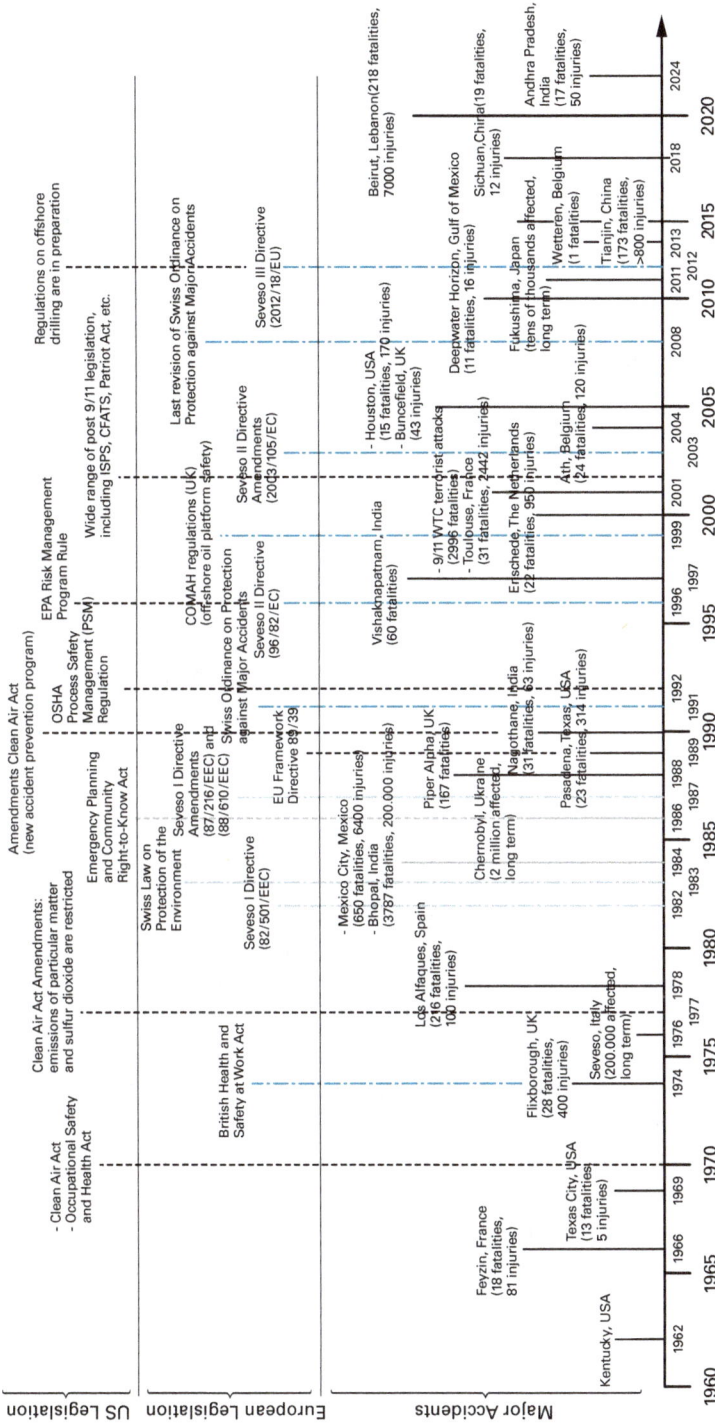

Fig. 7.1: A non-exhaustive number of major industrial accidents and their relation with European and US safety legislation.

- The release of flammable liquids or gases in tonnage quantities that may either burn to produce high levels of thermal radiation or form an explosive vapor cloud
- The explosion of unstable or reactive materials

Apart from routine safety and health provisions, special attention should be paid by competent authorities to major hazard installations by establishing a major hazard control system. This should be implemented at a speed and to an extent dependent on the national financial and technical resources available. The works management of each major hazard installation should strive to eliminate all major accidents by developing and implementing an integrated plan of safety management. Works management should develop and practice plans to mitigate the consequences of accidents that could occur. For a major hazard control system to be effective, there should be full cooperation and consultation, based on all relevant information, among competent authorities, works management and workers and their representatives.

7.2 Major industrial accidents: examples

Literature sources for the accidents include Wells [5], Reniers [6] and Paltrinieri [7].

7.2.1 Feyzin, France, January 1966

A tank farm including eight spheres containing propane and butane was undergoing a routine drainage operation. The wrong procedure was applied to a 1,200-m^3 sphere. The isolating valves became inoperable, and an uncontrolled leak from a propane sphere was ignited by a car on a nearby road, flashing back to burn like a flare directly under the sphere. Propane snow had also accumulated within the bund. The refinery fire brigade attempted to put out the fire but ran out of dry foam. The municipal fire company continued to fight the fire using fire water. After 30 min, the safety valve lifted, and 1 h later a boiling liquid expanding vapor explosion (BLEVE) ensued. The fire brigade had concentrated on cooling the other spheres and not the burning sphere on the assumption that the relief valve would provide protection. Approximately 340 m^3 of liquid propane was released and partially vaporized, producing a large fireball and an ascending mushroom cloud. The BLEVE killed and injured approximately 100 people in its vicinity.

One debris missile broke the legs of an adjacent sphere, which contained 857 m^3 of propane. A second piece tipped over another sphere containing 1,030 m^3 of butane. Another section traveled 240 m to the south and severed all the product piping connecting the refinery area to the storage area. One fragment broke piping near four floating roof tanks. Fires were initiated in this area. Extensive structural damage was caused in the village of Feyzin, about 500 m away. Some 2,000 people were evacuated from the surrounding area. A further BLEVE and other explosions occurred as the fire spread. Firefighting continued for a further 48 h until the three spheres that were still intact (and full of propane and butane) were cooled to an appropriate level.

7.2.2 Flixborough, UK, June 1974

In the month of March, a caprolactam plant, using six reactors in series, was shut down and reactor no. 5 was taken out of service. A 20-inch pipe was used to fabricate a dog-leg bypass pipe. The plant was started up on the first of April. On Wednesday, May 29, a leak was discovered in the bottom isolation valve on a sight glass fitted to a reactor. The plant was depressurized and cooled down, the leak repaired and the plant restarted. Normal operating conditions of 8.8 bar and 155 °C were achieved on June 1, with the plant on hold pending the arrival of high-pressure nitrogen needed for the commencement of oxidation. Shortly before 17:00 h, the bypass in place on no. 5 reactor became unstable with the result that the two bellow units attached to no. 4 and no. 6 reactors failed, and the bypass pipe collapsed. Hot cyclohexane was emitted with flash vaporization and massive entrainment. Two distinct clouds were observed: a larger elevated cloud and a base cloud.

A minor explosion took place in the control room some 10–25 s after the release. When the base cloud reached the hot hydrogen unit, part of it was carried up by the thermal draft and ignited by the open burners at the top. This occurred some 22 s before the explosion of the elevated cloud. Flames were seen moving back to the escape point from the hydrogen plant and control room area and probably caused the elevated cloud to ignite some 54 s after the escape started. The main aerial explosion then occurred, followed by a major fire with firestorm characteristics. For 20 min, the fire raged over an area of 180 × 250 m with flames over 100 m in height. At the time of deflagration, it was believed that the large aerial cloud contained about 45 tons of cyclohexane. Ninety percent of buildings on the site suffered damage, with blast being the primary factor. Fire extended the damage where the blast breached the containment of flammable inventories. The incident killed 28 people, all on the plant site. Over 400 people received treatment for injuries.

There has been much argument about the cause of failure of the bellows and much criticism of the way the bellows were installed. However, inventory levels were high. Each reactor had a capacity of 27 tons, which could empty in 10 min. The total process inventory was 400 tons of cyclohexane and cyclohexanone. Furthermore the pump rate through the reactors was large due to the low conversion in the reactors. This meant that a 10-min flow corresponded to a throughput of 43 tons. This figure is closest to the estimated size of the cloud based on general evaluation of the explosion.

7.2.3 Seveso, Italy, July 1976

On Saturday, July 10, 1976, an explosion occurred in a TCP (2,4,5-trichlorophenol) reactor belonging to the ICMESA chemical plant on the outskirts of Meda, a small town about 20 km north of Milan, Italy. A toxic cloud containing TCDD (2,3,7,8-tetrachlorodibenzo-*p*-dioxin), then widely believed to be one of the most toxic man-made chemicals, was accidentally released into the atmosphere. It was an exothermic decomposition that led to the release of dioxin-containing material into the atmosphere. The chemical reaction to produce the material had been performed earlier. Subsequently a number of plant activities had been carried out contrary to operating procedures. Operations only distilled off 15% instead of 50% of the total charge of ethylene; operators did not add water to cool the reaction mixture and did not remain with the unit until the target cooling temperature was reached. It is considered bad practice to discharge a liquid directly into the atmosphere. In this case, about 2 kg of dioxin was discharged. Also, there has been justifiable criticism that the emergency plan was implemented slowly. An area of some two square miles was declared contaminated, a figure that was later increased by a factor of five. The release killed off large areas of vegetation and about 3,300 animals were killed by the dioxin, while a further 70,000 were slaughtered to prevent dioxin from entering the food chain. People in the affected areas suffered from skin infections and other symptoms. At least 250 cases of the infection were identified, some 600 people were evacuated and the land was later decontaminated. Subsequent effects of the dioxin causing deaths in the long term are still debated. The disaster led to the Seveso Directive (legislation), which was issued by the European Community and imposed much harsher industrial regulations.

7.2.4 Los Alfaques, Spain, July 1978

In this disaster on a Spanish camp site, an articulated tank car carrying over 23 tons of liquefied petroleum developed a leak. The driver stopped beside a campsite frequented by holiday-makers. Gas was released into the ground and ignited, probably in the campsite disco. The tanker was engulfed in flames, and this may have been followed by a BLEVE. Others have suggested a flash fire and a gas explosion. Certainly the people in the vicinity, largely holiday-makers and tourists, were singularly ill-protected against heat radiation; many were sunbathing at the time of the accident and large numbers of them were photographed watching a pall of smoke rising from the tanker, which was hidden by light trees. More than 200 people died on-site and a similar number succumbed to injuries, as they were unaware that they should have been escaping from the area rather than spectating. The tank car had been overloaded; it had no relief valve and the steel tank was deteriorated, having been used for the transport of ammonia. Hence, a crack developed. The route selected by the driver was the coastal road, and it is presumed that no advice was given on what emergency procedures to follow in the event of a leak.

7.2.5 Mexico City, Mexico, November 1984

At approximately 05:35 h on November 19, 1984, a major fire and a series of catastrophic explosions occurred at the government owned and operated PEMEX LPG Terminal at San Juan Ixhuatepec, Mexico City.

Some 11,000 m^3 of liquefied petroleum gas (LPG) was stored in six 1,600 m^3 spheres and 48 horizontal cylindrical bullets, all in close proximity. The legs of the spheres were not fireproofed. It is believed that no fixed water sprays or deluge systems were fitted to the tanks. A leak of LPG from an unknown source formed a vapor cloud that was ignited by a plant flare. The storage area was bounded into 13 separate areas by walls of about 1-m high. A fierce fire developed, engulfing the spheres, which went up one after the other in a series of BLEVEs. Nine explosions were recorded. The accident is, to date, the most catastrophic accident in history involving domino effects. The series of LPG explosions at the distribution center resulted in 542 fatalities, and over 7,000 people were injured. Some 200,000 people were evacuated. The fireballs were up to 300 m in diameter and lasted as long as 20 s. Rain consisting of liquid droplets of cooled LPG fell over the housing area covering people and property, and were consequently set alight by the heat from the fireballs. Since the construction of the plant, some 100,000 people had settled in crowded housing on the valley floor and slopes. The local housing was mainly single story and built of brick supported by concrete pillars. LPG was used for heating and cooking, and each household had its own small bottles. Some 2,000 houses at 300 m were destroyed, and 1,800 were badly damaged. Windows were broken at 600 m, and debris missiles were thrown a considerable distance. One cylinder was thrown 1,200 m. The emergency plan functioned well considering the circumstances. This accident is also called "the forgotten accident" due to another, much worse, disaster that happened approximately a month after this accident: the Bhopal catastrophe (see Section 7.2.6).

7.2.6 Bhopal, India, December 1984

Ingress of water (because of pipe washing) in tank E610 initiated a runaway reaction that caused the release of some 25 tons of methyl isocyanate and probably hydrogen cyanide, causing 3,787 fatalities and 200,000 injuries instantaneously. In total, the death toll is estimated to have risen to a number as high as 20,000, which means that the Bhopal accident is considered to be the worst disaster ever to take place in the chemical industry. The cause was initially claimed by the company to have been sabotage. This was later denied by official investigators; the incident undoubtedly reached the proportions it did because of operating instrumentation; the safety interlock systems and mitigating systems were inaccurate, inoperable or undersized. The standard of maintenance was appalling, and the chemical plant should not have been operating under such conditions. The emergency plan was extremely poor with negligible communication to the public.

The material that leaked was an intermediate product that need not have been stored in such quantities. Alternative routes might have been adopted. As a result of the incident, Union Carbide, at that time one of the biggest chemical companies in the world, the owner of the plant, saw share ratings plummet on the US stock market, revealing how a major incident can also be a true financial disaster for any company.

7.2.7 Chernobyl, Ukraine, April 1986

An experiment was carried out to investigate whether or not a nuclear reactor could develop enough power to keep auxiliary equipment running while shutting down. The plant was operated below the power output at which a reactor remains stable. The design of the reactor made it liable to a positive void coefficient at power settings below 20% of maximum. At one point, the power dipped below 1% of maximum and slowly stabilized at 7% of maximum. Operators and engineers continued to improvise by gradually removing rods. The plant went "super-prompt-critical," and an explosion followed. When the temperature increased it rose 100-fold in 1 s. The fire and radiation release caused many deaths. Exact numbers are unclear, but about 2 million people are believed to have been affected. The incident caused the permanent evacuation of 600,000 people and vast contamination of the environment.

7.2.8 Piper Alpha, North Sea, July 1988

A condensate pump on an oil rig tripped. The duty of the condensate system exceeded the initial design, and such problems were not uncommon. Staff started up the spare, which had earlier been shut down for maintenance and during which the pressure relief cap had been removed and replaced by a cap that was not leakproof. Clearly there were failures in communication of information during the shift changeover on the evening of the incident. Gas escaped from the cap and ignited. The resulting explosion destroyed the fire control and communication systems and demolished a firewall. The incoming gas pipeline was ruptured upstream of the emergency isolation valve and the gas burned fiercely as it does in a blowtorch. A fireball engulfed the platform. The adjacent rigs continued to feed gas and oil to Piper Alpha for over an hour. Other pipelines ruptured, intensifying the fire and eventually most of the platform toppled into the water. The platform controller had tried to enact the emergency plan, which involved mustering in the galley followed by evacuation by helicopter. However, the explosions made escape by helicopter impossible. Some survivors escaped by jumping into the sea from a height of up to 50 m. One hundred and sixty-seven oil workers were killed, the platform was totally destroyed and UK hydrocarbon production dropped temporarily by 11%. Most of the fatalities were caused by smoke inhalation in the galley or accommodation areas.

7.2.9 Pasadena, Texas, USA, October 1989

Early afternoon on October 23, 1989, Phillips' 66 chemical complex at Pasadena, near Houston (USA), experienced a chemical release on the polyethylene plant. A flammable vapor cloud formed and subsequently ignited resulting in a massive vapor cloud explosion. Following this initial explosion, there was a series of further explosions and fires.

The consequences of the explosions resulted in 23 fatalities, and 314 people were injured. Extensive damage to the plant facilities occurred.

The day before the incident, scheduled maintenance work had begun to clear three of the six settling legs on a reactor. A specialist maintenance contractor was employed to carry out the work. A procedure was in place to isolate the leg to be worked on. During the clearing of no. 2 settling leg, a part of the plug remained lodged in the pipework. A member of the team went to the control room to seek assistance. Shortly afterward the release occurred. Approximately 2 min later the vapor cloud ignited.

7.2.10 Enschede, the Netherlands, May 2000

On May 13, 2000, a devastating explosion at a fireworks depot ripped through a residential district in the eastern Dutch town, leaving 22 people dead and 2,000 families homeless. A small fire at the factory triggered several massive explosions. In this incident, several important necessary safety precautions had not been observed. The fireworks were not stored properly. They had been put in sea containers, offering insufficient delay in fire inhibition. To exacerbate matters, the classification on the boxes of fireworks was incorrect, leading the authorities and fire brigade to believe that they were dealing with consumer fireworks instead of professional explosives. Statistics about heat radiation require consumer fireworks to be at a distance of 20 m from inhabited buildings. Professional fireworks, however, need to be hundreds of meters away from residential areas.

7.2.11 Toulouse, France, September 2001

On September 21, 2001, an explosion in Shed 221 of the AZF (Azote de France) plant killed 31 people and injured 2,442 people. The catastrophe cost the French government 228 million euros and TotalFinaElf, owner of AZF, more than 2 billion euros.

Although there remain some uncertainties and unsolved questions about the disaster, it has been assumed that the explosion was caused by a human handling error. A worker from a subcontracted company mistook a 500-kg sack of a chlorine compound (dichloroisocyanuric acid) for nitrate granules and poured it onto the stock of ammonium nitrate in Shed 221 15 min before the explosion. The mixture is said to have produced trichloroamine, an unstable gas that explodes at normal temperatures.

7.2.12 Ath, Belgium, July 2004

A huge gas explosion occurred at about 09:00 h on July 30, 2004, in the small Belgian town of Ghislenghien just outside Ath, 40 km south of Brussels. It sent a wall of flame into the air, triggering a chain of explosions. A leak was reported on the pipeline, which runs from the Belgian port of Zeebrugge into northern France, 37 min before the explosion. Firefighters attempting to establish a security perimeter around the site were among those killed when the explosions destroyed two factories in the industrial park. The blast was heard several miles away. It melted or burned everything within a 400-m radius and left a large crater between the two factories. Bodies and debris were thrown 100 m into surrounding fields. Twenty-four people died in the accident, and 120 people were injured, half of them seriously.

7.2.13 Houston, Texas, USA, March 2005

At approximately 13:20 h on March 23, 2005, a series of explosions occurred at the BP Texas City refinery during the restarting of a hydrocarbon isomerization unit. Fifteen workers were killed and about 180 others were injured. Many of the victims were in or around work trailers located near an atmospheric vent stack. Investigators reported that the explosions occurred when a 170-feet distillation tower flooded with hydrocarbons was overpressurized, causing a geyser-like release from the vent stack. As a result of this mistake, a mixture of liquid and gas flowed out of the gas line at the top of the column, traveled through emergency overflow piping and was discharged from a tall vent that was located hundreds of feet away from the distillation column. A vapor cloud accumulated at or near ground level. The cloud was further ignited by a vehicle that had been left in the area with its engine idling. Finally, a number of mobile offices that had been located far too close to the plant were destroyed by the explosion, killing and injuring their occupants.

7.2.14 St. Louis, Missouri, USA, June 2005

St. Louis was experiencing a heat wave, with bright sunlight and temperatures exceeding 35 °C on June 24, 2005. At Praxair, a gas repackaging plant, operations proceeded normally during the morning and early afternoon; however, in the afternoon a security camera video from the facility showed the release and ignition of gas from a cylinder in the propylene return area. As workers and customers evacuated, the fire spread to adjacent cylinders. The video shows nearby cylinders igniting in the first minute. At 2 min, cylinders began exploding, flying into other areas of the facility, and spreading the fire. After 4 min, the fire covered most of the facility's flammable gas cylinder area and explosions were frequent. Fire swept through thousands of flammable gas cylinders and dozens of exploding cylinders were launched into the surrounding community and struck nearby homes, buildings and cars, causing extensive damage and several small fires.

7.2.15 Buncefield, UK, December 2005

From around 18:50 h on Saturday December 10, 2005, a delivery of unleaded petrol was being pumped down the T/K pipeline from Coryton Oil Refinery into tank 912 (situated within bund "A"). The automatic tank gauging system, which records and displays the level in the tanks, had stopped indicating any rise in tank 912's fuel level from around 03:00 h on Sunday, 11 December. At about 05:40 h on Sunday morning, tank 912 started to overflow from the top. The safety systems that were designed to shut off the supply of petrol to prevent overfilling, failed to operate. Petrol cascaded down the side of the tank, collecting in bund A. As overfilling continued, a vapor cloud that was formed by the mixture of petrol and air, flowed over the bund wall, dispersed and flowed off the site and toward the Maylands industrial estate. Up to 190 tons of petrol escaped from the tank, about 10% of which turned to vapor that mixed with the cold air, eventually reaching concentrations capable of supporting combustion. The release of fuel and vapor is considered to be the initiating event for the explosion and subsequent fire.

At 06:01 h on December 11, 2005, the first of a series of explosions took place. The main explosion was massive and appears to have been centered on the Maylands Estate car parks. These explosions caused a huge fire, which engulfed more than 20 large storage tanks over a large part of the Buncefield depot. The fire burned for 5 days and a plume of black smoke from the burning fuel rose high into the atmosphere. There were only 43 injuries, but the human death toll could have been very large if the accident had happened during a regular day of the week instead of during the night at the weekend. The damage was estimated at several billion euros.

7.2.16 Port Wenworth, Georgia, USA, February 2008

On February 7, 2008, a huge explosion and fire occurred at the Imperial Sugar refinery northwest of Savannah, Georgia, causing 14 deaths and injuring 38 others, including 14 with serious and life-threatening burns. The explosion was fueled by massive accumulations of combustible sugar dust throughout the packaging building. The investigation report issued by US Chemical Safety and Hazard Investigation Board concluded that the initial blast ignited inside a conveyor belt that carried sugar from the refinery's silos to a vast packaging plant where workers bagged sugar under the Dixie Crystals brand.

7.2.17 Deepwater Horizon, Gulf of Mexico, April 2010

A fire aboard the oil rig Deepwater Horizon started at 9:56 p.m. on April 20, 2010. At the time, there were 126 crew on board. Suddenly, two strong vibrations were felt by the employees. A bubble of methane gas escaped from the well and shot up the drill column, expanding quickly as it burst through several seals and barriers before exploding. The event was basically a blowout, and indeed a number of significant problems have been identified with the blowout preventer. Survivors described the incident as a sudden explosion, which gave them less than five minutes to escape as the alarm went off. The explosion was followed by a fire that engulfed the platform. After burning for more than a day, Deepwater Horizon sank on April 22, 2010, at approximately 10:21 h. As a result of the accident, 11 workers died, and approximately 5 million barrels of oil (790,000 m^3) were spilled into the Gulf of Mexico. Besides the huge immediate consequences for all life at sea, in October 2011 dolphins and whales continued to die at twice the normal rate. In April 2012, 2 years after the accident, scientists reported finding alarming numbers of mutated crab, shrimp and fish they believe to be the result of chemicals released during the oil spill.

7.2.18 Fukushima, Japan, March 2011

Following a major earthquake, a 15-m tsunami disabled the power supply and cooling of three Fukushima Daiichi reactors, causing a nuclear accident on March 11, 2011. All three cores largely melted in the first three days. High radioactive releases were measured. After two weeks, the three reactors were stable with water addition but no proper heat sink for the removal of decay heat from the fuel. By July 2011, they were being cooled with recycled water from a new treatment plant. Reactor temperatures had fallen to below 80 °C at the end of October 2011, and an official "cold shutdown condition" was announced in mid-December 2011, although it would take decades to decontaminate the surrounding areas and to decommission the plant altogether.

Apart from cooling, the basic ongoing task is to prevent release of radioactive materials, particularly in contaminated water leaked from the three units. A few of the plant's workers were severely injured or killed by the disaster conditions resulting from the earthquake. There were no immediate deaths due to direct radiation exposures, but at least six workers have exceeded lifetime legal limits for radiation, and more than 300 have received significant radiation doses.

7.2.19 West, Texas, USA, April 2013

On April 17, 2013, an explosion at the West Fertilizer Company in West, Texas killed 15 people, injured approximately 160–200 and damaged 150 homes. The explosion occurred after a fire broke out in the plant. Investigators from the Texas Department of Insurance and State Fire Marshall's Office concluded ammonium nitrate was the cause of the explosion. The explosion at West Fertilizer resulted from an intense fire in a wooden warehouse building that led to the detonation of approximately 30 tons of ammonium nitrate stored inside in wooden bins. Not only were the warehouse and bins combustible, but the building also contained significant amounts of combustible seeds, which likely contributed to the intensity of the fire. According to available seismic data, the explosion was a very powerful event. In May 2016, the Bureau of Alcohol, Tobacco, Firearms and Explosives stated that they had determined that the fire leading to the disaster had been deliberately set. However, this finding is widely discussed. Legal and forensic experts have criticized the investigation, which remains disputed.

7.2.20 La Porte, Texas, USA, November 2014

On November 15, 2014, nearly 24,000 pounds of methyl mercaptan was released inside the Lannate® unit at the E. I. du Pont de Nemours chemical manufacturing facility in La Porte, Texas. The release resulted in the fatalities of three operators and a shift supervisor inside the Lannate® manufacturing building. The four DuPont employees died from a combination of asphyxia and acute exposure to toxic chemicals including methyl mercaptan. All four victims were located inside the manufacturing building – three on the third floor and one descending the stairs between the third and second floor. At the time of the incident, the manufacturing building ventilation fan for the portion of the unit where the methyl mercaptan was released (wet end fan) was not operating despite an "urgent" maintenance work order written on October 20, 2014, nearly a month prior to the incident. As a result of the release, the manufacturing building stairways were contaminated with highly toxic and highly flammable methyl mercaptan. The stairways were not a safe location for workers. However, these stairways provide the primary means to access the equipment or exit the building in the event of an emergency.

The accident followed a series of mistakes, which began days earlier with the inadvertent introduction of water into a methyl mercaptan storage tank. The water, methyl mercaptan and cold temperatures combined to form a hydrate that blocked the tank's feed line. Workers warmed pipes to break up the hydrate. They opened and closed valves and vents to redirect the methyl mercaptan while they worked to reduce the hydrate. Eventually operators succeeded, and the piped material began to flow. Meanwhile, in another part of the production line, two workers began a routine mission to drain a vent in a poorly ventilated manufacturing building. The vent piping contained methyl mercaptan because of a jerry-rigged configuration to reduce the hydrate. The toxic chemical was released and vaporized, exposing the unprepared workers. The stricken workers immediately called for help. Two others, brothers, responded. All four died [8].

The accident is still (end of 2015) under investigation by the US Chemical Safety Board to discover what really happened.

7.2.21 Tianjin, China, August 2015

A series of explosions that killed at least 173 people and injured hundreds of others occurred at a container storage station at the Port of Tianjin on August 12, 2015. The first two explosions occurred within 30 s of each other. The second explosion was far larger and involved the detonation of about 800 tons of ammonium nitrate. Fires caused by the initial explosions continued to burn uncontrolled throughout the weekend, repeatedly causing secondary and higher-order explosions. The exact cause of the explosions was not immediately known, but Chinese state media reported that at least the initial blast was from unknown hazardous materials in shipping containers at a plant warehouse. Poor coverage of the event and the emergency response to it received criticism. As of 12 September 2015, the official casualty report was 173 deaths [9], 8 missing and 797 nonfatal injuries.

7.2.22 Cambria, USA, May 2017

At approximately 11:00 PM on May 31, 2017, explosion(s) at the Didion Milling (Didion) facility in Cambria, Wisconsin, resulted in 5 worker deaths and injured an additional 14 workers. Because the event occurred at night, only 19 employees were working within the facility at the time of the incident. Shortly before the explosion(s) at Didion, workers saw or smelled smoke on the first floor of one of the mill buildings. In trying to find its source, workers focused on a piece of equipment called a gap mill. While inspecting the equipment, workers witnessed a filter connected to an air intake line for the mill blow-off, resulting in corn dust filling the air, and flames shooting from the air intake line, followed by one or more explosions.

7.2.23 Tangerang, Indonesia, October 2017

An accidental ignition of stored fireworks occurred in Tangerang, Indonesia on October 26, 2017, at the PT Panca Buana Cahaya Sukses, a fireworks manufacturing factory in Kosambi, Tangerang. The explosion occurred in a warehouse connected to the factory and ignited a massive fire. A total of 103 people were working at the factory at the time of the explosion. At least 49 people were killed and 46 others were injured in the accident.

7.2.24 Sichuan, China, July 2018

At least 19 people died and 12 were injured when the explosion ripped through the Yibin Hengda Technology Co. plant in Jiang'an County on the evening of July 12, just as a shift change was underway. Following the initial massive explosion and several smaller ones, the plant was rapidly engulfed in flames, which sent a thick plume of black smoke into the sky. Firefighters said the blaze appeared to have been fueled by methanol. The initial investigation discovered that the factory had been producing a range of chemicals that had not been approved or declared during an earlier safety inspection. Moreover, management and staff lacked the skills and experience to properly and safely manufacture the chemicals being produced [China Labor Bulletin, 16/07/2018].

7.2.25 Yancheng, China, March 2019

On March 21, 2019, a major explosion occurred at a chemical plant in Chenjiagang Chemical Industry Park, Chenjiagang, Xiangshui County, Yancheng, Jiangsu, China. According to reports published on March 25, 78 people were killed and 617 injured. According to the Beijing News, the State Administration of Work Safety, which regulates occupational safety in China, had reported problems with the body responsible for monitoring safety at the Tianjiayi plant, citing insufficient analysis and identification of risk factors and hazards, and a lack of targeted measures to rectify issues (Reuter press, published 28 March 2019).

7.2.26 Chicago, USA, May 2019

On May 3, 2019, a silicone manufacturing process generated a flammable gas inside an enclosed production building at the AB Specialty Silicones (AB Specialty) facility in Waukegan, Illinois. At approximately 9:30 p.m., the flammable vapor cloud found an ignition source and ignited, causing an explosion and fire. The flammable vapor originated from the area where AB Specialty was making a silicon hydride emulsion. The explosion fatally injured four AB Specialty employees and caused serious injury to another AB Specialty employee. At the time of the incident there were nine AB Specialty employees onsite. The explosion heavily damaged AB Specialty's production building. Additionally, the force from the explosion was felt up to 20 miles away in the surrounding communities, and some nearby businesses sustained damage from the blast. Post-incident, AB Specialty has resumed some of its operations at another location [13].

7.2.27 Abqaiq–Khurais attack, Saudi Arabia, September 2019

On September 14, 2019, drones were used to attack the state-owned Saudi Aramco oil processing facilities at Abqaiq and Khurais in eastern Saudi Arabia. The attack by 10 drones caused fires at a major oil processing facility and a nearby oil field. Houthi, movement in Yemen, have claimed responsibility. The drone attack targeted Abqaiq, the world's biggest oil processing facility, and the Khurais oil field, which produces around one million barrels of crude oil a day. Saudi Aramco said that around half of Saudi Arabia's daily oil production had been suspended as a result. In such attacks, the drones are typically packed with explosives and flown at speed into their targets. While the group used to use standard, off-the-shelf hobbyist drones with a limited range, later attacks have used more sophisticated models with an estimated range in excess of 900 miles, the UN reported in January [14].

7.2.28 Gas leak, Visakhapatnam, India, May 2020

On May 7, 2020, an uncontrolled release of styrene vapor occurred at LG Polymers in RR Venkatapuram, Visakhapatnam, from a styrene storage tank (M6 Tank). This was the first such incident of styrene vapor release from a storage tank into the atmosphere in India. The accident resulted in the immediate loss of 12 lives and required 585 people to receive hospital treatment. It also caused significant damage to livestock and vegetation. Known widely as the "Vizag Gas Leak," this incident stands as one of the most significant styrene vapor releases from a bulk storage tank anywhere in the world [16].

7.2.29 Beirut Port Explosion, Lebanon, August 2020

On August 4, 2020, a substantial storage of ammonium nitrate at the port in Beirut, Lebanon, exploded, causing at least 218 deaths and approximately 7,000 injuries. An amount of 2,750 tons of the substance (equivalent to around 1.1 kton of TNT) had been stored in a warehouse without proper safety measures from 2013 on, after having been confiscated by Lebanese authorities from the abandoned ship MV Rhosus. The confiscated cargo of the ship remained in the warehouse despite multiple safety warnings from Lebanese customs to high-ranking officials. Ammonium nitrate, a compound used in fertilizers, is explosive and can undergo decomposition when heated in a confined space. On the day of the explosion, a fire broke out in the port area near the warehouse where the ammonium nitrate was stored, and, just as customs officials had often warned, the ammonium nitrate detonated.

The explosion caused severe economic and environmental damage, with much of the port area being destroyed, impacting Lebanon's economy heavily. This disaster led to an international outcry, prompting investigations and calls for accountability concerning the storage of hazardous materials. The tragedy highlighted serious safety and regulatory failures, leading to significant reforms in port management and hazardous materials storage worldwide [17].

The blast was so powerful that it was not only felt throughout Lebanon, but also in Turkey, Syria, Palestine, Jordan, Israel and parts of Europe, and was heard in Cyprus, more than 240 km away. To this date, the blast was the largest single-fired ammonium nitrate explosion in history.

7.2.30 Explosion, Samut Prakan, Thailand, July 2021

On July 5, 2021, a catastrophic explosion occurred at the Ming Dih Chemical Co. factory in Samut Prakan province, near Bangkok, Thailand. The facility, which produced plastic foam and pellets, was engulfed in flames following the blast, leading to extensive damage within a 1-km radius. The incident resulted in the death of a volunteer firefighter and injuries to at least 29 individuals, including emergency responders. In response to the potential release of toxic fumes and the risk of further explosions, authorities evacuated residents within a 5-km radius of the factory. The fire, fueled by hazardous chemicals such as styrene monomer – a flammable substance used in foam production – proved challenging to control and reignited multiple times over a 24-h period.

The explosion caused significant damage to surrounding communities, affecting approximately 80,000 people across four subdistricts in Bang Phli district. Over 1,200 residents filed complaints regarding property damage, with initial estimates of losses reaching 423 million baht [18].

7.2.31 Chemical explosion, Leverkusen, Germany, July 2021

On July 27, 2021, an explosion occurred at the Chempark chemical complex in Leverkusen, Germany. The blast took place at a hazardous waste incineration plant operated by Currenta within the Chempark, one of Europe's largest chemical parks. The explosion and resulting fire led to the deaths of at least seven people, injured over 30, and left several others unaccounted for.

The incident released toxic fumes into the air, prompting officials to declare "extreme danger" and advise residents nearby to stay indoors, keep windows closed and avoid consuming homegrown produce due to potential contamination. Investigations pointed to issues in the storage and handling of hazardous substances, raising concerns about safety protocols and emergency response practices at chemical plants [19].

7.2.32 Coal mine explosion, Amasra, Turkey, October 2022

On October 14, 2022 the explosion happened at a coal mine in Amasra, Bartın Province, Turkey,. The blast occurred deep underground in the mine and was believed to have been caused by a buildup of methane gas. The explosion resulted in the tragic deaths of at least 41 miners and injured several others. Rescue operations were challenging due to the remote location and the dangerous conditions within the mine, leading to a delay in retrieving some of the miners [20].

The Turkish government launched an investigation into the explosion, and the tragedy sparked public discussions around worker safety standards, especially in Turkey's mining sector, which has seen other fatal accidents in recent years.

7.2.33 Container fire, Sitakunda, Bangladesh, June 2022

Fire broke out at a container facility on the night of June 4, 2022, at BM Container Depot in Sitakunda, Chittagong, Bangladesh. The fire triggered multiple explosions, killing at least 41 people, including 12 firefighters, and injuring at least 300 others. The explosions were so intense that they shattered windows in nearby buildings and caused structural damage within a large radius [21].

7.2.34 Explosion, West Reading, Pennsylvania, USA, March 2023

On March 24, 2023, an explosion happened at the RM Palmer candy factory in West Reading, Pennsylvania, USA, The explosion, suspected to be caused by a natural gas leak, led to a massive fire that destroyed part of the facility. Tragically, seven employees were killed, 11 people were injured, three households were displaced from a neighboring apartment building and many more people were evacuated from the area [22].

The explosion prompted a large-scale response from emergency services, who conducted extensive search-and-rescue operations for missing workers. Investigations pointed to potential issues with gas line maintenance, and the incident raised concerns about the need for regular inspection and monitoring of gas systems in industrial plants.

7.2.35 Dairy farm explosion, Dimmitt, Texas, USA, April 2023

On April 10, 2023, a catastrophic explosion occurred at the South Fork Dairy Farm near Dimmitt, Texas, resulting in the deaths of approximately 18,000 cows and critically injuring one worker. This incident is considered one of the deadliest barn fires recorded in the United States.

Investigations by the Texas State Fire Marshal's Office determined that the explosion originated from a failure in a piece of equipment used daily within the dairy. Specifically, an engine fire in a manure vacuum truck ignited flammable materials, leading to a rapid expansion of gases and a subsequent explosion [23].

7.2.36 Battery factory fire, Hwaseong city, South Korea, June 2024

On June 24, 2024, a devastating fire erupted at a lithium battery manufacturing plant in Hwaseong, Gyeonggi Province, South Korea. The incident resulted in the deaths of at least 22 workers, including 19 foreign nationals, and left eight others injured. The blaze was triggered by explosions of flammable batteries on the second floor of the three-story factory building. Firefighters managed to control the fire after approximately five hours.

The factory, operated by Aricell, housed over 35,000 lithium batteries. The rapid spread of the fire was exacerbated by the flammable components within the batteries, leading to multiple explosions and significant structural damage.

In response to the tragedy, South Korean President Yoon Suk Yeol ordered a comprehensive investigation into the cause of the fire and emphasized the need for stringent safety measures in industrial settings. Aricell's CEO, Park Soon-kwan, issued a public apology and pledged full cooperation with the investigation. Subsequent inquiries revealed potential safety violations, leading to the arrest of the CEO and his son on charges related to industrial safety law breaches [24].

7.2.37 Explosion, Andhra Pradesh, India, August 2024

On August 21, 2024, a catastrophic explosion occurred at the Escientia Advanced Sciences Private Limited facility in the Atchutapuram Special Economic Zone (SEZ) of Anakapalli district, Andhra Pradesh, India. The incident resulted in the deaths of at least 17 workers and injuries to over 50 others. The explosion took place around 2:15 p.m. during the lunch break, which likely prevented a higher casualty count. The blast was so powerful that it caused the first-floor slab of the four-story building to collapse, trapping several workers on the upper floors. Thick smoke and intense flames engulfed the area, complicating rescue operations. Initial investigations suggest that the tragedy was caused by the leakage of methyl tertiary butyl ether solvent from a pipeline flange, followed by an explosion.

The Andhra Pradesh government ordered a high-level inquiry into the incident. Chief Minister N. Chandrababu Naidu announced the formation of a committee to investigate the incident and directed officials to use air ambulances to shift the injured for better treatment [25].

7.3 Learning from accidents

Many disasters have occurred because organizations have ignored the warning signs of precursor incidents or have failed to learn from the lessons of the past. Normal accident theory suggests that disasters are the unwanted, but inevitable output of complex sociotechnical systems, while high reliability theory sees disasters as preventable by certain characteristics or response systems of the organization (see Section 3.10.2). Industrial accidents and hazards have become the order of the day, with new technologies evolving every day and few people knowing how to use them. Disasters have at least one thing in common: the inability of the organization involved to effectively synthesize and share the information from separate "precursor" incidents with the relevant people across the organization, so that appropriate action can be taken to reduce the risk of disaster. Kletz [10] reports several examples in the chemical industry of the same accident occurring multiple times in the same organization. We could then suppose that it is not natural for organizations to learn from safety incidents. Even if ad hoc learning is occurring, it is not enough.

It is then essential that incidents and accidents are properly reported. The investigation of any accident will never progress unless it is first properly reported within an organization. A formal policy requiring the consistent and adequate reporting of all incidents and accidents is one of the most important principles of any accident investigation program. Much of what is known today about accident prevention and loss control was, in fact, learned from loss incidents that were properly reported and investigated.

The accident investigation should be objective and complete. It is all too common that accident investigations fail to identify the reasons for the occurrence of the accident because the investigator focused on assigning blame rather than determining the underlying causes. This is the reason that people are reluctant to report any event that might reflect unfavorably on their own performance or that of their department. However, it is obvious that without complete reporting of incidents, near-misses, accidents, losses, etc., followed by a comprehensive and "honest" investigation into the causes, an organization and its management will never know the extent and nature of the conditions that no doubt will have downgraded the efficiency of the company.

To efficiently investigate accidents, management should determine appropriate parameters within which the investigation is to be carried out. Parameters may include the types of occurrences that will require reporting and investigating, to what extent the investigation has to be conducted, how incidents and accidents shall be reported and what information should be given, what use shall be made of the information reported, etc.

Vincoli [11] indicates that the fact that an accident occurs is a strong indication that a bad and/or erroneous decision was made by management within the organization somewhere. The nature of accident investigation requires an analysis of all possible reasons for such decisions. It is obvious, also from the examples above, that faulty communications, lack of adequate information, improper training and improper behavior are examples of accident causal factors that are often repeated.

There are nine essential elements of a successful accident investigation program:
1. Consistent reporting to management
2. Interview of witnesses and examination of evidence at scene
3. Determination of immediate causal factors
4. Study of all evidence and formulation of interpretations
5. Determination of basis or root causal factors
6. Reconstruction of accident (if required)
7. Analysis of basic causal factors and management involvement
8. Implementation of corrective actions
9. Follow-up of planned and implemented actions

Specific plans should be developed to organize and manage the carrying out of an accident investigation in anticipation of any possible future incident or accident. Adequate planning ensures that sufficient resources will be available to guarantee a proper investigation if an incident or accident should occur. Individual or team investigation, team member composition depending on the characteristics of the incident or accident, coordination requirements, collaboration specificities between investigators, required background information and data, exchange of information, sharing ideas and brainstorming sessions, training, expertise, know-how, information, resources, etc. should all be thought of in advance. Hence, preparation is the key to the accurate determination of causal factors for incidents and accidents, in the shortest possible time and with the lowest expenditure of resources.

What does it take to learn?
- Opportunity (to learn) – learning situations (cases) must be frequent enough for a learning practice to develop.
- Comparable/similar – learning situations must have enough in common to allow for generalization.
- Opportunity (to verify) – it must be possible to verify that the learning was "correct" (feedback).

The purpose of learning (from accidents, etc.) is to change behavior so that certain outcomes become more likely and other outcomes less likely. According to Kletz, there are several *myths on accident investigation* to note (see [12] for supplementary information) and to develop with your own culture, education and goal. The myths are:

1. Most accidents are caused by human error. (The authors of this book would suggest using/interpreting "most" into "all" to avoid possible misunderstandings. Another possibility is to explicitly indicate that by "accidents," Kletz means "major accidents.")
2. Natural accidents are "Acts of God" and are unavoidable.
3. The purpose of accident reports is to establish blame.
4. If someone's action has resulted in an accident, the extent of the damage or injuries is a measure of his or her guilt.
5. We have reached the limit of what can be done by engineering to reduce accidents. We must now concentrate on human factors.
6. Most industrial accidents occur because managers put costs and output before safety.
7. Most industrial accidents occur because workers fail to follow instructions.
8. Most industrial accidents occur because workers are unaware of the hazards.
9. Two competent and experienced investigating teams will find the same causes for an accident.
10. When senior people do not know what is going on, this is because of the failures of their subordinates to tell them.
11. When everything has run smoothly for a long time, and accidents are few, we know we have got everything under control.

Once you have debated these myths, these comments can be raised:
- As individuals, we do learn from the experience of accidents and rarely let the same happen again, but when we leave a company we take our memories with us. We need to find ways of passing on our knowledge and building a "company memory."
- The next best thing to experience is not reading about an accident, or listening to a talk on it, but discussing it.
- When we investigate an accident, and particularly if we are involved in some way, it is hard for us to be dispassionate and our mindsets and self-interests can

easily take over. Without making a conscious decision to do so, we look for short-comings in other departments rather than our own and tend to blame those below us rather than those above us.

7.4 Finding problems

Many managers at all levels tell their people that they hate surprises. They encourage people to tell them "the bad news," if there is any. Still, due to many reasons, problems often remain concealed in organizations. Bad news and simple problems do not tend to rise upward a lot. Therefore, warning signals, precursors, near-misses, bad behavior and what have you should be looked for. Managers should not only learn adequately (single loop and double loop, see Chapter 6) from past accidents and catastrophes to prevent gray rhinos from happening, but they should also become hunters who venture out in search of the problems that might lead to disasters for their organizations. The sooner problems are identified and surfaced, the more likely a major accident can be avoided. After all, engineering risk managers are not merely "problem solvers," but they are also and foremost, *"potential problem* solvers," hence, problem finders and proactive fixers.

Roberto [15] has identified seven sets of skills and capabilities that managers, also engineering risk managers and middle and top managers, should master for effective problem-finding:

1. *Circumvent the gatekeepers*: Remove the filters at times, and go directly to the source to see, experience and hear the raw data. Listen aggressively to the people actually doing the work. Keep in touch with what is happening at the periphery of the organization, not just simply at the core.
2. *Become an ethnographer*: Many anthropologists observe people in natural settings, which is known as ethnographic research. As a manager dealing with risks, the same approach should be embraced. Do not simply ask people how things are going. Do not depend solely on data from surveys and focus groups. Do not simply listen to what people say; watch what they do: go out and observe how employees, contractors, visitors and what have you actually behave in the organization. Effective problem-finders become especially adept at observing the unexpected without allowing preconceptions to cloud what they are seeing.
3. *Hunt for patterns*: Reflect on and refine your individual and collective pattern-recognition capability. Focus on the efficacy of your personal and organizational processes for drawing analogies to past experiences. Search deliberately for patterns amidst disparate data points in the organization.
4. *Connect the dots*: Recognize that large-scale failures are preceded by small problems that occur in different parts of the organization. Foster improved sharing of information, and build mechanisms to help people integrate critical data and knowledge.

5. *Encourage useful failures*: Encourage people to come forward when mistakes are made. Reduce the fear of failure in the organization. Help your people understand the difference between bona fide/excusable and mala fide/inexcusable mistakes.
6. *Teach how to talk and listen*: Employees should be able to surface and discuss problems and concerns in an effective manner. They should learn and be able to speak up; and managers need to handle all comments and concerns appropriately.
7. *Watch the game film*: Reflect systematically on your organization's conduct and performance, as well as on the behavior and performance of competitors. Create opportunities for individuals and teams to practice desired behaviors so as to enhance their performance.

7.5 Conclusions

Accidents happened. Accidents happen. Accidents will happen. It is obvious that a zero-risk or zero-accident organization does not exist. Industrial activities go hand-in-hand with risks and regretfully with incidents and accidents. This does not mean, however, that nothing can be done to prevent most incidents and accidents, or to prevent the consequences of accidents from aggravating. Accidents will happen within organizations, but the number of accidents should be very low, and the consequences of accidents should be none to minor, when applying the engineering risk management and its principles, methods, concepts, etc., explained and discussed in this book. We need especially to learn from minor and major accidents and insert the lessons learned of any disaster worldwide into the DNA and the memory of all organizations. On the bright side, the vast majority of possible incidents and accidents never happen, thanks to adequate engineering risk management. On the dark side, disasters keep on taking place somewhere worldwide, and they really are preventable, either by preventing the initiating causes of the disaster, or by preventing the accident's consequences from becoming disastrous.

References

[1] Taleb, N.N. (2007). The Black Swan. The Impact of the Highly Improbable. New York, NY: Random House.
[2] Wucker, M. (2016). The Gray Rhino. How to Recognize and Act on the Obvious Dangers We Ignore. New York, NY: St, Martin's Press.
[3] Kahneman, D., Tversky, A. (1979). Prospect theory: An analysis of decisions under risk. Econometrica. 47: 263–291.
[4] Sutherland, V., Making, P., Cox, C. (2000). The Management of Safety. London, UK: Sage Publications.

[5]   Wells, G. (1997). Major Hazards and Their Management. Rugby, UK: Institution of Chemical Engineers.

[6]   Reniers, G. (2006). Shaping an Integrated Cluster Safety Culture in the Chemical Process Industry, PhD dissertation. Antwerpen, Belgium: University of Antwerp.

[7]   Paltrinieri, N. (2012). Development of Advanced Tools and Methods for the Assessment and Management of Risk Due to Atypical Major Accident Scenarios, PhD dissertation. Bologna, Italy: University of Bologna.

[8]   Johnson, J. (2015). Fatal accident at Dupont probed. Chem. Eng. News. 93(6): 6.

[9]   The Guardian. (2015) Tianjin explosion: China sets final death toll at 173, ending search for survivors. The Guardian. Associated Press. Retrieved 14 September 2015.

[10] Kletz, T.A. (1993). Lessons from Disaster – How Organizations Have No Memory and Accidents Recur. Houston, TX: Gulf Publishing Co.

[11] Vincoli, J.W. (1994). Basic Guide to Accident Investigation and Loss Control. New York: John Wiley & Sons.

[12] Kletz, T.A. (2001). Learning from Accidents. Oxford, UK: Butterworth-Heinemann.

[13] U.S. Chemical Safety and Hazard Investigation Board (2019) Explosion and Fire at AB Specialty Silicones Facility, Report No. 2019-03-I-IL, 1–22.

[14] United Nations Security Council, Letter dated 25 January 2019 from the Panel of Experts on Yemen addressed to the President of the Security Council, S/2019/83.

[15] Roberto, M.A. (2009). Know what you don't know. How great companies prevent problems before they happen. Upper Saddle River: NJ. ISBN: 9780134177014.

[16] Tammineni, Y., Dakuri, T. (2020). Vizag gas leak-a case study on the uncontrolled styrene vapour release for the first time in india. EPRA International Journal of Research and Development. 5(8).

[17] El Sayed, M.J. (2022). Beirut ammonium nitrate explosion: A man-made disaster in times of the COVID-19 pandemic. Disaster Med. Public Health Prep. 16(3): 1203–1207.

[18] Natt Leelawat, N., Vilaivan, T. (2021). A polystyrene foam factory fire in a Bangkok satellite city: Incident and lessons learned. ACS Chem. Health Saf. 28(6): 394–396.

[19] Ferl, L. et al,. (2023). Doing Trust and Crisis Communication. Narratives of the2021 Explosion in the Chempark Leverkusen. DIEGESIS. InterdisciplinaryE-Journal for Narrative Research / Interdisziplinäres E-Journal für Erzählforschung 12(1): 4–27.

[20] Bilginsoy, Z. (2022) Family mourns coal miner's death in Turkey, demands punishment for explosion. PBS News, retrieved December 2024.

[21] Acaps (2022) BANGLADESH, Industrial accident: Depot fire in Sitakunda. Acaps report.

[22] National transportation Safety Board (2024) Pipeline Investigation – 151 Docket Items – PLD23LR002, Public Release Date & Time: 02/22/2024.

[23] Baio, A. (2023) Fiery explosion at Texas dairy farm kills 18,000 cows. The Independent. Retrieved April 14, 2023.

[24] Hoey, I.(2024) Catastrophic fire at battery plant in South Korea results in 22 deaths. International Fire and Safety Journal, online, June 2024.

[25] Sanjay, K. (2024) 17 dead in solvent explosion at Indian plant. Chemistry World. Retrieved 23 august 2024.

8 Crisis management

We often hear different statements after a major event happened, such as
- "Crises only happen to others."
- "If that should happen, we will improvise."
- "A crisis is unpredictable, so why should we get organized?"
- "People always exaggerate the severity and the consequences of a crisis."
- "We are in a crisis every day and we always get out of it."

Disasters have served to guide and shape history, and many great civilizations have been destroyed by the effects of disasters.

So, would it not make sense to identify the first signs of a crisis, so that when one happens, we know what to do, or when it is over that we know how to recover? Those are the several points that this chapter will try to highlight.

All crisis management techniques, methods and plans are based on the question, "What's the worst that can happen?" While many people do not want to think of worst-case scenarios and only hope for the best, looking at the world through rose-colored glasses does not help individuals, teams, companies, corporations or governments to plan or prepare to meet an emergency. Some crises happen immediately, while others may take longer to develop and form. Being able to recognize various stages of a crisis is essential for crisis management teams, managers and supervisors.

Swartz et al. [1] identified at least three stages in managers' mindsets regarding crisis management. The authors have been suggesting a potential evolutionary path for crisis management since the 1970s, from the initial focus on technological activities and the predominant concern with respect to hardware, to a growing interest in the value of business continuity management. They stated that there is no suggestion that these mindsets are restricted to specific periods, but rather they suggest that each represents the dominant paradigm for a particular decade:

technology mindset (1970s) → auditing mindset (1980s) → value mindset (1990s) → normalization of business continuity management mindset (2000s) → resilience mindset (2020s)

Crisis management, as a corporate activity, has the fundamental strategic objectives of ensuring corporate survival and economic viability when business profits and/or continuity are threatened by external or internal potentially destructive events.

Three characteristics can be expressed as separating crises from other unpleasant occurrences:
- *Surprise* – even naturally occurring events, such as floods or earthquakes, do not escalate to the level of crisis unless they come at a time or a level of intensity beyond everyone's expectations.

https://doi.org/10.1515/9783111493633-008

- *Threat* – all crises create threatening circumstances that reach beyond the typical problems organizations face.
- *Short response time* – the threatening nature of crises means that they must be addressed quickly. This urgency is compounded by the fact that crises come as a *surprise* and introduce *extreme threat* into the situation.

In this chapter, we will consider a crisis as a specific, unexpected and nonroutine event or series of events that creates high levels of uncertainty and that is threatening. **!**

8.1 Introduction

Damages, following an accident, can lead to a degraded situation by weakening the target affected by the accident. This target is then exposed to the emergence of other accidents, against which it is usually unprotected. The degradation situation then becomes a crisis situation. A breach in the defense system occurs, and it is the whole system that is at threat. This phenomenon is highly dependent on time, as illustrated in Fig. 8.1. The crisis is multifaceted and multitemporal. It can be

- The result of a succession of past events, leading to a creeping degradation
- The result of a major event, a sort of sudden cataclysm
- The outcome of future events that will be linked together in cascade

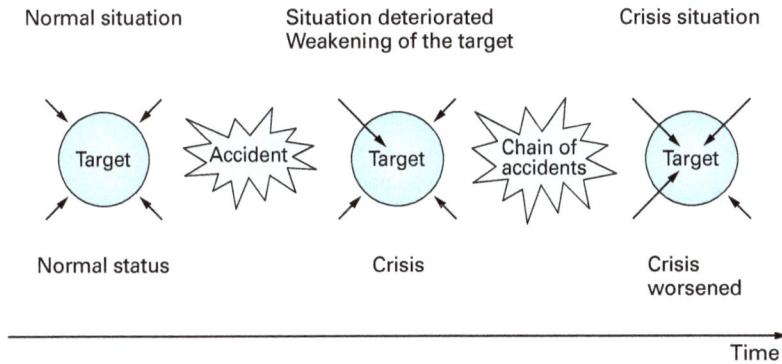

Fig. 8.1: Physical modeling of the time evolution effect on a crisis.

As observed, at the beginning, threats (represented by arrows) are not reaching the target (the vulnerable resource, process, human, etc.). This represents the normal status; the target is not harmed and the safety barriers are fulfilling their purpose. As the situation deteriorates with time, the target can be reached by the threats, and then damages and/or harms appear. The immediate consequence is a weakening of the target as the safety barriers are either deteriorated or are not completely performing their duty. In the end, a chain of accidents allows multiple threats on a weakened target and the safety barriers, indicating that the crisis has worsened and that the target will suffer multiple and more important damages and/or harms [2].

This implies that, following an accident, one must not only repair the noticed damage, but at the same time, be more careful to try to prevent other risks, whose likelihood of occurrence momentarily increases drastically.

A *crisis* is often defined "as an acute situation, hard to manage, with important and lasting consequences (sometimes harmful)." It can result from an accident or from the normal evolution of a situation. This last point can lead us to think that it would be more correct to consider a crisis as a passing process from one situation to another, for example, from a critical situation to a catastrophic situation. A crisis assumes a decision-making process and an immediate action to get out of it. *A crisis is an unusual situation, characterized by its instability, forcing a specific governance to be applied in order to get back to a normal status.*

We must not confuse a crisis with an exceptional event or an emergency situation. A crisis is a period of intense instability, more or less long, not anticipated and not always visible or known, which causes negative consequences on the environment and its actors. Figure 8.2 provides an overview of the differences between emergency, crisis and business continuity management.

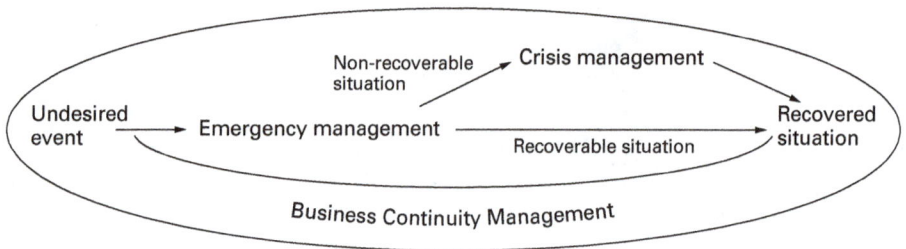

Fig. 8.2: Emergency, crisis and business continuity management.

The components of a crisis management system are composed of
- Pre-event risk and vulnerability assessment
- Mitigation and loss control
- Planning
- Response management
- Recovery functions

The fundamental strategic objective of any organization is to ensure its long-term survival and economic success. Crisis management is a strategic function that links functions such as risk management, safety management, environmental management, security, contingency planning, business recovery and emergency response [3].

When an event such as an industrial accident, a toxic release or a major oil spill occurs, management often finds itself simultaneously involved in emergency management (the specialized response of emergency forces), including disaster management (the management of the incident and the management and support of the response organization); crisis management (the management of the crises situations that occur as a result of the accident); and business recovery and continuity (the entire management process to deal with the recovery and continuation of the profitable delivery of products and services). In addition, crisis management often includes a strong focus on public relations to recover any damage to public image and assure stakeholders that recovery is underway.

A crisis is often defined as an acute situation, hard to manage, with important and lasting consequences (sometimes harmful). It can result from a triggering accident or from the normal evolution of a situation. It has the following rules:

- Being unprepared is no excuse.
- You know the threats, get ready for them.
- Know in advance before you are asked.
- Admit that you are wing-it-challenged.
- Adopt a short key message communication type.
- Beware of the court of public opinion.
- The first 72 h of any crisis are crunch time.
- Do not forget that in a crisis situation, time flies.
- Get every help or support you may need.
- Every crisis is an opportunity.

Murphy's law and expressions such as "bad things happen in threes" and "waiting for the other shoe to drop," suggest the possibility of a single crisis expanding into something like a knock-on effect. Because it is extremely important for individuals responsible for establishing control to take action during a crisis, it is extremely important to choose members of a crisis management team that are efficient, effective and able to give their best in such situations.

8.2 The steps of crisis management

In reality, managing is predicting. Anticipating and preparing a protocol and crisis management organization is a very big responsibility for the company's management. Managing a crisis and containing the crisis so that it does not extend or deteriorate is a high-priority objective of any organization.

In a crisis situation, it is difficult to predict human behavior with regard to the undergone stress and emotional load that can overwhelm an individual. *Anticipation,* the master word of crisis management, is
- Imagining and studying the different past scenarios in the company and integrating with the known catastrophic scenarios.
- Equipping ourselves with "thinking" and "acting" in a structured way when the moment comes, known as the "crisis cell."
- Preparing a minimum of the functioning logistics and procedures.
- Preparing crisis communication so as not to add an aggravating factor, in case of mediatization.
- Identifying in advance the right people – those who have the knowledge and expertise, those who are united and can work in teams, those who will not "flinch" when there are decisions to be made and finally those whose legitimacy would not be questioned.

Anticipating and preparing for a crisis is answering the following questions:
- How should we react and be organized to face an unpredictable disaster?
- Which crisis management device should be deployed?
- Which collaborators should be involved? What roles should be assigned to them and what would their responsibilities be?
- Which logistic processes/activities should be planned?
- Which crisis communication should be put in place? And for which addressees?
- Which procedures and which operations should be implemented to reduce the crisis impact and proceed with the restarting of activities?

There are the risks that are known, have been integrated and that are now assumed, and then there are all the other risks, the new ones, the unknown unknowns, the unexpected, which play tricks on our emotions and lead to psychological issues, anger, irrationality, etc. This constitutes one of the main difficulties in crisis communication. To make sure the words are not transformed into ills, every used term must be chosen with care and precision by the public powers and political, economic and media actors.

And it is precisely, this mixture of uncertainty, anxiety and acceleration that multiplies tenfold the evolution risk of a "sensitive" situation (or emergency situation) into a "crisis" situation. This crisis situation is always characterized by its instability, which forces a specific governance to be adopted in order to go back to a normal way of life. By crisis management, we mean a governance mode. In a crisis situation, we are heavily engaged in the future, and in a way that is necessarily more closed than usual. As an extreme example, it is not when an airline pilot approaches critical takeoff speed and he notices a motor failure that he will start consultations and negotiations to decide which maneuver to make or whether to perform an emergency brake.

What is done at that moment does not translate to instantaneous goodwill of the individual: it is the anterior decisions that play a large role.

8.2.1 What to do when a disruption occurs

No matter what the quality of the risk management system is, an unwanted and unexpected event will happen sooner or later, as "zero risk" does not exist. As long as the risk is there, identified and quantified, no matter how well-managed, we know the accident may happen. But the moment when it will happen remains, of course, unknown. The same goes for a crisis: it surprises and, under Murphy's law, some say it always happens at the wrong moment [4].

Mythology is ripe with tragically sad stories. The story of Cassandra is such a mythological tale. Cassandra was the princess of the legendary city of Troy; she was the daughter of King Priam and Queen Hecuba. She was a precocious and charming child who easily caught the attention of mortals and gods alike.

As was very common in Greek mythology, the God Apollo lusted after the beautiful young mortal woman and intended to make her his own. To convince her to give into his advances, he promised to bestow upon her the gift of prophecy.

While Cassandra was obviously flattered that an Olympian God sought her favors, she was not at all convinced that she wanted to take him as lover. Still, unable to resist the gift he offered, she eventually relented.

Apollo took Cassandra under his wing and taught her how to use her prophecies. Once her mentorship was finished, however, Cassandra refused to give her body to Apollo as promised.

Furious at being rejected by a mere mortal, Apollo decided to punish her. However, he could not take back the gift he had already given her. So he leveled a terrible curse upon her head. While Cassandra would still be able to foresee the future, the curse ensured that no one would believe her. Worse than that, they would believe that she was purposely telling lies. True to his word, Cassandra was able to foresee the future for herself and those around her. But every attempt she made to warn people in advance of an impending doom, it was ignored or, worse yet, labeled as an outright lie.

The careful reader immediately sees the analogy with crisis and disaster managers: before anything happens, they are often regarded by others as fantastic doomsayers, who do not have to be believed. "Nothing ever happened, so why would something happen now?" is often the reasoning. Hence, in the proactive phase, crisis and disaster managers are often treated exactly the same way as Cassandra.

The major steps of crisis management can be expressed as three phases (see Fig. 8.3):

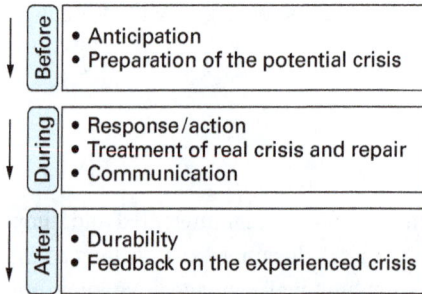

Fig. 8.3: Major steps of crisis management.

1. Preparation, *before*, is composed of
   - Major risk identification – linked to strategic impact risks; necessitates risk quantification; focuses on critical consequences)
   - Prospective crisis organization – crisis scenarios and continuity plans, crisis levels, ethics and image. In case of a crisis, we can be confronted with ethical problems. For example, it is possible that some resources have to be sacrificed in order to save others.

     The main actions to be considered by a company, prior to a crisis situation, are
   - Appoint someone within the organization – who can also represent company's top management – as the responsible person for company crisis management and the person in charge, in case of a crisis situation (the "crisis manager"). Think also about a backup person.
   - Make a draft of a crisis procedure. Who carries out which tasks in what order and under what circumstances? Compose a crisis team comprising at least the following functions: coordinator of crisis team, secretary (responsible for the log), person responsible for communication, person knowledgeable about safety and technology. Develop training schedules with yearly exercises.
   - List the organizations/authorities/emergency services that need to be informed/averted in case of a crisis situation. Inform these parties about the crisis procedure of your company.
   - Conceptualize a crisis center inside and/or outside the premises of the organization, equipped with the necessary communication means. Use only a single point of contact (in case of crisis), that is, the e-mail/phone/social media specificities of the person responsible for communication.
   - Anticipate contacts with the press. Make a list of press contacts (e-mail addresses, telephone numbers and social media specificities).
2. Treatment and repair, *during*, is composed of
   - Triggering of crisis management. The importance of wisely triggering the crisis management process cannot be overstated. If it is done too early, the system is destabilized, credibility is lost and we risk being no longer capable of

mobilization if the situation really requires it. Too late, and the consequences will be dramatic.

- Crisis cell. The establishment of a crisis cell cannot be improvised at the last moment, but should be prepared; the members of the cell having previously participated in one or more simulations. The crisis cell has the mission of evaluating the disaster, and taking immediate protection measures for the affected people and facilities. It must also inform the board of directors and the local authorities with assured communication, and determine a strategy to get back to a normal situation. The crisis cell is an emergency organization of the service leader to identify a problem, put a contingency plan into action and return as quickly as possible to a normal situation.
- Crisis communication. This should be prepared: the possible attitudes, the choice of speech and the tone of the message to send out are predetermined in the crisis scenario. It establishes a relationship with the media. Note that the media are quick – simplifying, aggravating and amplifying.
- Repair. Once the crisis reaches its peak and is in its descending phase, repairs can begin. Temporary solutions were put into place in the period preceding the crisis. Repairs do not stop at these solutions; their mission is a return-to-normal operations.
- End of the crisis situation. The crisis situation has ended; it is thus necessary to record it. Recording enables an end to the emergency and to surveillance devices that have been put in place in parallel with the crisis. There is a subtle balance between taking the devices away too fast and leaving them in place for too long.

The main actions to be considered by a company during a crisis situation are:

- No two crises are completely similar. Hence, anticipate improvisation and at the same time, be rational and keep a cool head.
- Capture all the information in a log (oral, telephonic, social media and written contacts).
- Show people – internal as well as external to the company – that you are managing the crisis situation in a professional way.
- The setting of the first news coverage often determines the further course of the crisis. Therefore, prepare the first coverage very well.
- Show empathy with those affected by the crisis (victims, their families, employees, surrounding communities, etc.). Remember that installations, buildings, machines and any kind of material losses are easier to replace than people. Avoid jargon.
- Be honest, transparent and clear in your messages. Show competence and engagement while trying to mitigate losses and lower nuisance. Limit the information to sheer facts and avoid reacting to assumptions, speculations, etc.
- Be realistic (honest) and admit that the situation is not yet fully recovered, or that information is still not fully known, when this is the case.

- Disseminate an internal message, as promptly as feasible, containing all known information at that time. Inform the families of victims (if there are) in the first instance. While drafting a press release/message for external use, answer the questions who, what, when, where and why.
- Look after journalists and accommodate them in a specific press room.
- Organize a press conference when you are ready and with at least the company management, the crisis manager and the person responsible for communication present. Depending on the crisis, the authorities and rescue services may be a part of this press conference.

3. Memorization, *after*, is composed of
   - Learning from the situation as well as documenting and analyzing the different steps and sequences
   - Seeking feedback to look for ways to improve future responses
   - Updating continuity plans with knowledge gained from the crisis

When disruptions occur, they are handled in three steps:
1. Response (incident response involves the deployment of teams, plans, measures and arrangements).
2. Continuation of critical services (ensure that all time-sensitive critical services or products are continuously delivered or are not disrupted for longer than is permissible).
3. Recovery and restoration (the goal of recovery and restoration operations is to recover the facility or operation and maintain critical service or product delivery).

The main actions to be considered by a company after a crisis situation are
- A thorough investigation, searching for the causes, the exact losses/damages/detriment and the time necessary for full recovery.
- After completion of the investigation, the press to be informed of relevant findings.
- Evaluate the crisis based on the log and the press releases. Learn from what went wrong and adapt/improve the crisis procedure, if necessary.

Crises and causes of crises can be adverse weather, bribery, blackmail, computer breakdown and/or failure, fire, flood, pandemic, IT problems (computer virus), liability issues, loss of resources, loss of staff, major accident, energy supply problems, reputation problem, sabotage, supplier problem, internet failure, etc. Mitroff et al. [5] developed a matrix for classifying types of resources, according to where the crisis is generated and which systems are the primary causes. This matrix provides a starting point for a creative brainstorming session that provides a means of identifying the range of disruptions an organization might experience; see Fig. 8.4.

Technical/Economic

Major accidents	Natural disasters
Product recalls	Aggressive takeover
IT problems	Global economic crisis

Internal ←————————————————→ External

Sabotage	Terrorism
Occupational health disease	Bomb alarm
Major fraud	Product tampering

Human/Organizational/Social

Fig. 8.4: Generic crisis typology matrix.

Hence, different types of crises are, for example,
– Internal crises
  – Product problems, supply chain rupture, process problems, absence of key people, fire, hostility, financial problems, etc.
– External crises
  – Product/services, consumers, suppliers, community, competition, regulation, fire, hostility, stock exchange movement, natural disasters, etc.

> Managing a crisis involves three steps: before, during and after the crisis; the third, often being the most lasting and more demanding of time and resources. **!**

Rather than deal with individual incidents, "families" of crises may be clustered together and provide a focus for preparations.

8.2.2 Business continuity plan

When the crisis situation occurs, it is too late to ask the question, "What to do?" and too late to search for information – we must simply act. The principle consists of elaborating different scenarios, looking for failures that could follow an initiating event. *The purpose of developing a business continuity plan (BCP) is to ensure the continuation of the business during and following any critical incident, which results in disruption to the normal operational capability.* It must enable us to work in a degraded mode or in a major crisis situation. It is a strategic document, formalized and regularly updated, of planned reactions to catastrophes or severe disasters. Its main goal is to minimize the impact of a crisis or the effect a natural, technological or social catastrophe has on the activity (thus, the sustainability) of an enterprise, a government, an institution or a group [6].

The scenarios that are deemed possible will constitute the contingency plan, also called the continuity plan. It focuses on how to keep in service an activity, whose usual means of operation have been partially destroyed. It must also plan for measures to limit the damage and prevent it from spreading to other activities. Business continuity impact analysis identifies the effects resulting from the disruption of business functions and processes. It also uses information to make decisions about recovery priorities and strategies.

Disaster recovery and business continuity planning are processes that help organizations prepare for disruptive events. BCP is designed to protect personnel and assets and make sure they can function quickly when disaster strikes. Disaster recovery is the process by which you resume business after a disruptive event. The event might be something huge, like an earthquake, or something small, like malfunctioning software caused by a computer worm. Recovery strategies are alternate means to restore business operations to a minimum acceptable level, following a business disruption, and are prioritized by the recovery time objectives developed during the business impact analysis. Recovery strategies require resources, including people, facilities, equipment, materials and information technology (IT). An analysis of the resources required should be conducted to identify gaps. For example, if a machine fails but other machines are readily available to make up lost production, then there is no resource gap. However, if all machines are lost due to a flood, and insufficient undamaged inventory is available to meet customer demand until production is restored, production might be made up by machines at another facility – whether owned or contracted. Strategies may involve contracting with third parties, entering into partnership or reciprocal agreements or displacing other activities within the company. Staff with in-depth knowledge of business functions and processes are in the best position to determine what will work.

Nowadays, IT includes many components such as networks, servers, desktop and laptop computers and wireless devices. The ability to run both office productivity and enterprise software is critical. Therefore, recovery strategies for IT should be developed such that technology can be restored in time to meet the needs of the company. Manual workarounds should be part of the IT plan so that business can continue while computer systems are being restored.

Given the human tendency to look at the bright side, many business executives are prone to ignoring "disaster recovery" because disaster seems an unlikely event. BCP suggests a more comprehensive approach to making sure you can still do business, not only after a natural calamity but also in the event of smaller disruptions, including illness or departure of key staff, supply chain partner problems or other challenges that businesses face from time to time.

The following steps can be identified:

1. Size up the risks
   - Assess how potential risks to the business/process will impact the ability to deliver products and services.
   - Differentiate between critical (urgent) and noncritical (nonurgent) organization functions/activities.
   - Develop a list of possible scenarios that could seriously impede operations.
2. Identify vulnerable targets and critical services, and assess the risks
   - Determine who would be affected most by each possible crisis, and which areas of operations would be hit the hardest.
   - Critical services or products are those that must be delivered to ensure survival, avoid causing injury and meet legal or other obligations of an organization.
   - Compile a list of key contacts and their contact information when determining which external stakeholders would be most affected.
3. Formulate plans to respond to an emergency
   - Create plans to address each of the vulnerable targets identified in the previous step – plans, measures and arrangements for business continuity.
   - Develop a generic checklist of actions that may be appropriate when an emergency occurs.
   - Test the plan; when crisis occurs, it is too late to ask oneself if the BCP will work.
4. Readiness procedures
   - Determine critical core businesses or facilities.
   - Determine the core processes required to ensure the continuation of the business.
   - Prepare the response (team response, emergency facilities, how to react, etc.).
5. Return to business
   - Create plans to return to "business as usual;" the ultimate goal in any disaster is to bring operations back into line by rebuilding damaged facilities and reorganizing the work force to cover any losses.
6. Quality assurance techniques (exercises, maintenance and auditing).
   - Review of the BCP should assess the plan's accuracy, relevance and effectiveness. It should also uncover which aspects of a BCP need improvement. Continuous appraisal of the BCP is essential to maintaining its effectiveness. The appraisal can be performed by an internal review or by an external audit.

BCP is a proactive planning process that ensures critical services or products are delivered during a disruption.

Critical services or products are those that must be delivered to ensure survival, avoid causing injury or meet legal or other obligations of an organization. Business continuity planning is a proactive planning process that ensures critical services or products are delivered during a disruption. **!**

8.3 Crisis evolution

Similar to a fire, it is in the first few minutes that an exceptional event can become an emergency situation, which can then degenerate into a crisis.

When are we in a crisis?

– In principle, too late!
– Not always when we expect it.
– Never when we would like.
– When the director of the company knows or assumes that the risky event could be serious for the company.
– Following a non-mastered emergency situation.

A crisis, like every other dynamic process, does not evolve in a monotonous way in time [7]. We can schematize the different steps in time, as depicted in Fig. 8.5:

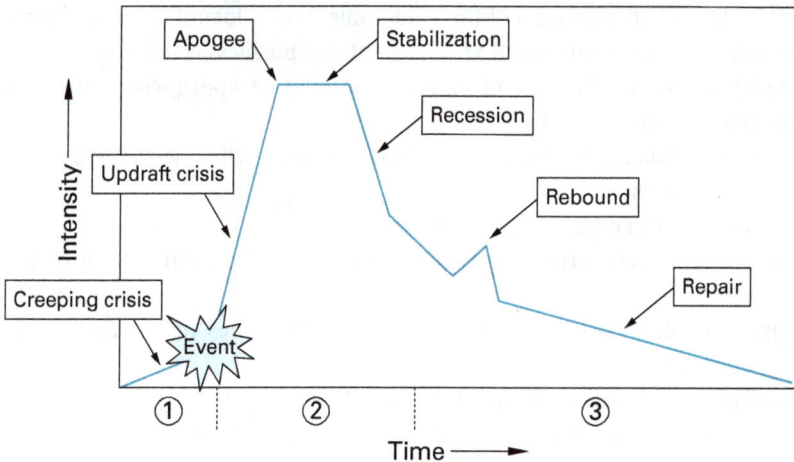

Fig. 8.5: Chronology of a crisis situation.

We could simplify this model by breaking up a crisis into three stages: pre-crisis stage, acute-crisis stage and post-crisis stage, identified by the circled numbers 1, 2 and 3.

8.3.1 The pre-crisis stage or creeping crisis

When someone in an organization discovers a critical situation, they usually bring it to the attention of their supervisors/managers. This is known as either the pre-crisis warning or precursor. At this point in time, the critical situation is known only inside the organization and is not yet visible to the general public. When managers are told of the critical situation, their job is to analyze it to determine if it has the potential to become serious. If managers are then comfortable with it and feel it will be resolved without any action on their part, they will not take any action. If, however, they see the critical situation as a serious problem requiring intervention, they will take action to mitigate it. They should then manage it and prevent it from moving into the acute-crisis stage. This is considered a time of opportunity to turn it from a negative situation into a positive one. The first issue then is to recognize the situation for what it is and what it might become.

8.3.2 The acute-crisis stage

A crisis moves from the pre-crisis to the acute stage when it becomes visible outside the organization. At this point in time, we have no choice but to address it. It is too late to take preventative actions, as any action taken now is more associated with "damage control." Once the problem moves to the "acute" stage, the crisis management team should be activated. The main actions to be taken are:
- Take charge of the situation quickly.
- Gather all the information you can about the crisis and attempt to establish the facts.
- Communicate the story to the appropriate groups that have vested interest in the organization, namely, the media, the general public, the customers, the shareholders, the vendors and the employees.
- Take the necessary remedial actions to fix the problem.

8.3.3 The post-crisis stage

A crisis moves from the acute-crisis stage to the post-crisis stage after it has been contained. This is when the organization will try to recoup its losses. The organization must show the customer, all shareholders, public authorities and the community that it cares about the problems the crisis has caused. During this stage, the organization must not forget to
- Recoup any losses – recovery phase
- Evaluate its performance during the crisis – challenging phase
- Make any changes that were identified during the crisis – learning phase

A crisis evolves with time. It is not a straight or smooth process; it can be split into three major time evolution stages: pre-crisis stage, acute-crisis stage and post-crisis stage. **!**

8.3.4 Illustrative example of a crisis evolution

A crisis is not static, it is dynamic. It evolves, following a cycle that is almost always the same, represented in Fig. 8.5.

Let us illustrate this chronology by taking the example of Hurricane Katrina in August 2005 (see Fig. 8.6), with a partial extract from the US Department of Health & Human Services and the National Institute of Environmental Health [8, 9].

Hurricane Katrina was most destructive hurricane ever to strike the USA. On August 28, 2005, Hurricane Katrina was in the Gulf of Mexico where it powered up to a category 5 storm on the Saffir-Simpson hurricane scale. This catastrophe illustrates perfectly the chronology of a crisis, as depicted in Fig. 8.5. We can divide the crisis evolution as follows:

Fig. 8.6: Satellite picture of Katrina Hurricane (with reprint permission of NOAA [10]).

1. Creeping crisis
   – Thursday, August 25, 2005: Tropical Storm Katrina threatens to gain power in the warm waters of the Atlantic, and grows to a category 3 hurricane, before striking the southeast coast of Florida.
2. Updraft crisis
   – Friday, August 26, 2005, 04:45: Hurricane Katrina reaches the southeast coast of Florida, where she kills at least two people and leaves hundreds of thousands of houses without power.
   – 21:50: Hurricane Katrina kills five people in south eastern Florida and knocks out power to more than two million people, before returning to the warm waters of the Gulf of Mexico and strengthening.

3. Apogee
   - Sunday, August 28, 2005, 09:25: Residents of New Orleans begin to evacuate the city threatened by Hurricane Katrina, which is approaching with renewed vigor.
   - 09:50: The merchants of the French Quarter (Vieux Carré) pile sandbags in front of their stores to protect them from flooding that may be caused by Hurricane Katrina.
   - 22:15: The winds of Hurricane Katrina have significantly increased steadily, reaching category 5, and blowing at 284 km/h, threatening to cause catastrophic damage.

4. Recession
   - Monday, August 29, 2005, 09:26: Katrina is downgraded to category 4, but it is still possible that it could strengthen to a category 5.
   - 12h54: In the eye of the hurricane, the Mississippi River overflows and floods Slidell, LA and Mobile, AL, up to 9–10 m.
   - 13h00: The hurricane is located 112 km from the city; it is moving at a speed of 25 km/h.

5. Rebound
   - Monday, August 29, 2005, 13:21: Conditions begin to improve in New Orleans, but degrade rapidly in Picayune and Springhill.
   - 13:38: There is water seepage in the Superdome, where 10,000 people are present because they could not be evacuated.
   - 14:04: Ochsner Hospital is flooded up to the first floor; patients are moved to the upper floors.
   - 15:05: The drinking water system is polluted and unfit for consumption.
   - 16:53: Hurricane Katrina is downgraded to a Category 3 hurricane.

6. Repairs
   - Tuesday, August 30, 2005, the extent of the major damage in several states causes the initial estimate to be over 25 billion dollars. Hurricane Katrina is the most expensive hurricane in history and one of the most deadly.
   - Wednesday, August 30, 2005, 13:00: Given the number of damaged or flooded dams, and with over 90% of homes flooded, the mayor of New Orleans orders the total evacuation of the last residents and refugees and orders the total closure of the city for at least 6–12 weeks; it is likely that the pumping out of flooded areas will take months.
   - Emergency services decide not to deal, at the moment, with the floating bodies, and to concentrate their efforts on the survivors.
   - Thursday, September 1, 2005, the evacuation of the refugees from the Superdome in New Orleans is suspended after shots are fired at military helicopters searching for them. The National Guard decides to send hundreds of military police to regain control.

Currently, in 2022, repairs of the damage caused by Katrina are not yet fully completed. The efforts to return to normality continue.

8.4 Proactive or reactive crisis management

As anyone who has been involved in a crisis knows, bad news travels alarmingly fast (especially in our era, where social media is ever more important in society and is used by more and more people). Proactive crisis management involves forecasting potential crises and planning how to deal with them, for example, how to recover if the computer system completely fails.

Reactive crisis management involves identifying the real nature of a current crisis, intervening to minimize damage and recovering from the crisis. When an organization is reactive in response to an incident, it shows the public that it is responsive and that the incident will be investigated so that it will not occur again in the future, as many lives are at stake. By being reactive, the company responds immediately. When a company is proactive, it does all it can to avoid such mistakes. A proactive approach focuses on eliminating problems before they have a chance to appear, and a reactive approach is based on responding to events after they have happened. The difference between these two approaches is the perspective each provides in assessing actions and events.

The more an organization denies its vulnerability, the more its activities will be directed to reactive crisis management. Hence, the more it will be engaged in cleanup efforts "after the fact." Conversely, the more crises an organization anticipates, the more it will engage in proactive behavior directed toward activities of proactive crisis management.

Among the most compelling benefits of proactive management is a greater sense of control. Rather than having crises and worker demands determine your schedule, you create a plan that allows you to control when and how you lead. Proactive management also identifies the best way to do things before a problem pops up with a poor system or process; freedom from firefighting allows for greater time to implement best practices. Risks are managed with careful planning and orchestrated execution. The proactive management also makes positive worker morale a priority and is always looking for better ways to lead and develop the organization.

No matter what, we should realize that nearly every potential crisis is thought to leave a repeated trail of early warning signals. Thus, if organizations could only learn to read these warning signals and to attend to them more effectively, then there is much they could do to prevent many crises from ever occurring.

8.5 Crisis communication

The Chinese character representing the idea of "crisis" means, at the same time, "danger" and "opportunity": whatever does not kill you makes you stronger. A crisis, in the end, is nothing but an opportunity to learn from past mistakes and to start over for an enterprise or an organization, on the condition it keeps its guard up, as the next crisis may not be far around the corner. A crucial matter in this respect is the way the crisis is communicated.

To make sure words do not become ills, every term used must be chosen with care and precision by the public powers and the political, economic and media actors. The most challenging part of crisis communication management is reacting – with the right response – quickly. This is because behavior always precedes communication. Non-behavior or inappropriate behavior leads to spin, not communication. In emergencies, it is the nonaction and the resulting spin that cause embarrassment, humiliation, prolonged visibility and unnecessary litigation [11].

Crisis management is primarily a matter of communication. When choosing a single spokesperson, be sure it is someone who speaks well and who knows how to step back and choose his/her register. The main objective is to restore or maintain trust or confidence. The best communication is therefore not to deny that something has happened, but to indicate that everything will be done to understand the facts and take the necessary measures so that the event will not recur. Crisis communication should be prepared and should evolve with the incoming information. Crisis communication refers to the technologies, systems and protocols that enable an organization to effectively communicate during an emergency situation.

The skills needed for good communication could be expressed as follows:
- Keep calm
- Try to stand back
- Competence for rapid synthesis and analysis
- Be reactive instead of passive
- Learn to decide quickly and well
- Be positive
- Use speaking skills
- Know how to inform

Every word counts in a crisis situation. A badly written press release, a defensive or arrogant speech, and the whole communication strategy collapses.

Fuller and Vassie [12] indicate that the effectiveness and efficiency of the communication process depends on the level of attention provided to the communicator by the receiver, the perceptual interpretation of the message by the receiver, the situational context in which the information is provided and the trust that the receiver has in the provider of the information. Key elements of the communication process are summarized in Fig. 8.7.

Communicator	Message	Recipient
Status/credibility Appeal Trust Presentation	Verbal/non-verbal Explicit/implicit Appeal/fear For and against arguments Presentation style	Attitude Education Entrenchment Persuasibility

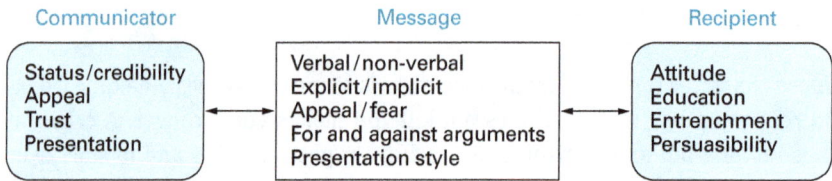

Fig. 8.7: Key elements of the communication process.

They also mention that "framing effects," which relate to the context in which information is presented or "framed," can lead to bias in the recipients' views of the risks. The most common framing effect is that of the *domain effect*, which involves changing the description of a risk from a negative to a positive description by, for example, identifying the benefits associated with the risk rather than the losses. Where a choice must be made between two undesirable options, both options can be framed in terms of their relative gains rather than the losses, so that those people affected by the decision would be left with a positive feeling that they had made a real gain, whichever decision was made. Framing effects are routinely employed in the advertising industry and by politicians.

The basic steps of effective crisis communications are not so difficult, but they require advance work in order to minimize damage. In order to be serious about crisis preparedness and response, one should follow and implement the following nine steps of crisis communications, the first six of which can and should be undertaken before any crisis occurs.

PRE-CRISIS
1. Anticipate crises: *Be proactive and prepare for crises, gather the Crisis Communications Team (CCT) for intensive brainstorming sessions on all the potential crises that could occur at the organization.*
2. Identify your CCT: *A small team of senior executives should be identified to serve as the organization's CCT. The team also needs to include other members: specific managers, those with special knowledge related to the current crisis, for example, subject-specific experts and/or important operational persons.*
3. Identify and train spokespersons: *The organization should ensure via appropriate policies and training that only authorized spokespersons speak for it. Each CCT should have people who have been pre-screened and trained to be the lead and/or backup spokespersons for different channels of communications. The training should include how to be prepared and how to be ready to respond in a way that optimizes the response of all stakeholders. The latter, internal and external, are just as capable of misunderstanding or misinterpreting information about the organization as the media. It is the senior management responsibility to minimize the chance of that happening.*
4. Establish notification and monitoring systems: *It is absolutely essential, in pre-crisis, to establish notification systems that will allow communication to rapidly reach the stakeholders using multiple modalities. If more than one modality is used*

to reach the stakeholders, the chances are much greater that the message will go through. Intelligence gathering is an essential component of both crisis prevention and crisis response. Knowing what is being said about the organization on social media, in traditional media, by the employees, customers and other stakeholders often allows to catch a negative "trend," which, if unchecked, turns into a crisis. Likewise, monitoring feedback from all stakeholders during a crisis situation allows one to accurately adapt the strategy and tactics. Both require monitoring systems to be established in advance.

5. Identify and know the stakeholders: *Who are the internal and external stakeholders that matter to the organization?*
6. Develop holding statements: *While full message development must await the outbreak of an actual crisis, "holding statements," messages designed for use immediately after a crisis breaks, can be developed in advance to be used for a wide variety of scenarios to which the organization is perceived to be vulnerable, based on the assessment conducted in Step 1.*

POST-CRISIS

7. Assess the crisis situation: *Assessing the crisis situation is, therefore, the first crisis-communications step that cannot be taken in advance. If the CCT is not prepared in advance, the reaction will be delayed by the time it takes the in-house staff or quickly hired consultants to run through steps 1 to 6.*
8. Finalize and adapt key messages: *With holding statements available as a starting point, the CCT must continue developing the crisis-specific messages required for any given situation. The team already knows, categorically, what type of information its stakeholders are looking for.*
9. Post-crisis analysis: *Once the crisis is over, the question must be asked, "What did we learn from this?" A formal analysis of what was done right, what was done wrong, what could be done better next time and how to improve various elements of crisis preparedness is another must-do activity for any CCT.*

The organization has to prepare a crisis communications plan, the common axiom "It can't happen to us" should be banned from our mind. Instead, we should be prepared for "It can happen to us."

In crisis situations, the pace of the conflict accelerates dramatically. This means that the parties have to react very quickly to changing conditions or risk, as their ability to protect their interests is substantially reduced. Crises are likely to be further complicated by the increased levels of fear, anger and hostility that are likely to be present. Often, in crises, communication gets distorted or cut off entirely. As a result, rumors often supplant real facts, and worst-case assumptions drive the escalation spiral. In addition, parties often try to keep their real interests, strategies, and tactics secret and use deceptive strategies to try to increase their relative power (Inspired by Conflict Research Consortium, University of Colorado, USA [13]).

8.6 Tips for implementing crisis management

You can never be too prepared when it comes to crisis management. "Surprise" will be the main word; the intensity of it will depend on the state of preparation. Many recent crises have demonstrated that the most important things one must do in any "crisis" situation are:
– Disseminate accurate information as quickly as possible (if you do not do it, it will be done through social medias or other means, without control)
– Respond to incorrect information that may be circulating (in a clear and prompt manner)
– Activate appropriate mechanisms to keep the public, media and stakeholders informed on an ongoing basis (use all communication channels available)

It is essential that the initial information is kept simple so that the message to the public and the media is as clear as possible. Once the initial phase of the crisis has passed, it is often important to provide more detailed background information to the stakeholders.

Recommended hints could be expressed as follows:
– Forecast the crisis; do not wait for it to put a crisis management plan together.
– Plan a crisis management team with a primary spokesperson (assign a backup person) to represent the organization throughout the crisis process.
– Set priorities according to the crisis and information evolution.
– Respond in a timely manner; the more you wait, the more damage can be done.
– Be honest and straight to the point, but be factual.
– Use different channel communications to reach the different stakeholders.
– Never say "no comment" (it implies guilt) or speak "off the record" (there is no such thing).
– Express empathy and concern when victims are involved (bring humanity).
– Do not hide bad news – it will get out anyway.
– Step down from the crisis when it is over.

8.7 Polycrisis

A polycrisis in risk management refers to a situation where multiple distinct crises occur simultaneously, interact with each other, and amplify the overall impact, creating complex challenges that are difficult to address through conventional methods. The key feature of a polycrisis is that the interconnections between the crises make their combined effect greater than the sum of their individual impacts [14].

The characteristics of a polycrisis can be expressed as:
– Convergence of crises: Multiple crises (e.g., economic, political, environmental) unfold at the same time.

- Interconnectedness: Crises are interconnected, meaning one crisis worsens or exacerbates the others.
- Amplified impact: The interactions between crises lead to compounded consequences, escalating their severity.
- Complexity: The interconnected nature makes it challenging to isolate and address any single crisis without influencing others.
- Unpredictability: The cascading effects of a polycrisis make outcomes harder to predict and control.

8.7.1 Risk management in polycrisis

Managing risks in a polycrisis requires a comprehensive, adaptive and systemic approach. Unlike isolated crises, polycrises involve complex interconnections and cascading effects, necessitating strategies that address both the immediate impacts and their broader implications. Addressing a polycrisis requires:

- Holistic risk assessment: Understanding how crises are interconnected and anticipating the cascading effects
- Resilience building: Strengthening the systems to withstand and adapt to shocks (see also Chapter 5, Section 5.7)
- Scenario planning: Preparing for multiple plausible outcomes to enhance flexibility in decision-making
- Adaptive governance: Implementing policies that can evolve as the situation changes

In essence, managing a polycrisis demands a systemic approach that recognizes and addresses the interplay of diverse risks rather than treating them in isolation.

8.7.2 Examples of a polycrisis

2008 GLOBAL FINANCIAL CRISIS AND POLITICAL INSTABILITY
- Economic crisis: The collapse of major financial institutions led to widespread economic downturns.
- Social unrest: Rising unemployment and austerity measures triggered protests, such as the Occupy Wall Street movement.
- Political polarization: The crisis eroded trust in governments and financial systems, fueling populism in many countries.
- Global trade disruptions: Interconnected financial systems amplified the crisis across borders.

COVID-19:
- Health crisis: Overwhelmed healthcare systems worldwide.
- Economic crisis: Lockdowns led to widespread unemployment, supply chain disruptions, and global recessions.
- Geopolitical tensions: Nations blamed one another for the pandemic's spread, worsening international relations.
- Environmental impacts: Resource diversion slowed climate initiatives and increased pollution from medical waste.
- Global financial crisis and Climate change: Economic instability undermining investments in renewable energy, exacerbating environmental issues.

CLIMATE CHANGE AND MIGRATION
- Environmental crisis: Rising sea levels, extreme weather, and desertification force people to leave their homes.
- Migration crisis: Mass displacement creates refugee challenges in neighboring and distant countries.
- Political tensions: Host countries face political instability and social backlash over immigration policies.
- Economic strain: Increased pressure on infrastructure, housing, and social services in host countries.

EXTREME WEATHER EVENTS AND FINANCIAL STRESS
- Climate crisis: Severe droughts, floods, and hurricanes devastate regions globally.
- Insurance crisis: Insurers struggle with the financial burden of compensating for frequent disasters.
- Economic disparities: Developing countries bear the brunt of climate impacts, despite contributing the least to global emissions.
- Political pressure: Governments face increasing calls to implement climate adaptation and mitigation strategies, but financial and political challenges hinder progress.

The list could be extensive, encompassing ongoing wars, global political and economic tensions, and severe climate disturbances causing economic, social, and societal upheavals. Addressing polycrises demands global collaboration to recognize interconnected risks, proactive policies to anticipate cascading effects, equity-focused solutions to protect vulnerable populations, and innovative, resilient frameworks to mitigate complex challenges. These examples emphasize the importance of systems thinking in managing an interconnected world.

8.8 Conclusions

Certain issues that relate to crisis management offer important lessons for practice. The issues center on the processes by which early warnings of vulnerability can be generated and acted upon, the processes by which we can communicate within and around the complex dynamics of crises, and the processes by which organizations learn from near-miss events and problems in other organizations, as well as from major crisis events across a range of organizations and activities. However, even when processes are known, there may be difficulties in implementing solutions. Nonetheless, when critical services and products cannot be delivered, consequences can be severe. All organizations are at risk and face potential disaster, if unprepared. Wherever possible, economically, socially and environmentally sustainable policies and measures must be set up, not only to moderate risk, avoid major accidents and crises, but also to continuously deliver products and services, despite disruptions that will/ might happen sooner or later. Adequate crisis management for both single crises and polycrises is simply a part of the resilience of an organization.

References

[1]   Swartz, E., Elliott, D., Herbane, B. (2003). Greater than the sum of its parts: Business continuity management in the UK finance sector. Risk Manag. 5: 65–80.

[2]   Le Ray, J. (2010). Gérer Les Risques Pourquoi? Comment? AFNOR. La Plaine Saint-Denis Cedex. France: Afnor Editions.

[3]   Mitroff, I.I., Pearson, C.M., Harrington, L.K. (1996). The Essential Guide to Managing Corporate Crises. New York: Oxford University Press.

[4]   Fink, S. (2000). Crisis Management – Planning for the Inevitable. Lincoln, NE: iUniverse, Inc.

[5]   Mitroff, I., Pauchant, T., Shrivastava, P. (1988). The structure of man-made organizational crisis. Technol. Forecast Soc. 33: 83–107.

[6]   Hiles, A. (2011). The Definitive Handbook of Business Continuity Management. Chichester, UK: Wiley & Sons.

[7]   Bhagwati, K. (2006). Managing Safety – A Guide for Executives. Weinheim, Germany: Wiley-VCH.

[8]   Department of Health & Human Services. USA (accessed in 2012). http://www.hhs.gov/.

[9]   National Institute of Environmental Health. USA (accessed in 2012). http://www.niehs.nih.gov/.

[10] National Oceanic and Atmospheric Administration. USA (accessed in 2012). http://www.noaa.gov/.

[11] Ulmer, R.R., Sellnow, T.L., Seeger, M.W. (2010). Effective Crisis Communication: Moving from Crisis to Opportunity. California, USA: SAGE Publications, Thousand Oaks.

[12] Fuller, C.W., Vassie, L.H. (2004). Health and Safety Management. Principles and Best Practice. Essex, UK: Prentice Hall.

[13] Conflict Research Consortium, University of Colorado, USA (accessed in 2012). http://conflict.colorado.edu/.

[14] Albert, M.J. (2024). Navigating the Polycrisis. USA: The MIT Press. ISBN: 9780262547758.

9 Economic issues of safety

Any organization tries to be profitable to be healthy in the long term. Adequate managerial decisions should be taken about uncertain outcomes to achieve this goal. Hence, uncertainties, whatever they are, should be managed well within organizations. In this regard, it is important that managers realize that the results of decisions are two-sided (as was already mentioned in Chapter 2): they may have a positive outcome, and they may have a negative outcome. Furthermore, basically, two fields of uncertainties are important for organizations, and decisions are made in those fields to make (or to continue) a company (being) profitable: financial uncertainties and operational uncertainties. Figure 9.1 provides an idea of the different fields where adequate or optimal managerial decisions matter for the long-term viability of any organization.

As shown in Fig. 9.1, operational risks should be seen as an essential domain of any organization to increase profitability. Similar to the fact that good decisions should be taken in the fields of financial uncertainties and profit-related operational uncertainties (such as production investments and innovation), good decisions should be taken to avoid operational losses. In other words, operational safety is actually a domain fully contributing to the profitability of a company by generating hypothetical benefits (through the avoidance of real losses via the use of adequate operational safety). Organizations, and managers within organizations, should genuinely look at operational safety this way. Operational safety is not part of the cost structure of a company; on the contrary, it is a domain that truly leads to making profits (in the short term and long term). Similar to financial uncertainties, one needs to be careful with high-uncertainty decisions (related to type II risks), and one needs to focus on making profits via low-uncertainty decisions (related to type I risks). In reality, the brain part corresponding to the financial positive and negative uncertainties/risks as well as the positive operational uncertainties/risks is usually much bigger than the part of the operational risks. Much more attention and efforts from the average manager's brain are devoted to financial risks and the increase of production than to the avoidance of losses due to operations.

It seems evident that risk management and safety management are essential to any manager. However, Perrow [1] indicates that there are indeed reasons why managers and decision-makers would not put safety first. The harm, the consequences, is not evenly distributed: the latency period may be longer than any decision-maker's career. Few managers are punished for not putting safety first even after an accident, but will quickly be punished for not putting profits, market share or prestige first. But in the long term, this approach is obviously not the best management solution for any organization. The economic issues of risks – playing a crucial role in the decision-making on safety management budgets and budget constraints, and also having a pro-

https://doi.org/10.1515/9783111493633-009

found impact on other organizational budgets such as production budgets, human resource budgets and maintenance budgets – are explained in this chapter.

One question increasingly on the mind of a lot of corporate senior executives
today is: "What are the risk/opportunity trade-offs of investing in safety?" Of course,
safety is not a monolith; it might be very heterogeneous. We could draw a parallel
with emerging technologies. The rate of growth in emerging technologies for the past
decades has been higher than the growth rate of classical technologies. This is why
there is increasing interest in emerging technologies by companies in advanced econ-
omies, where growth has been much slower. Importantly, these growth differentials
reflect a secular transformation in the structure of the global economy, not a cyclical
phenomenon occasioned by the current economic/financial crisis.

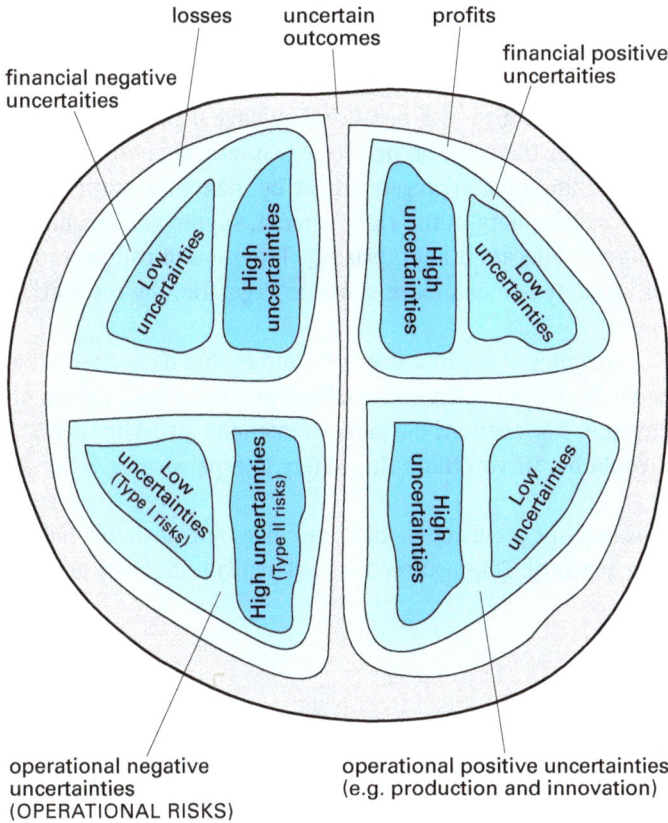

Fig. 9.1: The position of operational risks within the brain of the manager – how it ideally should be: in
balance (source: [9]).

Accidents and illness at work are matters of health, but they are also matters of economics, as they stem from work, and work is an economic activity. The economic per-

spective on safety and health encompasses both causes and consequences: the role of economic factors in search for causes and their effects with respect to the economic prospects for workers, enterprises, nations and the world as a whole. It is therefore a very broad perspective, but it is not complete, because neither the causation nor the human significance of safety and health can be reduced to their economic elements. Economics means one thing to the specialist and another to the general public.

For economics, a central concept is that of costs. On the one hand, we have the costs of improving the conditions of work in order to reduce the incidence of injury and disease. On the other hand, we have the costs of not doing these things. Hence, discussion of safety, however defined, involves a discussion of choice. Safety is largely the product of human action consequent upon prior choices. The economics of safety consists of pricing it, or compensating for its absence, so as to produce the economically optimal amount of safety at a socially optimal cost. Safety may be treated as a resource-absorbing commodity or service. Its production therefore involves the allocation of resources to the production of safety and the allocation of the resulting safety to those in need of it and entitled to it because they have paid for it in some fashion. These two allocative functions can be performed, at least supposititious, by the market, through other institutional arrangements or by their combination. The object of allocation is to produce safety in the right amount, supply this amount as cost-effectively as possible and allocate it appropriately. Provision of compensation against the incidence of unsafety is much more suited to regulation by the market than the allocation of safety.

Safety is not costless. Providing it absorbs scarce resources that have alternative uses; they constitute the visible cost of safety. We could always raise the question of, "Should it be worth to invest in stock options the money I intend to invest in safety?"

Again, one answer could be, "If you think that safety is expensive . . . . Try an accident!"

Indeed, the fact that safety – or prevention – has a cost does not mean that it does not have a benefit, on the contrary. The next sections will explain the costs and the benefits related to safety.

9.1 Accident costs and hypothetical benefits

Accidents do bear a cost – and often not a small one. An accident can indeed be linked to a variety of direct and indirect costs. Table 9.1 reveals a straightforward non-exhaustive list of potential socioeconomic costs that might accompany accidents.

Hence, by implementing a sound safety policy and by adequately applying engineering risk management, substantial costs can be avoided, namely *all costs related to accidents that have never occurred. We call them "hypothetical benefits."* However, in reality companies place little or no importance on "hypothetical benefit" because of its complexity.

Nonquantifiable costs are highly dependent on nongeneric data such as individual characteristics, a company's culture and/or the company as a whole. Rather, the costs assert themselves when the actual costs supersede the quantifiable costs. In economics (e.g., in environment-related predicaments), monetary evaluation techniques are often used to specify nonquantifiable costs, among them the contingent valuation method and the conjoint analysis or hedonic methods [3]. In the case of nonquantifiable accident costs, various studies demonstrate that these form a multiple of the quantifiable costs [4–8].

Tab. 9.1: Non-exhaustive list of quantifiable and nonquantifiable socioeconomic consequences of accidents (see also [2]).

Interested parties	Nonquantifiable consequences of accidents	Quantifiable consequences of accidents
Victim(s)	– Pain and suffering – Moral and psychic suffering – Loss of physical functioning – Loss of quality of life – Health and domestic problems – Reduced desire to work – Anxiety – Stress	– Loss of salary and bonuses – Limitation of professional skills – Time loss (medical treatment) – Financial loss – Extra costs
Colleagues	– Bad feelings – Anxiety or panic attacks – Reduced desire to work – Anxiety – Stress	– Time loss – Potential loss of bonuses – Heavier work load – Training and guidance of temporary employees
Organization	– Deterioration of social climate – Poor image, bad reputation	– Internal investigation – Transport costs – Medical costs – Lost time (informing authorities, insurance company, etc.) – Damage to property and material – Reduction in productivity – Reduction in quality – Personnel replacement – New training for staff – Technical interference – Organizational costs – Higher production costs – Higher insurance premiums – Sanctions imposed by parent company – Sanctions imposed by the government

Tab. 9.1 (continued)

Interested parties	Nonquantifiable consequences of accidents	Quantifiable consequences of accidents
		– Modernization costs (ventilation, lighting, etc.) after inspection
		– New accident indirectly caused by accident (due to personnel being tired, inattentive, etc.)
		– Loss of certification
		– Loss of customers or suppliers as a direct consequence of the accident
		– Variety of administrative costs
		– Loss of bonuses
		– Loss of interest on lost cash/profits
		– Loss of shareholder value

The quantifiable socioeconomic accident costs (see Tab. 9.1) can be divided into direct and indirect costs. Direct costs are visible and obvious, while indirect costs are hidden and not immediately evident. In a situation where no accidents occur, the direct costs result in direct hypothetical benefits, while the indirect costs result in indirect hypothetical benefits. Resulting indirect hypothetical benefits comprise, e.g., not having sick leave or absence from work, not having staff reductions, not experiencing labor inefficiency and not experiencing change in the working environment. Figure 9.2 illustrates this reasoning. Since Fig. 9.2 only relates to ONE accident, and the number of accidents happening in reality are much fewer than the number of accidents (accident scenarios) that never realize and are avoided due to safety measures (both in time and in frequency) (non-accidents), one might say that overall, the hypothetical benefits are arguably even much higher than the calculable costs of accidents that happened.

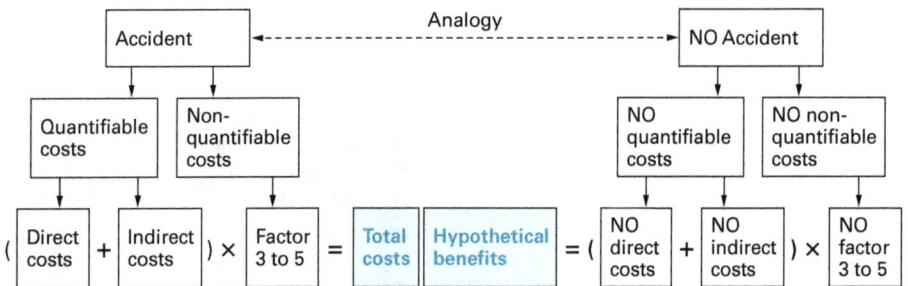

Fig. 9.2: Analogy between the costs and hypothetical benefits related to one accident, either happening or not happening.

Although hypothetical benefits seem to be rather theoretical and conceptual, they are nonetheless important to fully understand safety economics. Hypothetical benefits can be identified by the enterprise by asking questions such as "How much does the installation or a part of it cost?" or "What would be the higher price of the premium should a specific accident occurs?" Hypothetical benefits resulting from *non-occurring accidents* can be divided into five categories at the organizational level:

1. The first category concerns the non-loss of work time and includes the nonpayment of the employee, who at the time of the accident adds no further value to the company (if payments were to continue, this would be a pure cost).
2. The non-loss of short-term assets forms the second category and can include, e.g., the non-loss of raw materials.
3. The third category involves long-term assets, such as the non-loss of machines.
4. Various short-term benefits, such as non-transportation costs and non-fines, constitute the fourth category.
5. The fifth category consists of non-loss of income, non-signature of contracts or non-price reductions.

Clearly the visible or direct hypothetical benefits generated by the avoidance of costs resulting from non-occurring accidents only make up a small portion of the factors responsible for the total hypothetical benefits resulting from non-occurring accidents. Next to visible benefits, invisible benefits might exist. Invisible benefits may manifest themselves in different ways: e.g., non-deterioration of image, avoidance of lawsuits and the fact that an employee, thanks to the safety policy and the non-occurrence of accidents, does not leave the organization. If an employee does not leave the organization, the most significant benefit arises from work hours that other employees do not have to make up. Management time is consequently not given over to interviews and routines surrounding the termination of a contract. The costs of recruiting new employees can prove considerable if the company has to replace employees with experience and company-specific skills. Research into the reasons that form the basis of recruitment of new staff reveal a difference between, on the one hand, the recruitment of new employees because of the expansion of a company and, on the other hand, the replacement of staff who have resigned due to poor safety policies [7].

Costs of incidents and accidents that happened ("accident costs") and costs of incidents and accidents that were avoided and that never happened ("hypothetical benefits") are different in their nature. Nonetheless, their analogy is clear, and therefore they can easily be confused when making safety cost–benefit considerations.

9.1.1 Quick calculation example of (type 1) accident costs based on the number of serious accidents with certain consequence category

We can use the bird pyramid from Fig. 3.5 to make a rough estimate of the total yearly costs of occupational incidents and accidents within an organization, based on the number of serious accidents. Companies usually have a good understanding of the cost per type of accident. For example, the cost of a serious accident is calculated on a yearly basis by organizations, and equals €x. Similarly, the costs of the different types of incidents and accidents can be determined from Fig. 3.10. Figure 9.3 shows some theoretical ways of calculating the different costs.

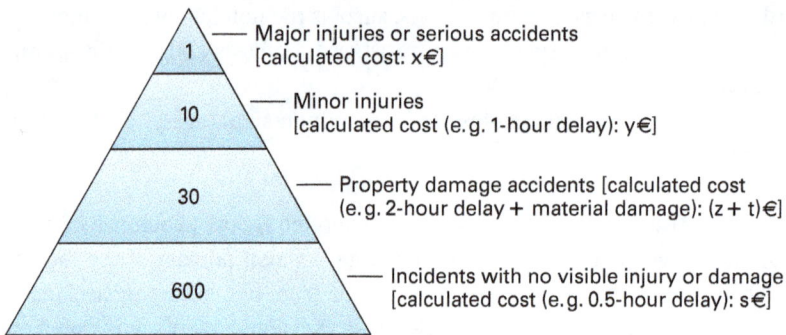

Fig. 9.3: The bird accident pyramid with costs per category of consequences.

Based on these costs, a rough estimate of the total yearly cost can be made. Legislation requires every company to know the number of serious accidents that happen per year, so let us assume that the number of serious accidents is N, and hence this rough calculation can be done for any company. Table 9.2 shows us how to calculate the total yearly accident costs.

Table 9.2 shows that, based on the ratio between serious accidents and other types of accidents, a rough estimate can be made of the total average accident costs for a given year if the number of incidents and accidents in that year is known, and if the average cost per type of accident is known. Table 9.2 does not consider that only 21% of all incidents lead to serious accidents, since this knowledge has no impact of the overall costs of all incidents happening within the organization. It does however impact the amount of costs avoided purely related to serious accidents, and should be taken into consideration when calculating those avoided costs.

A calculation example of hypothetical benefits is much more difficult to carry out, as there are no numbers available of the number of serious accidents avoided, or of the avoided cost per category of accident consequence. Future research has to be carried out to gain a better understanding of the hypothetical benefits and their calculation.

Tab. 9.2: Quick calculation of the total yearly costs based on the number of serious accidents.

Type of incident/ accident	Bird pyramid	Number of incidents/accidents	Cost per type of incident/accident	Cost
Serious	1	N	x	$N.x$
Minor injury	10	$10.N$	y	$10.N.y$
Property damage	30	$30.N$	$z + t$	$30.N.(z + t)$
Incident	600	$600.N$	s	$600.N.s$
			Total cost:	$N.(x + 10).y + 30.(z + t) + 600.s$

9.2 Prevention costs

In order to obtain an overview of the various kinds of prevention costs, it is appropriate to distinguish between fixed and variable prevention costs, on the one hand, and direct and indirect prevention costs, on the other hand.

Fixed prevention costs remain constant irrespective of changes in a company's activities. One example of a fixed cost is the purchase of a fire-proof door. This is a one-off purchase, and the related costs are not subject to variation in accordance with production. Variable costs, in contrast to fixed costs, vary proportionally in accordance with a company's activities. The purchase of safety gloves can be regarded as a variable cost because the gloves have to be replaced sooner when used more frequently or more intensively because of increased productivity levels.

Direct prevention costs have a direct link with production levels. Indirect prevention costs, on the contrary, are not directly linked to production levels. A safety report, e.g., will state where hazardous materials must be stored, but this will not have a direct effect on production. Hence, the development of company safety policy includes not only direct prevention costs such as the application and implementation of safety material, but also indirect prevention costs such as development and training of employees and maintenance of the company safety management system. A non-exhaustive list of prevention costs is given:
– Staffing costs of the company HSE department
– Staffing costs for the rest of the personnel (time needed to implement safety measures, time required to read working procedures, safety procedures, etc.)
– Procurement and maintenance costs of safety equipment (fire hoses, fire extinguishers, emergency lighting, cardiac defibrillators, pharmacy equipment, etc.)
– Costs related to training and education with regard to working safe
– Costs related to preventive audits and inspections
– Costs related to exercises, drills, simulations with regard to safety (e.g., evacuation exercises)
– A variety of administrative costs
– Prevention-related costs for early replacements of installation parts

- Maintenance of machine park, tools, etc.
- Good housekeeping
- Investigation of near-misses and incidents

In contrast to quantifying hypothetical benefits, companies are usually very experienced in calculating direct and indirect non-hypothetical costs of preventive measures, as listed earlier.

9.3 Prevention benefits

An accident that does not occur in an enterprise, thanks to the existence of an efficient safety policy within the company, was referred to in Section 9.1 as a *non-occurring accident*. Non-occurring accidents (in other words, the prevention of accidents) result in avoidance of a number of costs and thus create hypothetical benefits, as explained earlier. An estimation of the amount of input and output of the implementation of a safety policy is thus clearly anything but simple. It is impossible to specify the costs and benefits of one exceptional measure. The effectiveness (and the costs and benefits) of a safety policy must be regarded as a whole.

Hence, if an organization is interested in knowing the efficiency and the effectiveness of its safety policy and its prevention investments, in addition to identifying all kinds of prevention costs, it is also worth calculating the hypothetical benefits that result from non-occurring accidents. By taking all prevention costs and all hypothetical benefits into account, the true prevention benefits can be determined.

9.4 The degree of safety and the minimum total cost point

As Fuller and Vassie [8] indicate, the total cost of a company safety policy will be directly related to an organization's standards for health and safety, and thus its degree of safety. The higher the company's H&S standards and its degree of safety, the greater the prevention costs and the lower the accident costs within the company. Prevention costs will rise exponentially as the degree of safety increases, because of the law of diminishing returns, which describes the difficulties of trying to achieve the last small improvements in performance. If a company has already made huge prevention investments, and the company is thus performing very well on health and safety, it becomes evermore difficult to further improve its H&S performance. On the contrary, the accident costs will decrease exponentially as the degree of safety improves because, as the accident rate is reduced and hence accident costs are decreased, there is ever less potential for further (accident reduction) improvements. Figure 9.4 illustrates this.

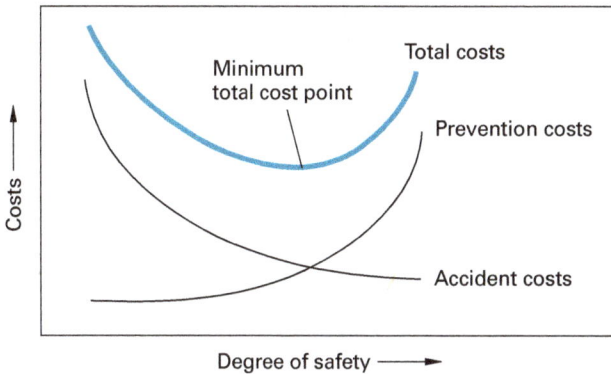

Fig. 9.4: Prevention costs and accident costs as a function of the degree of safety (qualitative figure).

From a purely microeconomic viewpoint, a cost-effective company will chose to establish a degree of safety that allows it to operate at the minimum total cost point (see Fig. 9.4). At this point, the prevention costs are balanced by the accident costs, and at first sight the "optimal safety policy" (from a microeconomic point of view) is realized. It is important for this exercise that the results should be as accurate as possible, the calculation of both prevention costs and accident costs should be as complete as possible, and all direct and indirect, visible and invisible costs should be taken into account. But the exercise results will still be largely improved, and the next paragraph and section will explain why and how.

We mention the phrase "from a microeconomic point of view" in the previous paragraph, because victim costs and societal costs should, in principle, also be taken into account in the total cost calculations. If only microeconomic factors are important for an organization, only accident costs related to the organization are considered by the company and not the victim costs and the societal costs. The human costs are obvious: loss in quality of life caused by, e.g., pain, stress and incapacity. Financial impacts on society arise from taxes that are used to provide medical services and social security payments for people injured at work, and prices of goods and services that are increased in order to recover the additional operating costs caused by accidents and ill-health at work. Hence, even if all direct and indirect microeconomic accident costs are determined by an organization, there will be an underestimation of the full economic costs because of individual costs and macroeconomic costs.

Furthermore, Fig. 9.4 is a qualitative figure that was drafted based on a large number of type I accidents. The "Minimum total cost point" can be situated more to the left or more to the right. This minimum cost point might even intersect the Y-axis or being at the point of 100% safety (to the far right). In the former case, investing in prevention is never recommended, while in the latter case it is always recommended. This is due to the fact that real accident costs, instead of hypothetical benefits, are considered in this figure. The fact that only real accidents (that happened) are employed in this figure, and not non-accidents, holds the problem that costs to avoid un-desired events are linked and compared with undesired events that already took place. So in fact, two aspects are being compared that should not be compared. When an accident happened, it is too late and no prevention cost was able to prevent it, nor will it be able to undo that accident. Furthermore, this figure should not be used, is not applicable, to type II accidents (or disasters).

> **!** True accident costs are composed of the organizational costs as well as the victim's costs and the society's costs. Moreover, to really take optimal decisions regarding prevention investments, organizations should make a distinction between the different types of risks (for the different types of risks, see Section 2.2), and between accident costs and hypothetical benefits.

9.5 Safety economics and the two different types of risks

The optimum degree of safety required to prevent losses is open to question, both from a financial and economic point of view and from a policy point of view. As explained earlier, on the one hand, developing and executing a sound prevention policy involves prevention costs, but, on the other hand, the avoidance of accidents and damage leads to hypothetical benefits. Consequently, in dealing with safety, an organization should try to establish an optimum between prevention costs and hypothetical benefits. Note that the real costs of actual accidents (such as displayed in Fig. 9.4) are not the same costs as those taken into consideration while determining the hypothetical benefits.

It is possible to further expand upon costs and benefits of accidents, in general, in terms of the degree of safety. The theoretical degree of safety can vary between $(0 + \varepsilon)\%$ and $(100 - \varepsilon)\%$, wherein ε assumes a (small) value, suggesting that "absolute risk" or "absolute safety/zero risk" in a company is, in reality, not possible. The economic break-even safety point, namely, the point at which the prevention costs equal the resulting hypothetical benefits, can be represented in graph form (see Fig. 9.5). The graph distinguishes between two cases: occupational accidents (or type I accidents, where a lot of information is available regarding the accidents) and major accidents (or type II accidents, where very scarce or no information is available regarding the accidents).

This distinction must be emphasized because there is a considerable difference in the costs and benefits relating to the two different types of accident. In the case of occupational accidents, the hypothetical benefits resulting from non-occurring accidents are considerably lower than in the case of major accidents. This means essentially that the company reaps greater potential financial benefits from investing in the prevention of major accidents than in the prevention of occupational accidents. When calculated, prevention costs related to major accidents are, in general, also considerably higher than those related to occupational accidents. Major accidents are mostly prevented by means of expensive technical studies (e.g., QRAs, see also Chapter 4) and the furnishing of (expensive) technical state-of-the-art equipment within the framework of process safety (e.g., SIL2, SIL3 and SIL4). Occupational accidents are therefore more associated with the protection of the individual, first aid, the daily management of safety codes, etc.

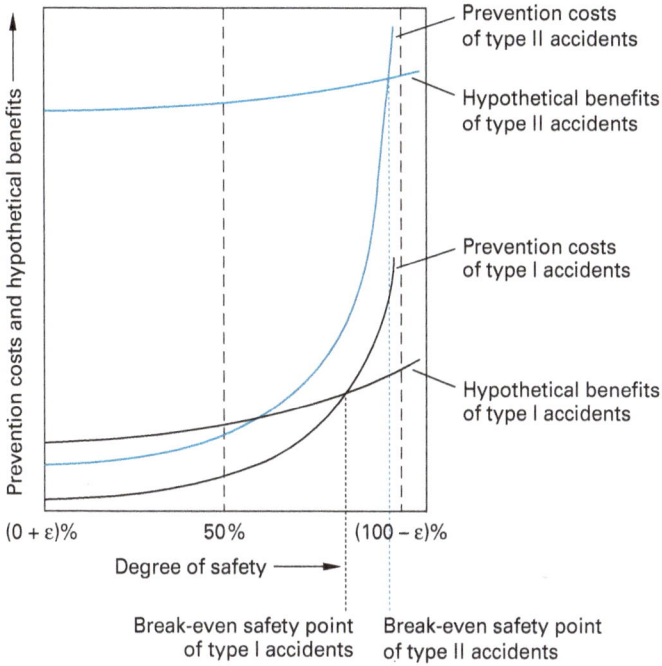

Fig. 9.5: Economic break-even safety points for the different types of accidents (qualitative figure).

As the company invests more in safety, the degree of safety will also increase. Moreover, the higher the degree of safety, the more difficult it becomes to improve upon this (i.e., to increase it), and the curve depicting investments in safety thereafter displays asymptotic characteristics. This was also explained in the text accompanying the discussion of Fig. 9.4 Moreover, as more financial resources are invested in safety from the point of $(0 + \varepsilon)$%, higher hypothetical benefits are obtained as a result of non-occurring accidents. These curves will display a much more leveled trajectory because marginal prevention investments do not produce large additional benefits in non-occurring accidents. It should be stressed that the curves displayed in Fig. 9.5 are qualitative and that the exact levels at which the curves are drawn are merely chosen for illustrative purposes and to increase the understanding of the theory explained for the readers. The hypothetical benefits curve and the prevention costs curve dissect at a break-even safety point. If there are greater prevention costs following this point, hypothetical benefits will no longer balance prevention costs. A different position for the costs and benefits curves is obtained for the different types of accidents. Figure 9.5 illustrates the qualitative benefits curves for the different types of accidents.

It is clear that hypothetical benefits and prevention costs relating to type I accidents are considerably lower than those relating to type II. The break-even safety point for type I accidents is likewise lower than that of type II and III accidents. Figure 9.5 also shows that in the case of type II and III accidents, where a company is subjected to extremely high financial damage, the hypothetical benefits are even higher and the break-even safety point must be located near the $(100 - \varepsilon)$% degree of safety limit. This supports the argument that, in the case of such types of accidents, almost all necessary prevention costs can be justified: the hypothetical benefits are certainly (nearly) always higher than the prevention costs. It should be noted that one also has to take into account that the uncertainty levels for type II risks are much higher than those of type I risks and, therefore, that the decision to allocate a prevention budget is not that straightforward.

Thus, for type II accidents it is possible to state that no real cost–benefit comparison is required but that a very high degree of safety should be guaranteed in spite of the cost of corresponding prevention measures. Of course, the uncertainties associated with these types of accidents are very high, and therefore, managers are not always convinced that such a major accident might happen in their organization. This is the main reason why the necessary precautions are not always taken, and why organizations are not always prepared.

In case of type I risks, regular cost–benefit analyses should be carried out, and the availability of sufficient data should lead to obtaining reliable results, based on which optimal precaution decisions can then be taken.

> **!** Risks possibly leading to occupational accidents (type I) are not to be confused with risks possibly leading to major accidents (type II) when making prevention investment decisions. Cost–benefit methods may yield reliable results for type I risks, whereas the hypothetical benefits of type II risks almost always outweigh the prevention costs for these types of risks. The uncertainties accompanying the risks should also be taken into account while making decisions. Uncertainties related to type II risks are much higher than those related to type I risks. Knowledge of type II risks is much lower than knowledge concerning type I risks. Making decisions about the two types of risks is thus fundamentally different, justifying two separate decision-making approaches.

9.6 Cost-effectiveness analysis and cost–benefit analysis

9.6.1 Cost-effectiveness analysis

Cost-effectiveness analysis is employed to maximize the return for a fixed budget. The technique can be used in the safety management field to compare the costs associated with a range of prevention measures and risk control measures that achieve similar benefits in order to identify the least cost option. The approach can be illustrated by two prevention options: having a different cost and avoiding different numbers of accidents (see Tab. 9.3).

Tab. 9.3: Cost-effectiveness analysis of two prevention options.

	Prevention costs	Number of similar accidents avoided	Average prevention cost per accident avoided
Prevention option 1	€5,000	25	200
Prevention option 2	€3,240	18	180

The average prevention cost per accident avoided is higher for prevention option 1 than for option 2 (see Tab. 9.3), indicating that the second option is more cost-effective than the first. We should remark that to use cost-effectiveness analysis to compare different options, they should share common outputs and be measured in similar terms.

9.6.2 Cost–benefit analysis

A cost–benefit analysis (CBA) is an economic evaluation in which all costs and consequences of a certain decision are expressed in the same units, usually money. Such an analysis may be employed in relation to operational safety, to aid normative decisions about safety investments (SIs). One should keep in mind, however, that CBAs cannot demonstrate whether one SI is intrinsically better than another. Nevertheless, a CBA allows decision-makers to improve their decisions by adding appropriate information on costs and benefits to certain prevention or mitigation investment decisions. Since decisions on SIs involve choices between different possible risk management options, CBAs can be very useful. Moreover, decisions may be straightforward in some cases, but this may not always be true. The decision-maker is thus recommended to use this approach with caution, as the available information is subject to varying levels of quality, detail, variability and uncertainty. Nevertheless, the tool is far from unusable and can provide meaningful information for aiding decision-making, especially if it takes the levels of variability and uncertainty into account and thus avoids misleading results.

The approach of CBAs can be used to determine whether an investment represents an efficient use of resources. An SI project represents an allocation of means (money, time, etc.) in the present that will result in a particular stream of nonevents, or expected hypothetical benefits, in the future. The role of a CBA is to provide information to the decision-maker; in this case, an employee or a manager who will appraise the SI project. The main purpose of the analysis is to obtain relevant information about the level and distribution of benefits and costs of the SI. Through this information an investment decision within the company can be guided and made in a more objective way. The role of analysis is thus to provide the possibility of a more objective evaluation and not to adopt an advocacy position either in favor of or against the SI, as there are also many other aspects that should be taken into account when deciding about SIs, such as social acceptability and regulatory affairs.

An SI project makes a difference, as the future will be different depending on whether the company decides to invest or not, or to invest in an alternative investment option. Thus, in the CBA, two hypothetical worlds are envisaged: one without the SI and another with the SI. During a CBA, a monetary value is put on the difference between the two hypothetical worlds. This process is shown in Fig. 9.6.

Since an SI project involves costs in the present and both costs and benefits in the future, the net benefit stream will be negative for a certain period of time and then will become positive at a certain point in time. This should be the case for both type I and type II risks, but the manner in which the calculations are performed differs. Therefore, a distinction between both types of risks is made later in this chapter.

CBA is used to evaluate the ratio between the benefits and the costs of certain measures. Fuller and Vassie [8] indicate that CBA requires financial terms to be assigned to both costs and benefits, to be able to calculate the "cost–benefit ratio." Several approaches are possible for determining such ratios, e.g., the value of averted losses divided by the prevention costs (over their lifetime), or the (liability of the original risk – liability of the residual risk) divided by prevention costs (over their lifetime). If the benefit–cost ratio is greater than 1, the safety management project could be accepted on the grounds that the benefits outweigh the costs, but if the ratio is less than 1, the project can be rejected on the grounds that the costs outweigh the benefits.

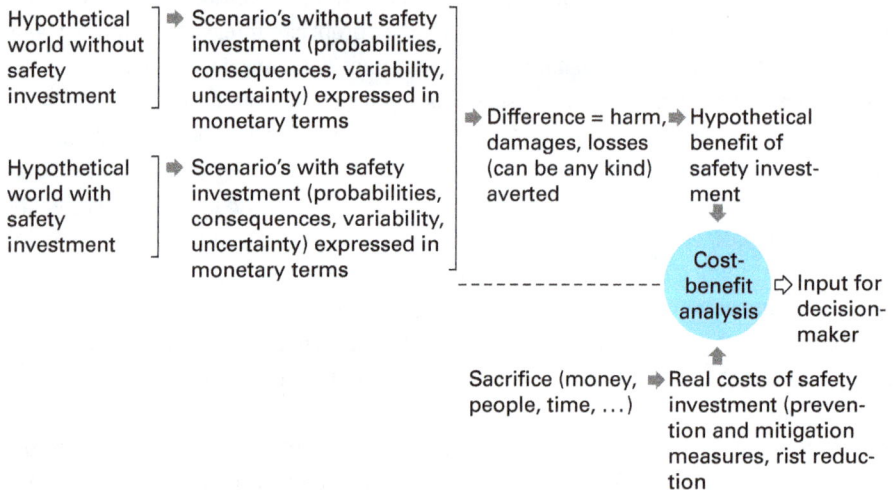

Hypothetical world without safety investment	➡ Scenario's without safety investment (probabilities, consequences, variability, uncertainty) expressed in monetary terms		
		➡ Difference = harm, damages, losses (can be any kind) averted	➡ Hypothetical benefit of safety investment ⬇
Hypothetical world with safety investment	➡ Scenario's with safety investment (probabilities, consequences, variability, uncertainty) expressed in monetary terms		

Cost-benefit analysis ▷ Input for decision-maker

Sacrifice (money, people, time, ...) ➡ Real costs of safety investment (prevention and mitigation measures, rist reduction)

Fig. 9.6: Cost–benefit analysis process for safety investments (source: [9]).

Tab. 9.4: Cost–benefit analysis of two prevention options.

	Prevention costs	Number of similar accidents avoided	Total benefit	Net benefit	Benefit–cost ratio
Prevention option 1	€5,000	25	€6,250	€1,250+	1.25
Prevention option 2	€3,240	18	€4,500	€1,260+	1.38

Table 9.4 illustrates the CBA approach, taking the numbers of Tab. 9.3 and adding the assumption that the financial benefit attributable to each accident avoided is €250.

The CBA from Tab. 9.4 shows that prevention option 1 has a benefit–cost ratio of 1.25, whereas option 2 has a ratio of 1.39. The second option is thus clearly preferred if a CBA would be used.

9.6.2.1 Decision rule, present values and discount rate

If a company uses a CBA, the recommendation whether to accept or to reject an investment project is usually based on the following process:
1. Identification of costs and benefits
2. Calculation of the present values (PVs) of all costs and benefits
3. Comparison of the total PV of costs and total PV of benefits

In order to compare the total costs and the total benefits, composed out of costs and benefits occurring at different points in time, one needs to take a discount rate into account in the calculation to obtain the PVs. Thus, during a CBA, all cash flows, from both costs and benefits in the future, need to be converted to values in the present. This conversion is carried out by discounting the cash flows by a discount rate. The discount rate represents the rate at which people (or companies) are willing to give up consumption in the present in exchange for additional consumption in the future. Another definition is that in a multiperiod model, people value future experiences to a lesser degree than present ones, as they are sure about present events and not sure about future events, which are subject to the environment. Thus the higher the discount rate they choose, the lower the PVs of the future cash flows [10].

An investment project is recommended when the total net PV (NPV) of all cash flows is positive, and an investment project is usually rejected when the NPV is negative. To calculate the NPV related to project management, all cash flows are determined, and future cash flows are recalculated to today's value of money by discounting them by the discount rate. The formula usually mentioned to calculate the NPV is

$$NPV = \sum_{t=0}^{T} \frac{X_t}{(1+r)^t}$$

where X_t represents the cash flow in year t, T is the time period considered (usually expressed in years) and r is the discount rate.

Applied to operational safety, the NPV of a project expresses the difference between the total discounted PV of the benefits and the total discounted PV of the costs. A positive NPV for a given SI indicates that the project benefits are larger than its costs:

$$\text{NPV} = \text{present value (benefits)} - \text{present value (costs)}$$

If NPV $\geq$ 0, recommend safety investment

If NPV $<$ 0, recommend rejecting safety investment

It is evident that the cash flows, i.e., prevention costs and certainly expected hypothetical benefits (due to nonevents), may be uncertain. Different approaches can be used in this regard. The cash flows can, e.g., be expressed as expected values, taking the uncertainties in the form of probabilities into consideration and also increasing the discount rate to outweigh the possibilities for unfavorable outcomes. In case of type II risks, it is recommended to use scenario analyses, determining expected cash flows for different scenario cases (e.g., worst-case and most credible case) and using a disproportion factor (DF) (see Section 9.6.2.2).

There can be different categories of costs related to an SI, e.g., initial costs, installation costs, operating costs, maintenance costs and inspection. These costs are evidently represented by negative cash flows. Some costs (e.g., initial costs and installation costs) occur in the present and thus do not have to be discounted, while other costs (e.g., operating, maintenance and inspection costs) occur throughout the whole remaining lifetime of the facility and thus will have to be discounted to the present. There may also be different categories of benefits linked to an SI, such as supply chain benefits, damage benefits, legal benefits, insurance benefits, human and environmental benefits, intervention benefits, reputation benefits and other benefits. The benefits represent positive cash flows, which all occur throughout the whole remaining lifetime of the facility and thus will all have to be discounted to the present.

In order to clarify the discount rate principle, all cash flows (for both costs and benefits) are assumed to occur on an arbitrarily chosen date, which can, e.g., be chosen to be the last day of the calendar year in which they occur. This assumption converts the continuous cash flows to a discrete range of cash flows, occurring at the end of each year. Then the cash flows at the end of each year have to be discounted to a PV, using a discount factor. As stated before, cash flows occurring in the current year do not have to be discounted. Therefore, the current year is called "year 0," and the following years "year 1," "year 2," . . ., "year n." Costs and benefits occurring in year 1 are discounted back one period, those occurring in year 2 are discounted back two periods and those occurring in year n are discounted back n periods. The implicit assumption is made that the discount rate remains the same throughout the entire remaining lifetime of the facility [10].

Thus, for calculating the PV of a benefit occurring in year 1, it needs to be discounted for one period to come to a PV in year 0. Similar to the calculation of a benefit occurring in year 1, the PV of benefits occurring in year 2 and year 3 are obtained by discounting them 2 and 3 periods, respectively. Similar to the previous calculations, the PV of a benefit occurring in year n is obtained by discounting it n periods. These calculations can be found in the following range:

$$\text{PV of a benefit in year } 1 = \frac{\text{Benefit}}{(1+r)}$$

$$\text{PV of a benefit in year } 2 = \frac{\text{Benefit}}{(1+r)^2}$$

$$\vdots$$

$$\text{PV of a benefit in year } n = \frac{\text{Benefit}}{(1+r)^n}$$

Now that the concept and method of discounting future cash flows is clarified, suppose an SI project has a cost in year 0 and then the same level of costs and benefits at the end of each and every subsequent year for the whole remaining lifetime of the facility. This means that the costs in year i are the same for all i, i.e., $C_i = C$; likewise, the benefits in year i are the same for all i, i.e., $B_i = B$. This concept is called an "annuity." The PV of such an annuity is given by the following formula, with n being the remaining lifetime of the facility:

$$\text{PV(annuity of a cost)} = C + \frac{C}{(1+r)} + \frac{C}{(1+r)^2} + \cdots + \frac{C}{(1+r)^n}$$

$$\text{PV(annuity of a benefit)} = B + \frac{B}{(1+r)} + \frac{B}{(1+r)^2} + \cdots + \frac{B}{(1+r)^n}$$

C and B are the equal annual costs (cost categories where costs are made in the future) and benefits (all benefit categories), respectively, that occur at the end of each year and are assumed to remain constant. This assumption is valid as long as inflation is omitted from the calculations and as long as the annual costs are assumed not to increase over time due to aging. These assumptions can be made to keep it rather simple while explaining the cost–benefit approach. Each term in the formula above is formed by multiplying the previous term by "$1/(1 + r)$." As the above formulas can become very long, the formula for calculating the PV of annuities can be rewritten by way of the series solution as follows:

$$\text{PV(annuity)} = A + \frac{A}{r} - \frac{A}{r(1+r)^n}$$

where A is the yearly cost or benefit of a cost/benefit category. Note that this general annuity goes to $(n + 1)A$ as the discount ratio r goes to zero. The term

$$\frac{1}{r} - \frac{1}{r(1+r)^n} = \frac{(1+r)^n - 1}{r(1+r)^n}$$

of the series solution is called "the annuity (discount) factor" and is applicable whenever the annuity starts from year 1.

Using this model, the benefits and costs in the future are assumed to be constant, and inflation is not included into the future costs and benefits, as already mentioned. Inflation is the process that results in a rise of the nominal prices of goods and services over time. Therefore in this (simplified) model, the real rate of interest[1] should be used as the discount rate instead of the money rate of interest.[2] Since the money rate of interest m includes two components, the real rate of interest r and the anticipated rate of inflation i:

$$m = r + i,$$

the anticipated rate of inflation is built into the money rate of interest. As the inflation is not being included into the numerator of the formula for calculating the PV of annuities (as the costs and benefits are constant throughout the whole remaining lifetime), it can also not be included in the denominator.

9.6.2.2 Disproportion factor

Type II accidents are related to extremely low frequencies and a high level of uncertainty. To take this into account, CBA preferably involves a so-called DF in order to reflect an intended bias in favor of safety above costs. This safety mechanism is vital in the calculation to determine the adequate level of investment in prevention measures, as, on the one hand, the probability influences the hypothetical benefits substantially through the number of years over which the total accident costs can be spread out, and on the other hand, the uncertainty regarding the consequences is high (see [11] for more details).

Traditional CBAs discourage SIs when the costs are higher than the benefits. If, however, a DF is included, an investment in safety is reasonably practicable unless its costs are grossly disproportionate to the benefits. If the following equation is true, then the safety measure under consideration is not reasonably practicable, as the costs of the safety measure are disproportionate to its benefits [11]:

$$\text{Costs/benefits} > \text{DF} \rightarrow \text{costs} > \text{benefits} \times \text{DF}$$

In order to determine the size of the DF, some guidelines and rules of thumb are available. They state that DFs are rarely greater than 10 and that the higher the risk, the

---

1 Real rate of interest (r) does not include the anticipated rate of inflation (i).
2 Money rate of interest (m) includes two components such as the real rate of interest (r) and the anticipated rate of inflation (i): $m = r + i$.

higher the DF must be to emphasize the magnitude of those risks in the CBA. There-fore, if the risk of accident is very high it might be acceptable to use a DF greater than 10 [11]. However, Rushton [12] strongly advises not to use a DF greater than 30. See also Reniers and Van Erp [9] for more detailed information on how to calculate and use DFs.

9.6.2.3 Different cost–benefit ratios

Several approaches are possible for presenting the cost–benefit principle, and differ-ent cost–benefit ratios can be calculated. Remark that sometimes benefits are divided by costs, then a benefit–cost ratio is obtained, and sometimes costs are divided by benefits and a cost–benefit ratio is obtained. In case of a benefit–cost ratio, the ratio should ideally be higher than 1, and as high as possible, while in case of a cost–benefit ratio, it should ideally be lower than 1, and as low as possible. The following ratios are mentioned by Fuller and Vassie [8]:

– Value of an averted loss:

$$\text{Benefit} - \text{cost ratio} = \text{value of averted losses} (= \text{hypothetical benefits}) / \\ \text{safety measures' costs over their lifetime}$$

– Value of equivalent life:

$$\text{Benefit} - \text{cost ratio} = \text{value of equivalent lives saved over the lifetime} \\ \textit{of the safety } \text{measures} / \text{safety measures' costs over their lifetime}$$

– Value of risk reduction:

$$\text{Benefit} - \text{cost ratio} = [(\text{liability of the original risk}) - (\text{liability of the} \\ \textit{residual risk})] / \text{Safety measures' costs over their lifetime}$$

9.6.2.4 Cost–benefit analysis for safety measures

It is possible to determine in a simple way whether the costs of a safety measure out-weighs – or not – its benefits. In general, the idea is simple: compare the cost of the safety measure with its benefits. The cost of a safety measure is easy to determine, but the bene-fit is much more difficult to calculate. The benefit can be expressed as the "reduced risk," taking into account the costs of accidents with and without the safety measure implemen-tation. The following equation may be used for this exercise [13]:

$$\{(C_{without} \cdot F_{without}) - (C_{with} \cdot F_{with})\}.\text{Pr}_{control} > \text{Safety measure cost}$$

Or, if sufficient information is not available for the frequencies of the initiating events to use the previous equation:

$$(C_{without} - C_{with}) \cdot F_{accident} \cdot \text{Pr}_{control} > \text{Safety measure cost}$$

where C_{without} is the cost of accident without safety measure, C_{with} is the cost of accident with safety measure, F_{without} is the statistical frequency of initiating event if the safety measure is not implemented, F_{with} is the statistical frequency of initiating event if the safety measure is implemented, F_{accident} is the statistical frequency of the accident and $\text{Pr}_{\text{control}}$ is the probability that the safety measure will perform as required.

The formulae show immediately why CBAs may only be carried out for risks where sufficient data is available: if not, the required "statistical frequencies" are not known, the probabilities may not be known, and rough estimates (more or less *guesses*) should be used, leading to unreliable results. If sufficient information is available, results from using these equations for determining the cost–benefit of a safety measure are reliable.

9.6.3 Risk acceptability

Risk acceptability is an extremely difficult and complex issue. The problem starts with the viewpoint: acceptable for whom? From whose perspective? In other words, who suffers the consequences when something goes awfully wrong and disaster strikes on the one hand, and who gains the benefits when all goes well and profits are made on the other hand? Are the risks equally spread? Are the risks justified/defensible and, even more important, just? Moral aspects always start to emerge when discussing the acceptability of risk, or in other words, when asking the question, "how safe is safe enough?"

Independent of the personalities of people, the discussion and argumentation about what risks and levels of risk are acceptable. Some people argue that one fatality is one too many, and other people interpret the prevailing accident rate in areas of voluntary risk-taking as a measure of the level of risk that society as a whole finds acceptable [14]. Obviously, both viewpoints have their merits. No one will argue against the fact that one death is one too many. But it also seems reasonable to assume that voluntary risk levels can be used as some sort of guideline for involuntary risk levels. However, both perspectives are far from usable in the real industrial world. On the one hand, lethal accidents do happen and should not be economically treated as if they are not possible. So being part of industrial practice and real-life circumstances, fatalities should be taken into account in economic analyses. On the other hand, research indicates that voluntary risk cannot be compared with involuntary risk. People are much less willing to suffer involuntary risks than voluntary risks, and the risk perception people have about involuntary risks is much higher compared to voluntary risks (see, e.g., [15]). Hence, these two sorts of risks, i.e., voluntary and involuntary risks, should not be used interchangeably in economic analyses, and risk criteria should not be based on them.

In this chapter on economic issues, *risks* and *dealing with risks* are viewed from a microeconomic decision-making perspective. Hence, the focus is on the optimal operational safety decision, based on the information available and thereby taking economic issues into account. Risks are relative, and making the best resource allocation decision for operational safety in a company means avoiding as much possible loss as possible within a certain safety budget. However, due to the "acceptability" aspect of risks, this seems easier than it is in reality. One of the best known and most used principles in this regard is the proactive concept of "as low as reasonably practicable" (ALARP). The term implies that the risk must be insignificant in relation to the sacrifice in terms of money, time or trouble required to avert it. Hence, ALARP means that risks should be averted unless there is a gross disproportion between the costs and the benefits of doing so. The difference between ALARP and other terms, such as BAT ("best available technology"), BACT ("best available control technology") and ALARA ("as low as reasonably achievable"), can best be described as the observation that to follow the ALARP approach, not only should the technology be available, but the costs of prevention should also be reasonable. BAT and ALARA demand to do work, research, engineering, etc. to make the prevention work, irrespective of the costs. The term "SFAIRP" ("so far as is reasonably practicable") is used in health and safety regulations in the UK and should not be confused with ALARP that has no legislative connotation. The SFAIRP term has been defined in UK courts. The formal definition given by Redgrave [16] is:

> "Reasonably practicable" is a narrower term than "physically possible," and implies that a computation must be made in which the quantum of [operational] risk is placed in one scale and the sacrifice involved in the measures necessary for averting the risk (whether in money, time or trouble) is placed in the other, and that, if it be shown that there is a gross disproportion between them – the risk being insignificant in relation to the sacrifice – the defendants [persons on whom the duty is laid] discharge the onus upon them [of proving that compliance was not reasonably practicable]. Moreover, this computation falls to be made by the owner at a point of time anterior to the accident.

An acronym similar to ALARP (but with slightly different meaning as regards the word "reasonable") is BATNEEC ("best available technology not entailing excessive costs"). Other acronyms focusing on the legislative (and reactive) part are CATNIP ("cheapest available technology not invoking prosecution") and ALARA (second meaning) ("as large as regulators allow").

It is clear that there is a diversity of acceptability approaches out there, but the most well-known and the most used is undoubtedly ALARP. ALARP is also the most useful on an organizational level, since the risk level being ALARP is influenced by international guidelines and directives, the progress of societal knowledge of risk, the continuous evolution of people's perception of what "risk" constitutes, political preferences and other forms of regulations in different industries. Hence, the ALARP principle allows a different risk judgment depending on industrial sectors, risk groups, risk-prone areas and even organizations.

The relationship between the ALARP principle and the terms "intolerable," "tolerable," "acceptable" and "negligible risk" is illustrated in Fig. 9.7.

Figure 9.7 shows that it is possible to make a difference between "tolerable" and "acceptable." The subtle distinction between both terms is used to indicate that a level of risk is never accepted, but it is tolerated in return for other benefits which are obtained through activities, processes, etc., generating the risk [17]. Figure 9.7 illustrates that activities characterized by high levels of risk are considered intolerable under any normal circumstances. At the lower end, there is a level below which risk is so low that it can be ignored – i.e. it is not worth the cost of actively managing it. There is a region between these two limits, called the ALARP region, in which it is necessary to make a trade-off between the risk and the cost of further risk reduction.

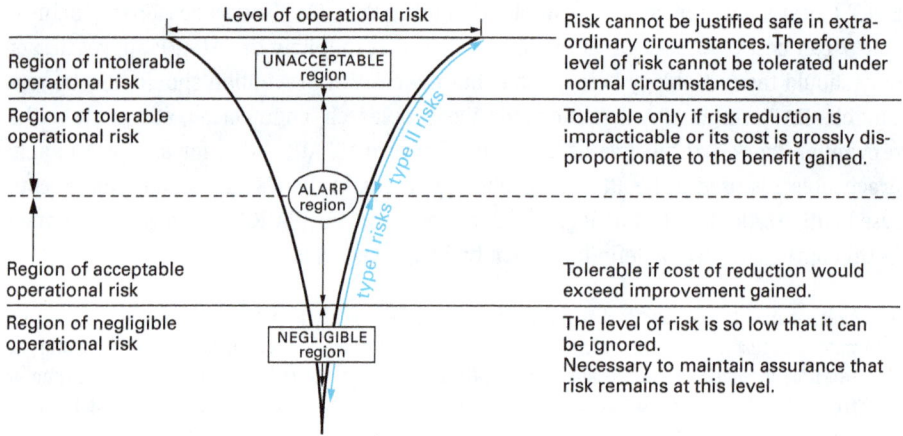

Fig. 9.7: The "ALARP" principle and its relationship with the terms "intolerable," "tolerable," "acceptable" and "negligible" (source: [9]).

The practical use of ALARP within one single organization can be demonstrated by using the risk assessment decision matrix (see Chapter 4). The goal of all reduction methods is summarized in Fig. 9.8, where one can observe that risks which are considered unacceptable (with the visual help of the risk matrix) have to be brought down to an acceptable or at least to a tolerable level.

In the middle zone of the matrix, two intermediate regions "2" and "3" between the unacceptable zone "1" (needing immediate action) and the negligible zone "4" can be noted in diagonal. These zones are the ones where the ALARP principle occurs. Using the risk matrix, it is thus easily possible for an organization to make a distinction between the tolerable region "2" and the acceptable region "3" as mentioned in Fig. 9.8, and to link certain actions to it, specifically designed by and for the company.

There are a number of disadvantages or issues concerning the using of economic (e.g., cost–benefit) analyses for ALARP. For example, there is a difference between the willingness to pay and willingness to accept evaluations (see also [9]), and both approaches are often used to put values on (un)safety. There are also problems of understanding "gross disproportionality" and of putting a value on human life. Furthermore, sometimes, risks are only replaced (exported) to another facility not falling under the same (or under any) ALARP calculation.

Likelihood

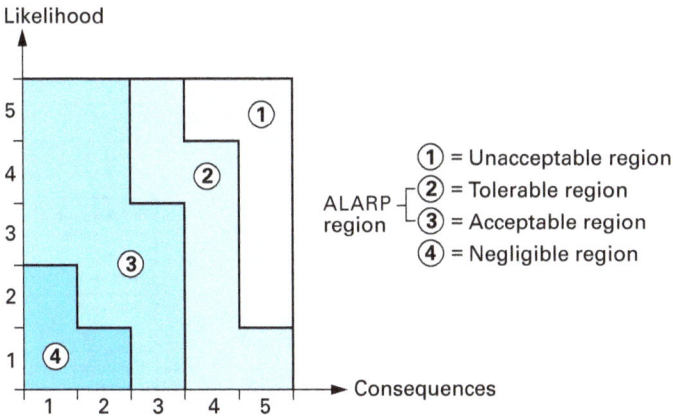

Fig. 9.8: The ALARP principle and the effect of risk reduction is represented in the risk matrix (source: [9]).

In any case, the ALARP principle needs a decision rule (or criteria) specifying what is "reasonable" for the person, group of persons or organization using it. The decision will ultimately depend on a combination of physical limits, economic constraints and moral aspects. The advantage is that the principle works in practice, but the downside is that it will sometimes lead to inconsistencies in effective valuation of various risks and have inefficiencies in the sense that it may be possible in some cases to save more lives at the same cost, or the same number of lives at a lower cost [18].

9.6.4 Application of the event tree for safety investments

As also explained in Reniers and Van Erp [9], both in case of type I and type II risks, it is possible to draft and use a decision analysis tree if a number of alternative outcomes are possible, as was discussed in Chapter 4. The probabilities of safety measures in case of certain events need to be taken into account and the expected values need to be calculated. This way, the value of an improbable outcome is weighed against the likelihood that the event will occur. Using the event tree analysis approach (for more information, see Chapter 4), a graph, such as in Fig. 9.9, is obtained, which is an illustrative example of a decision analysis tree for a runaway reaction event. All costs and probabilities are assumed to be yearly.

An analysis of SI decisions can be made. The costs of the safety measures can be collected and displayed on the figure (e.g., in the case of this illustrative runaway reaction event, a total prevention cost of €250,000 is obtained), as well as the expected total costs of the event. The expected total costs (which can be seen as the expected hypothetical benefits) can then be compared with the prevention costs. In this illustrative example, the expected total costs equal $P_1 \cdot [P_2 \cdot n + (1 - P_2) \cdot C_F]$, eventually to be multiplied with a DF in the case that the accident scenario is assessed to be a type II risk scenario

Event originator: failure of cooling system (Cost = 150,000€) | Temperature measure in the reactor (Cost = 100€) | Alarm signifying the rise in temperature to an operator (T>T_{max}) (Cost = 5000€) | Restoration of cooling system by an operator (Cost = 80,000€) | Automatic inhibition of the reaction at T>$T_{crit.}$ (Cost = 14,900€) | Sequences leading to:

P_4

Back to normal

$P_{A'} = P_1 \cdot P_2 \cdot P_3 \cdot P_4$
$C_A = 0€$

P_3

$\boxed{P_4 \cdot C_A + (1-P_4) \cdot k} = I$

P_5

Safekeeping of the installations (stopping of operations)

$P_B = P_1 \cdot P_2 \cdot P_3 \cdot (1-P_4) \cdot P_5$
$C_B = 100,000€$

$1-P_4$

$\boxed{P_5 \cdot C_B + (1-P_5) \cdot C_C} = k$

Success (YES)

P_2

$1-P_5$

Runaway reaction

$P_C = P_1 \cdot P_2 \cdot P_3 \cdot (1-P_4) \cdot (1-P_5)$
$C_C = 800,000€$

$\boxed{P_3 \cdot I + (1-P_3) \cdot m} = n$

Failure (NO)

P_5

Safekeeping of the installations (stopping of operations)

P_1

$\boxed{P_2 \cdot n + (1-P_2) \cdot C_F}$

$1-P_3$

$\boxed{P_5 \cdot C_D + (1-P_5) \cdot C_E} = m$

$P_D = P_1 \cdot P_2 \cdot (1-P_3) \cdot P_5$
$C_D = 100,000€$

$1-P_5$

Runaway reaction

$P_E = P_1 \cdot P_2 \cdot (1-P_3) \cdot (1-P_5)$
$C_E = 800,000€$

$1-P_2$

Runaway reaction

$P_F = P_1 \cdot (1-P_2)$
$C_F = 800,000€$

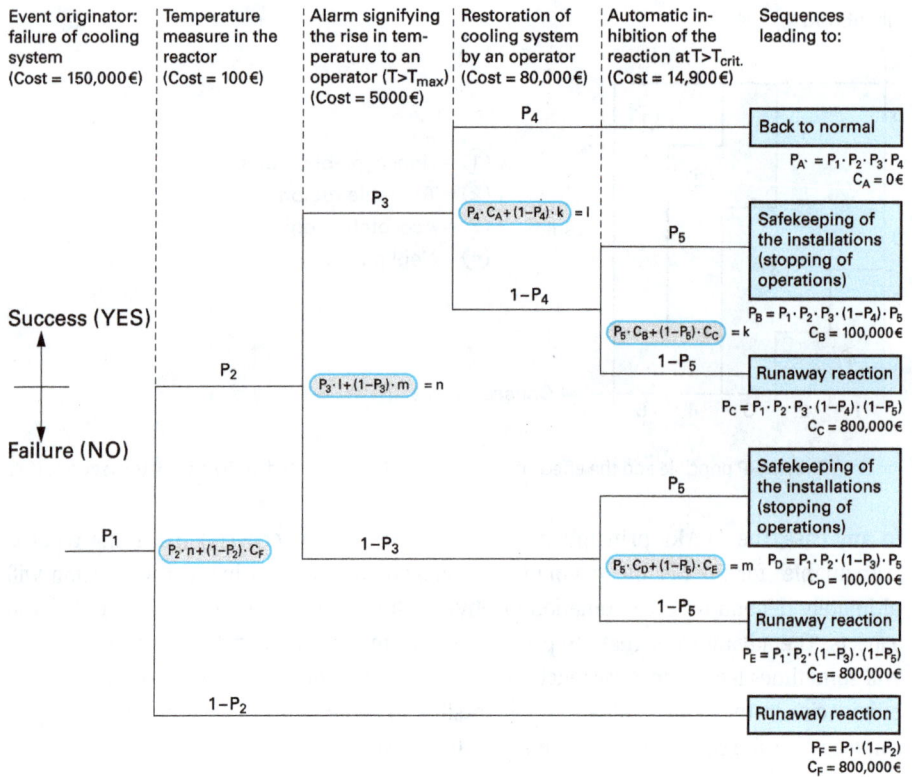

Fig. 9.9: Illustrative decision analysis tree for runaway reaction event (source: [9]).

(since the expected total costs can be seen as the hypothetical benefits which can be avoided by doing an investment), and should be compared with €250,000 (which is the accumulated cost of the safety measures, or the investment that has to be made to avoid the possible accident consequences) for formulating a decision recommendation. This way, a decision can be made regarding this SI portfolio composed of three barriers (i.e., the cooling system (itself consisting of three safety functions), the operator manual intervention and the automatic inhibition intervention).

9.6.5 Internal rate of return

The internal rate of return (IRR) can be defined as the discount rate at which the PV of all future cash flows (or monetized expected hypothetical benefits) is equal to the initial investment, or in other words, it is the rate at which an investment breaks even. Generally speaking, the higher an investment's IRR, the more desirable it is to carry on with the investment. As such, the IRR can be used to rank several possible investment options an organization is considering. Assuming that all other factors are equal among the various investments, the SI with the highest IRR would then be recommended to have priority. Remark that the IRR is sometimes referred to as "economic rate of return."

An organization should, in theory, undertake all SIs available with IRRs that exceed a minimum acceptable rate of return predetermined by the company. Investments may of course be limited by the availability of funds or safety budget to the company.

Because the IRR is a rate quantity, it is an indicator of the efficiency, quality or yield of an investment. This is in contrast with the NPV, which is an indicator of the value or magnitude of an investment.

A rate of return for which the NPV, expressed in function of the rate of return, is zero is the IRR r^*. This can be expressed as follows:

$$\mathrm{NPV}(r^*) = \sum_{n=0}^{N} \frac{C_n}{(1+r^*)^n} = 0$$

In cases where a first SI displays a lower IRR but a higher NPV over a second SI, the first investment should be accepted over the second investment. Furthermore, remark that the IRR should not be used to compare investments of different durations. For example, the NPV added by an investment with longer duration but lower IRR could be greater than that of an investment of similar size in terms of total net cash flows, but with shorter duration and higher IRR.

As a simple illustrative example, the following problem can be given for calculating an IRR. Assume that an SI is given by the sequence of cash flows (initial investment costs and yearly hypothetical benefits) as displayed in Tab. 9.5.

Tab. 9.5: Example of the calculation of the internal rate of return (IRR).

Year (n)	Cash flow (C_n)
0	−€123,400
1	€36,200
2	€54,800
3	€48,100

From Tab. 9.6 and the above equation, the IRR r^* can be determined by solving the following equation:

$$\mathrm{NPV}(r^*) = -123,400 + \frac{36,200}{(1+r^*)^1} + \frac{54,800}{(1+r^*)^2} + \frac{48,100}{(1+r^*)^3} = 0$$

In this case, the answer is 5.96% (in the calculation, i.e., $r^* = 0.0596$).

9.6.6 Payback period

The payback period is calculated by counting the time (usually expressed in a number of years) it will take to recover an investment. Hence, a break-even point of investment is determined in terms of time. The payback period of a certain SI for type I risks is a possible determinant of whether to go ahead with the safety project or not, as longer payback periods are typically not desirable for some companies. It should be noted that the PBP ignores any benefits that occur after the determined payback period and, therefore, does not measure profitability. Moreover, the time value of money is not taken into account in the concept, and neither is the opportunity cost considered. The PBP may be calculated as the cost of SI divided by the annual benefit inflows.

> As an illustrative example, let us assume that a company invests €400,000 in safety equipment. The expected (hypothetical) benefits from the new equipment is expected to be €100,000 per year for 10 years. The payback period is 4 years (€400,000 divided by €100,000 per year).
>
> A second safety project requires an investment of €200,000, and it generates expected hypothetical benefits as follows: €20,000 in year 1; €60,000 in year 2; €80,000 in year 3; €100,000 in year 4; €70,000 in year 5. The payback period is 3.4 years (€20,000 + €60,000 + €80,000 = €160,000 in the first 3 years + €40,000 of the €100,000 occurring in year 4).

Note that the payback calculation uses cash flows, not net income. The payback period simply computes how fast a company will recover its cash investment.

9.6.7 Application of investment analysis for type I risks

In order to illustrate the application of a CBA and parameters such as the payback period and the IRR, an extensive example is elaborated in this section.

Assume that a firm has decided to improve the working conditions defining a safety budget equal to €500,000. Two investment options are analyzed and aimed at to improve the safety levels within the company, increase the risk awareness and spread good safe practices in the working environment. The features of these alternative investment options are presented in the following sections.

9.6.7.1 Safety investment option 1

The duration of the investment is assumed to be 10 years. The beneficial effect of a product of an improved quality has been also quantified in additional profits of €25,000 per year. Moreover, a team responsible for the risk assessment has estimated yearly saving in the production operations equal to €135,000 due to waste reduction and a reduction of raw materials supply. On the investments side, the initial purchase cost of the machine is €280,000. Additional costs due to training, redesign of the layout in the production area and feasibility studies are to be initially sustained. The company has conducted an economic analysis associated with the investment quantifying the main benefits and costs associated with the investment as shown in Tabs. 9.6 and 9.7.

Tab. 9.6: List of costs associated with safety investment option 1.

Categories of costs	Subcategories of costs	Value
Initial costs	Investigation and preliminary study (€)	15,400
	Machine purchase costs (€)	280,000
	Initial training (€)	25,000
	Changing three layouts and production operations (€)	110,500
Installation costs	Machine configuration and testing (€)	5,500
	Equipment costs (€)	15,400
	Installation team costs (€)	25,000
Operating costs	Energy costs (€/year)	38,500
Maintenance costs	Material costs (€/year)	15,000
	Maintenance team costs (€/year)	7,750
Inspection costs	Inspection team costs (€/year)	2,500
Other safety costs	Other safety costs (€/year)	2,500

Tab. 9.7: List of hypothetical benefits associated with safety investment option 1.

Type of benefits	Subcategory	Value
Supply chain benefits	Production savings (€/year)	135,000
	Expected additional profits due to increased sales (€/year)	25,000
Damage benefits	Damage to own material/property (€/year)	2,500
Legal benefits	Fines (€/year)	10,000
Insurance benefits	Insurance premium (€/year)	20,000
Human and environmental benefits	Yearly reduction of days of illness (€/year)	2,500
Other benefits	Cleaning (€/year)	4,500

The costs and the hypothetical benefits summarized in Tabs. 9.6 and 9.7 can be further distinguished either in initial costs/benefits or yearly recurring costs/benefits. Initial costs/benefits are supposed to be sustained in year 0 (in which the evaluation is made). Yearly recurring costs/benefits are sustained during the time horizon in which the investment is evaluated. Analyzing the nature of costs and benefits, it can be observed that the total initial investment required to the company is equal to 0.476 M€.

The formula from Section 9.6.2.1 has been used to compute the NPV associated with the investment, assuming for each year t the previously mentioned negative cash flows (C_t) and positive cash flows (B_t) and the discount factor $I = 3\%$:

$$\text{NPV} = \sum_{t=0}^{10} \frac{(B_t - C_t)}{(1+i)^t}$$

The total NPV associated with the investment in this case is equal to €702,501. Since this value is greater than 0, the investment is profitable for the company. The payback period is equal to 3.33. This means that after 3 years and 4 months, the investment will cover the costs, and it will start producing profits for the firm. Another indication that is often used by decision-makers to assess the profitability of an investment is the IRR, as also indicated before. This measure represents the discount factor that makes the NPV equal to zero or in other words the value of interest rate at which an investment reaches and breaks an even point. In this case, the IRR is equal to 24.93%.

9.6.7.2 Safety investment option 2
After a thorough assessment, a team of experts has estimated benefits and costs associated with the investment, as summarized in Tabs. 9.8 and 9.9, respectively.

Tab. 9.8: List of costs associated with safety investment option 2.

Categories of costs	Subcategories of costs	Value
Initial costs	Investigation costs (€)	8,300
	Selection and design costs (€)	10,200
	Material costs (€)	85,500
	Training costs (€)	4,500
	Changing guidelines and informing costs (€)	6,500
	Purchase costs (€)	195,500
Installation costs	Start-up costs (€)	23,500
	Equipment costs (€)	58,500
	Installation costs (€)	86,500
Operating costs	Energy consumption costs (€/year)	10,000
Maintenance costs	Material costs (€/year)	3,500
	Maintenance team costs (€/year)	1,500
Inspection costs	Inspection team costs (€/year)	1,500
Other safety costs	Other safety costs (€/year)	1,400

Next, costs and benefits have been analyzed deeper to distinguish between the initial investments from the yearly recurring costs/benefits as shown in Tab. 9.10.

As done for the SI option 1 described in the previous section, the NPV is computed considering a time horizon of 10 years. Since the time horizon is the same of the one used to evaluate the investment option 1, the comparison between the alternative investments will be simplified as it will be done using the same criteria. In addition, the amount of money requested by the initial investment (0.479 M€) is comparable with the previous investment option.

Tab. 9.9: List of hypothetical benefits associated with safety investment option 2.

Type of benefits	Subcategory	Value
Supply chain benefits	Production loss (€/year)	8,700
	Waste recycling (€/year)	5,000
	Expected additional sales due to better product quality (€/year)	5,500
	Saving in personal protective equipment	4,000
Damage benefits	Damage to own material/property (€/year)	41,000
Legal benefits	Fines (€/year)	15,000
Insurance benefits	Insurance premium (€/year)	15,000
Human and environmental benefits	Injured employees (€/year)	5,000
	Environmental damage (€/year)	5,000
Other benefits	Manager working time (€/year)	2,400
	Cleaning (€/year)	25,000

Tab. 9.10: Initial investment and yearly benefits and costs associated with safety investment option 2.

Description	Value
Initial costs (€)	–€479,000
Yearly costs (€/year)	–€17,900
Yearly benefits (€/year)	€131,600

In this case, the NPV is equal to €490,884, which is lower than €659,850 associated with the previous investment option. In addition, the payback period of about 4.5 years is slightly longer than 3.3 years. Therefore, the investment option 2 seems to be dominated by the investment option 1. Given the limited safety budget available for the company, SI option 1 should be preferred.

If additional resources would be found, investment option 2 should also be considered and implemented since its profitability on the medium long term is guaranteed by a positive NPV and a relatively limited payback period.

The IRR associated with the second SI option is computed as being equal to 18.89%. Therefore, also in this case, investment option 2 is clearly the worst option.

9.6.8 Decision analysis tree cost-variable approach

It is possible to draft a tree where the cost of each decision is presented as a variable named "cost." The cost is a running total of the costs of each event in a given event pathway. At each box in the tree, the running total is increased by the cost assigned to the box. Figure 9.10 presents an illustrative example.

For instance, if companies decide to proactively collaborate with respect to domino effects prevention, e.g., by elaborating and maintaining a software for warning and help- ing neighbors, it would cost them each €10,000. If nothing happens, the total cost for a company would be €10,000 (for the collaboration) + €0 (nothing happens) = €10,000. If a large-scale fire happens within a company and it has severe cross-border potential de- spite collaboration efforts, the company also incurs the initial costs of the large-scale fire (€12,000 to deal with the fire). If the fire accident then does not affect other companies in a severe way, or it rests internally, no extra costs are incurred for the company, and the final value of the variable cost at the end of the bottom pathway, in Fig. 9.10, is €22,000. An analogous reasoning can be made for all other pathways, and Fig. 9.11 is obtained. From Fig. 9.11, it follows that, for the probabilities and costs that were as- sumed, the overall cost of domino effects collaboration would be €10,705 and that of no collaboration would be €13,200. Hence, overall, "no collaboration" would be some €2,500 more expensive in expected costs than "collaboration" in this illustrative example.

9.6.9 The Borda algorithm approach

The Borda algorithm is mainly used in voting problems (see also [9]). The Borda rule assigns linearly decreasing points to consecutive positions, e.g., for three alternatives the points would be 3 for the first place, 2 for the second place and 1 for the third place. The algorithm is employed to develop an ordinal ranking of preferences. The Borda rule can also be employed in a risk management context. In the context of operational safety de- cision-making with respect to economics, the Borda algorithm can be employed to de- velop an ordinal ranking of SI options, thereby using several SI criteria.

In the operational SI context, the algorithm can, e.g., work as follows. All SI options are ranked by a number of criteria. In case of type I risks, criteria can, e.g., be the abso- lute cost of safety (investment amount), the expected hypothetical benefit of safety (ex- pected avoided accident cost), the cumulative probability of the accident scenarios avoided, the payback period of the SI and the internal rate of investment. In case of type II risks, criteria can, e.g., be the cost of the SI, the hypothetical benefit based on a worst- case scenario, the variability related to the accident scenarios avoided, the information

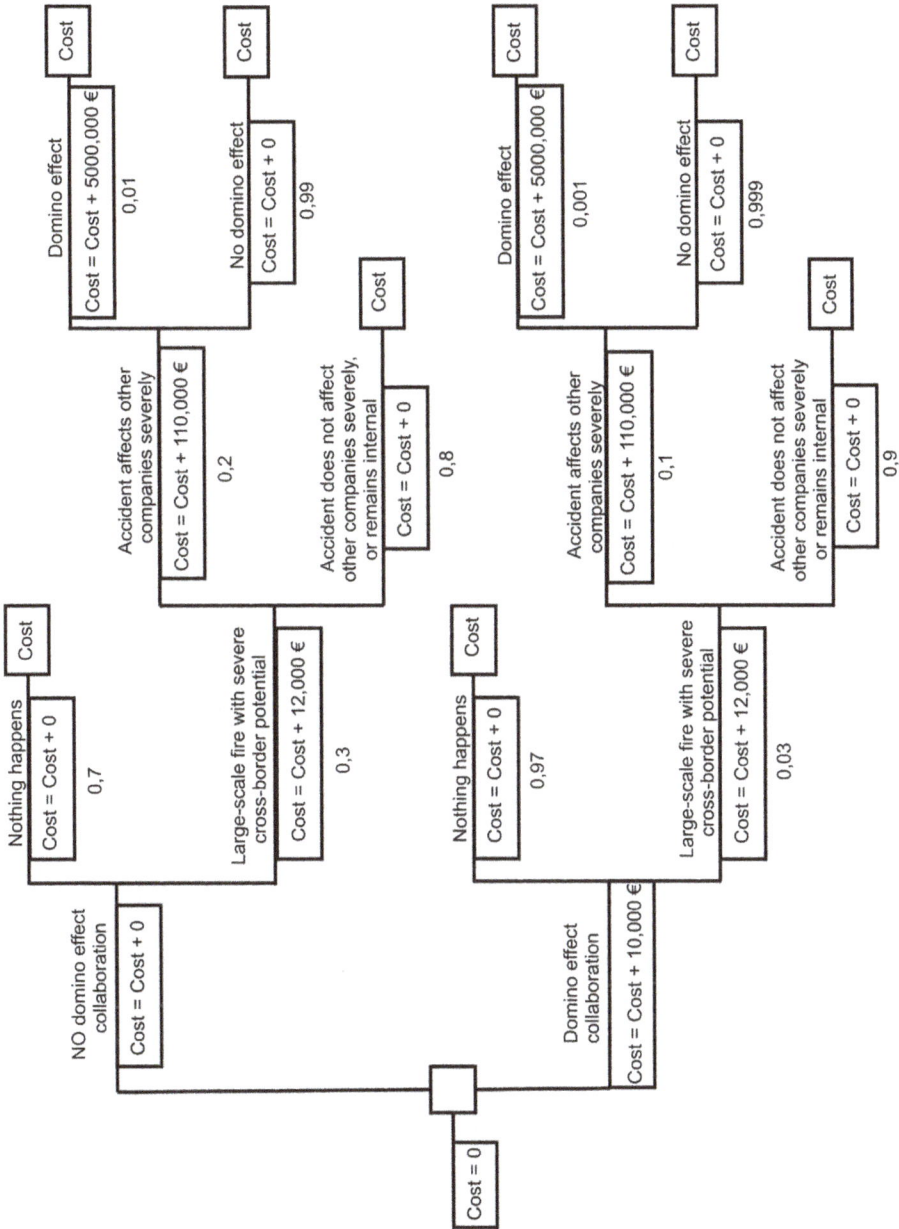

Fig. 9.10: Illustrative event pathway for domino effects prevention.

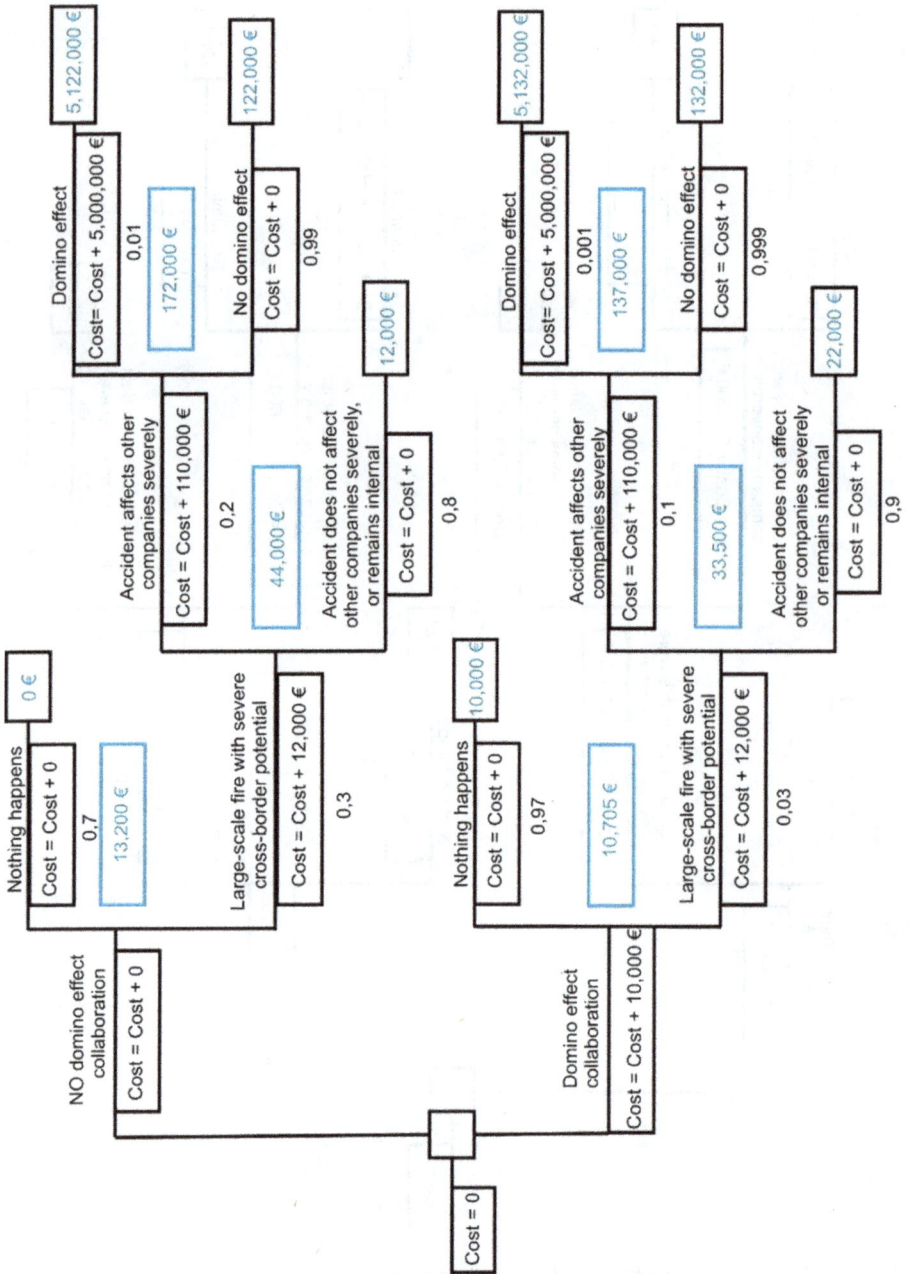

Fig. 9.11: Decision analysis tree rolled back to reveal the total cost of each strategy in the domino effects prevention illustrative example.

availability related to the SI, the equity principle and the fairness principle. Let us, e.g., explain it for type I risks and their SI. If there are n SI options to be compared, then the first-place option (e.g., according to the absolute cost of safety) receives $(n-1)$ points, the second-place option receives $(n-2)$ points and so forth, until the last-place option receives zero points. The same rule is used for assigning points according to the expected hypothetical benefit of safety, the cumulative probability, the PBP and the IRR. All the points obtained for the five criteria are summed for all installations and the option with the most points is ranked first, second, etc.

Let us explain this concept, e.g., in case of four SI options: SI1, SI2, SI3 and SI4. Suppose that the rank-order positions are as follows:

Absolute cost of safety (investment amount): $SI3 \succ SI2 = SI1 \succ SI4$
Expected hypothetical benefit: $SI1 \succ SI3 = SI2 = SI4$
Cumulative probability of accident scenarios avoided: $SI4 = SI1 \succ SI2 = SI3$
Payback period: $SI1 \succ SI3 \succ SI2 = SI4$
Internal rate of investment: $SI2 \succ SI4 \succ SI1 \succ SI3$

When ties occur, e.g., in case of the absolute cost of safety, SI2 and SI1 are tied, and points allocated to these positions are derived from the average; i.e., SI2 and SI1 will each receive $((n-2)+(n-3))/2$. In case of the expected hypothetical benefit, I3, I2 and I4 will each receive $((n-2)+(n-3)+(n-4))/3$.

The resulting point distribution is summarized in Tab. 9.11.

Tab. 9.11: Ranking investment options for type I risks using the Borda algorithm for a four-option illustrative example.

Criteria	Safety investment options			
	SI1	SI2	SI3	SI4
1. Absolute cost of safety	1.5	1.5	3	0
2. Expected hypothetical benefit	3	1	1	1
3. Cumulative probability of accident scenarios avoided	2.5	0.5	0.5	2.5
4. Payback period	3	0.5	2	0.5
5. Internal rate of return	1	3	0	2
Total Borda index for four safety investment options	**11**	**6.5**	**6.5**	**6**

From Tab. 9.11, it can be concluded that SI option 1 (SI1) has the highest Borda count and, therefore, ranks first and is the best SI according to the criteria used. The overall rank order of all four SI options employing the five criteria is as follows: $SI1 \succ SI2 = SI3 \succ SI4$.

The sole concern of the developed approach is the investigation of an SI option's position relative to other SI options if one looks simultaneously at the five criteria for, in the case of the illustrative example, the type I risks. This ranking information may lead to optimizing the allocation of safety budget resources within an organization.

9.6.10 Advantages and disadvantages of analyses based on costs and benefits

It should be remembered that the lack of accuracy associated with cost-effectiveness analysis and CBAs can give rise to significantly different outcomes in assessments of the same issues by different people. In addition, it is often much easier to assess all kinds of costs than to identify and evaluate the benefits.

CBA should only be used for type I risks, where a sufficient amount of information and data are available to be able to draw sufficiently precise and sufficiently reliable conclusions. If it is used for type II risks, it creates an image of accuracy and precision that it does not have. In that case, many of the valuations used for the costs and benefits reflect the perceptions of the person carrying out the analysis rather than their real values. More research is needed to develop reliable CBAs or cost-effectiveness analyses for type II risks.

Moreover, it is often difficult to incorporate realistic calculations of the NPV of future costs and benefits into the analyses. NPV calculations are widely used in business and economics to provide a means of comparing cash flows at different times on a meaningful "like by like" basis. Discounted values reflect the reality that a sum of money is worth more today than the same sum of money at some time in the future. As already expounded in a previous section, prevention costs incurred today should be compared with hypothetical benefits obtained at some time in the future, but equated to today's values. To achieve a benefit equal to €B in N years' time, one must have a benefit with a current NPV of €B/(1 + interest rate)^N.

Furthermore, Frick [19] claims that the cost–benefit approach underestimates the benefits and overestimates the costs of health and safety improvement programs.

Techniques based on costs and benefits that are used to make prevention decisions have one major undeniable strength: if used correctly, they allow us to allocate limited financial resources efficiently for occupational (type I) risks, and, if used with much caution, they provide us with some background knowledge for allocating financial resources aimed to deal with major accident (type II) risks.

9.7 Optimal allocation strategy for the safety budget

Safety measures show a diminishing marginal rate of return on investment: further increases in the number of a type of safety measure become ever less cost-effective, and the improvements in safety benefits per extra safety measure exhibit a decreasing marginal development [8]. In other words, the first safety measure of type "technology" provides the most safety benefit, the second safety measure provides less safety benefit than the first and so on. Hence, there should be safety measures chosen from different types (procedures, people and technology) to be most efficient. Of course, there may be differences in the safety benefit curves for the different types of safety measures.

Figure 9.12 shows the increased safety benefits from choosing a variety of safety measures.

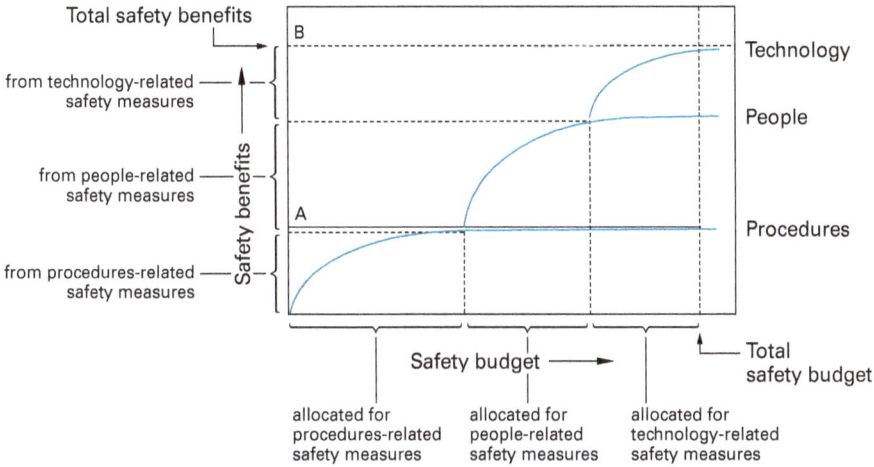

Fig. 9.12: Allocation strategy for the safety budget.

Figure 9.10 shows that, if the total budget available is spread over a range of proce- dure-related, people-related and technology-related safety measures, the overall safety benefit can be raised from point A (only investment in procedure-related safety measures) to point B (investment in P-, P-, T-related safety measures).

> Spreading the safety budget over different types of safety measures is always more efficient and effec- tive than only focusing on one type of safety measure. !

9.8 Loss aversion and safety investments: safety as economic value

Because of the psychological principle of "loss aversion" [20] and the fact that people hate to lose, especially precautionary investments to deal with highly unlikely events, are not at all evident. Risk managers, being human beings like all other people, also may let their decision judgment be influenced by this psychological principle.

To have a clear idea of "loss aversion," the following example can be given. Sup- pose you are offered two options: (A) you receive €5,000 from me (with certainty) or (B) we toss a coin and you receive €10,000 from me if it is heads, but if it is tails you receive nothing.

What will you chose? By far most of the people will choose option (A). They go for the certainty and prefer €5,000 for certain than to gamble and to have nothing.

Let us now consider two different options: (C) You have to pay me €5,000 (with certainty); or (D) We toss a coin and you need to pay me €10,000 if the coin turns up heads, but in case of tails you do not need to pay me anything.

What option will you prefer this time? By far most people in this case will prefer option (D). Hence, they go for taking the gamble and risking paying €10,000 with a level of uncertainty (there is a 50% probability that they will not have to pay anything) instead of paying €5,000 for certain.

From this example, it is clear that people hate to lose and that they love certain gains. People are more inclined to take risks to avoid certain losses than they are inclined to take risks to gain uncertain gains.

Translating this psychological principle into safety terminology, it is clear that company management would be more inclined to invest in production ("certain gains") than to invest in prevention ("uncertain gains"). Also, management is more inclined to risk highly improbable accidents ("uncertain losses") than to make large investments ("certain losses") in dealing with such accidents.

Therefore, management should be well aware of this basic psychological principle, and when making prevention investment decisions they should take this into account. The fact that we, as human beings, are prejudiced, and that we have some predetermined preferences in our minds, should thus really be considered in the decision-making process of risk managers.

Furthermore, safety and accident risk are not adequately incorporated into the economic planning and decision processes. What are the business incentives for investing in safety? There is a need for demonstrating that safety measures have a value in an economic sense. To what extent is it true that businesses would not invest in higher safety if such values cannot be demonstrated?

An overinvestment in safety measures is very likely if we ignore the fact that there is access to an insurance market, while an underinvestment in safety measures is very likely if we purchase insurance without paying attention to that. The probability and consequences can be reduced by safety measures.

Abrhamsen and Asche [21] state that the final decision with respect to how much resource should be spent on safety measures and insurance may be very different depending on what type of risks are considered. It makes a difference if they are risks we voluntarily assume for ourselves, or if they are risks imposed by others. Clearly, there is more reason for society to enforce standards in the latter case. However, the decision criterion is itself independent of the type of risk: an expected utility maximizer should combine insurance and investment in safety measures and directly take the costs of an accident, such that the marginal utility of the different actions are the same.

> **!** The fact that decision-makers have an in-built psychological preference to gamble with losses should be consciously considered by decision-makers when making precaution investment decisions.

9.9 Comparing risks with each other: the "micro-morts" concept

A micromort is a unit of risk that represents a one-in-a-million chance of death. It is used in risk analysis to quantify and compare the likelihood of fatal outcomes from various activities. Micromorts provide a standardized way to measure and compare the risk of death from different activities. Since humans often struggle to intuitively assess risk, micromorts help by assigning numerical values to different actions, making it easier to compare them.

Micromorts are for instance useful in activities where risk assessment is critical, such as healthcare, energy, manufacturing, transportation and public health. As an example, policymakers might use micromorts to assess the safety of a new medical procedure compared to existing treatments, while individuals can use them to understand the relative dangers of their daily activities. Governments and businesses might use them to communicate about risks, or to evaluate whether safety measures are cost-effective.

A Concrete Example: Deciding Between Two Travel Options

Imagine you need to travel 800 km, and you're deciding whether to drive or fly. You could use micromorts to compare the risk:

– *Driving has a risk of 1 micromort per 160 km, meaning a 800 km drive would be 5 micromorts.*
– *Flying has a risk of 0.1 micromort per flight, so a one-way flight would be 0.1 micromort.*

This comparison suggests that, from a purely statistical perspective, flying is about 50 times safer than driving over the same distance. While other factors like cost and convenience matter, micromorts provide a valuable, data-driven way to evaluate risk.

Another Example: Safety in the Energy Sector

Consider a government deciding between different energy sources. Assume the risk of fatal accidents per terawatt-hour (TWh) of energy produced is:

– *Nuclear power: 0.1 micromorts per TWh*
– *Wind power: 0.15 micromorts per TWh*
– *Coal power: 100 micromorts per TWh*

This suggests that coal is far riskier in terms of "per TWh of energy produced" than nuclear or wind energy in terms of fatalities. If a government applies safety economics, it may justify stricter regulations on coal plants or increased investment in nuclear safety to reduce risks.

Examples of Micromort Risks

Activity	Estimated Micromorts (per event or unit)
Skydiving (one jump)	8
Scuba diving (one dive)	5
Marathon running	7
Smoking 1 cigarette	0.1
Driving 100 miles (U.S.)	1
General anesthesia (for a routine surgery)	10 per operation
Working in offshore oil drilling (per year)	300
Commercial airline flight (per flight)	0.1
Construction work (per year)	50
Living near a nuclear power plant (per year)	0.01
Working in coal mining (per year)	1,000

By using micromorts, individuals and organizations can make more informed decisions about risk, helping to balance safety with the desire for adventure or efficiency. Furthermore, companies and governments can communicate about various risks and may evaluate whether investing in additional safety measures is justified based on the lives they might save versus their cost.

9.10 Conclusions

Economic issues of safety constitute much more than calculating the costs of accidents, or determining the costs of prevention. Hypothetical benefits, the benefits gained from accidents that have never occurred, should be considered, and the two types of risk should be taken into account when dealing with prevention investment choices. Decisions concerning safety investments make for a complex decision problem, where opportunity costs, perception and human psychology, budget allocation strategies, the choice of economic methodologies, etc., all play important roles.

References

[1]   Perrow, C. (2006). The Limits of Safety: The Enhancement of a Theory of Accidents. In: Key Readings in Crisis Management. Systems and Structure for Prevention and Recovery. Smith, Elliott, editors. Abingdon, UK: Routledge.

[2]   Reniers, G., Audenaert, A. (2009). Chemical plant innovative safety investments decision-support methodology. J. Saf. Res. 40: 411–419.

[3]   Hanley, N., Spash, C.L. (1993). Cost-Benefit Analysis and the Environment. Cheltenham, UK: Edward Elgar Publishing.

[4]   Mossink, J., Licher, F. (1997) Costs and Benefits of Occupational Safety and Health, Proceedings of the European Conference on Costs and Benefits of Occupational Safety and Health 1997, The Hague, The Netherlands.

[5]   Rikhardson, P.M., Impgaard, M. (2004). Corporate cost of occupational accidents: An activity-based analysis. Accid. Anal. Prevent.. 36: 173–182.

[6]   Leigh, J.P., Waehrer, G., Miller, T.R., Keenan, C. (2004). Costs of occupational injury and illnesses across industries. Scand. J. Work Environ. Health. 30: 199–205.

[7]   Chartered Institute of Personnel and Development. (2005). Recruitment, Retention and Turnover. London, UK: CIPD.

[8]   Fuller, C.W., Vassie, L.H. (2004). Health and Safety Management. Principles and Best Practice. Essex, UK: Prentice Hall.

[9]   Reniers, G., Van Erp, N. (2016). Operational Safety Economics. A Practical Approach Focused on the Chemical and Process Industries. London: Wiley-Blackwell.

[10] Campbell, H.F., Brown, R.P.C. (2003). Benefit-Cost Analysis. Financial and Economic Appraisal Using Spreadsheets. New York: Cambridge University Press.

[11] Goose, M.H. (2006) Gross Disproportion, Step by Step – A Possible Approach to Evaluating Additional Measures at COMAH Sites. In: Institution of Chemical Engineers Symposium Series, vol. 151, 952. Institution of Chemical Engineers; 1999, 2006.

[12] Rushton, A. (2006, April 4). CBA, ALARP and Industrial Safety in the United Kingdom. London, UK: Crown copyright.

[13] International Association of Oil and gas Producers. (2000). Fire System Integrity Assurance, Report No. 6.85/304. London, UK: OGP.

[14] Adams, J.G.U. (1995). Risk. London: UCL Press.

[15] Slovic, P. (2000). The Perception of Risk. London, UK: VA: Earthscan.

[16] Fife, I., Machin, E.A. (1976). Redgrave's Health and Safety in Factories. London: Butterworth.

[17] Melnick, E.L., Everitt, B.S. (2008). Encyclopedia of Quantitative Risk Analysis and Assessment. Chichester, UK: John Wiley & Sons, Inc.

[18] Wilson, R., Crouch, E.A.C. (2001). Risk-benefit Analysis. 2nd edn. Newton, MA: Harvard University Press.

[19] Frick, K. (2000). Uses and Abuses. In: The Role of CBA in Decision-making. Bilbao, Spain: European Agency for Safety and Health at Work, 12–13.

[20] Tversky, A., Kahneman, D. (2004). Loss Aversion in Riskless Choice: A Reference-dependent Model. In: Preference, Belief, and Similarity: Selected Writings. Shafir, Editor. Cambridge, MA: MIT Press.

[21] Abrahamsen, E.B., Asche, F. (2010). The insurance market's influence on investments in safety measures. Safety Sci. 48: 1279–1285.

10 Risk governance

Most of us are taught to think about the long-term consequences of our actions, but it is a life lesson that is easily forgotten – both at an individual and an organizational level. This is why, each year, the World Economic Forum poses the question, "What risks should the world's leaders be addressing over the next 10 years?"

10.1 Introduction

We all know the phrase "It's the economy, stupid!," which James Carville coined as a campaign strategist for Bill Clinton's successful 1992 presidential campaign. We can easily adapt this phrase for any organization going bankrupt, having lots of losses or having major financial problems, into "It's the risk governance, stupid!" *Risk governance can be defined* [1] *as the totality of actors, rules, conventions, processes and mechanisms concerned with how relevant risk information is collected, analyzed and communicated, and how management decisions are taken.*

Renn [1] provides several reasons why risk governance is crucial in today's organizations. First, risk plays a major role in our society: Ulrich Beck, the famous social scientist, called his book on reflexive modernity, *The Risk Society*. In his book, Beck argues that risk is an essential part of modern society, and hence, also the governance of risks. People, including customers, politicians and regulators, thus expect organizations to adequately govern their risks. Second, risk has a direct impact upon our life. People, especially employees of organizations, die, suffer, become ill or experience minor and major losses because they have ignored or misjudged risks, miscalculated the uncertainties or had too much confidence in their ability to master dangerous situations. Governing risks in a solid way, by everyone being part of an organization, would be a solution to this problem. Third, risk is a truly interdisciplinary phenomenon: risk, or domains or factors of it, are studied in natural, medical, engineering, social, cultural, psychological, legal and other disciplines. None of these science disciplines are able to grasp the holistic substance of risk; only if they combine forces, a truly adequate approach to understanding and managing risks is possible. Risk governance can be employed to ensure this. Fourth, risk is a concept that links the professional with the private person. If someone gets more experienced in dealing with risks in his professional life, in whatever way, they will also make better decisions in their personal life. An understanding of risk and risk governance directly contributes to this fact.

Risk governance starts with good corporate governance and integrated board management. Hilb [2] indicates that integrated board management has four precondi-

https://doi.org/10.1515/9783111493633-010

tions for being successful in developing, implementing and controlling fortunate organizations:
1. Diversity: strategically targeted composition of the board team
2. Trust: constructive and open-minded board culture
3. Network: efficient board structure
4. Vision: stakeholder-oriented board's measures of success

Although these preconditions of success have been proven in a variety of studies, they seem to be very hard to achieve by organizations. In the light of the recent economic crises, Peter Senge asks: "How can a team of committed board members with individual IQs above 120 have a collective IQ of 60?" [2].

As Fuller and Vassie [3] explain, one view of business is that the directors of a company are merely agents acting on behalf of the shareholders and, as such, their sole responsibility is to maximize the return on the investments of these owners: this is referred to as the "principal agent theory." This shareholder model is referred to as the "Anglo Saxon model." Corporate social responsibility, however, is a concept derived from a wider perspective that businesses have responsibilities to a range of people, in addition to shareholders: this is referred to as the "stakeholder theory," and the stakeholder model is referred to as the "Rijnland model."

Whether applying one model or the other for corporate governance, adequate risk governance, as part of good corporate governance, is absolutely necessary for any organization to be healthy in the long term.

As stated by the International Risk Governance Council (IRGC) in [4], "risk governance" involves the "translation" of the substance and core principles of governance to the context of risk and risk-related decision-making. Risk governance includes the totality of actors, rules, conventions, processes and mechanisms concerned with how relevant risk information is collected, analyzed and communicated and management decisions are taken.

Encompassing the combined risk-relevant decisions and actions of both governmental and private actors, risk governance is of particular importance in, but not restricted to, situations where there is no single authority to take a binding risk management decision, but where, instead, the nature of the risk requires the collaboration of, and coordination between, a range of different stakeholders. Risk governance, however, not only includes a multifaceted, multi-actor risk process, but also calls for the consideration of contextual factors, such as institutional arrangements (e.g., the regulatory and legal framework that determines the relationship, roles and responsibilities of the actors and coordination mechanisms such as markets, incentives or self-imposed norms) and political culture, including different perceptions of risk.

Risk governance therefore requires dealing with risks and uncertainties in a very holistic and general way, with the necessary procedures in place at high, strategic levels, in addition to the obvious operational and tactical levels, at different levels of the organization and even interorganizational, with collaboration as much as needed but as low as required (to avoid complexity), with a diversity of people, experts and disciplines, and above all, with an open mind. We will discuss a system and its requirements (Section 10.2), a framework (Section 10.3) and a model (Section 10.4) that together lead to effective risk governance, on an operational, tactical and strategic level.

A term that is often used in relation to risk governance is risk appraisal. Risk appraisal is an important part of risk governance and concerns the process of gathering all knowledge elements necessary for risk characterization, evaluation and management. This includes not only the results of (scientific) risk assessment, but also information about, for instance, risk perceptions and the economic and social implications of the consequences of risk. In this regard, governance includes matters of institutional design, technical methodology, administrative consultation, legislative procedures and political accountability on the part of public bodies, and social or corporate responsibility on the part of public enterprises. Two major challenges of risk governance exist: on the one hand, generating and collecting knowledge about the risk (risk appraisal), and, on the other hand, making decisions about how to handle and treat the risk (risk reduction/treatment). Risk communication forms the glue that brings both challenges together. All these areas of science belong to the encompassing field of organizational risk management and (on a higher level) risk governance.

10.2 Risk management system

Management systems for the safe operation of organizations require a system of structures, responsibilities, procedures and the availability of appropriate resources and technological know-how. Risks can be managed at different levels: at factory, at plant, at multiplant and at cluster level. Figure 10.1 illustrates the difference between these terms, and differentiates between the different levels. Plant B, e.g., consists of two installations within a single factory. Furthermore, plant B belongs to the multiplant area consisting of plants A and B, and is part of the larger cluster (or, in other words, industrial area) composed of plants A, B, C, D, E and F.

Factory-level risk management includes topics such as working procedures, work packages, installation-specific training, personal protective equipment and quality inspection. Plant-level risk management includes defining acceptable risk levels and documenting plant-specific guidelines for implementing and achieving these levels for every facility situated on the premises of the plant. It is current industrial practice to draft an organizational *safety management system (SMS)* to meet these goals (see further in this section). Multiplant or cluster-level risk management topics include defining risk-level standards by different plants, in collaboration; defining risk management cooperation levels; defining acceptable risks involving more than one plant; joint workforce planning and joint emergency planning in the event of cross-plant accidents. The latter multiplant-related topics can be documented in a type of cross-plant SMS dealing with these issues.

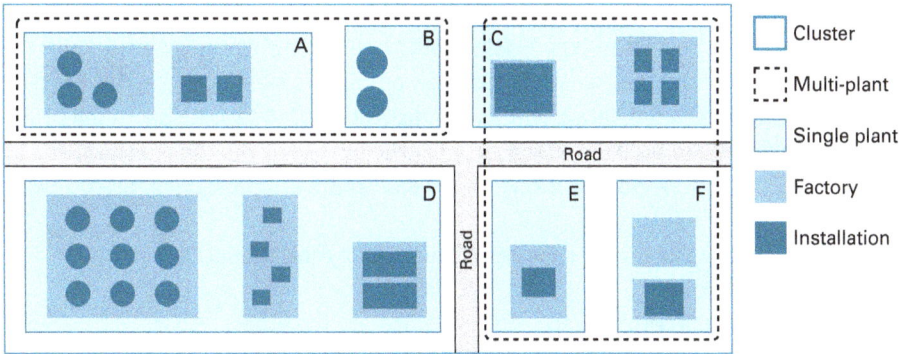

Fig. 10.1: Comparisons of installation, factory, plant, multiplant and cluster (inspired by Reniers [5]).

In the nuclear industry or in the chemical industry, e.g., at the factory level and at plant level, safety documents, guidelines and instructions, technical as well as meta-technical, are usually very well elaborated. An adapted version of the *plan-do-check-act* process is often used for continuously improving risk management efforts and safety (see also Chapter 3). To optimize risk management, the circular continuous improvement process can be used at all levels within the industry: from top (the cluster level) to bottom (the installation level). To be able to perform the process at all levels, a risk management cycle structure can be established at each level and provided with communication and cooperation links between the different levels. These links are necessary for further optimization of the different levels of looping risk management and for the prevention of double elaboration of certain multi-leveled risk management topics. Such a framework, characterized by loop-level risk management, can be arranged as illustrated in Fig. 10.2.

Many organizations already follow the plan-do-check-act loop because of their acquired know-how of internationally accepted business standards, e.g., ISO 9001, ISO 14001 and/or OHSAS 18001 (see also Chapter 1), addressing quality, environmental and safety management, respectively, and continuously improving performance concerning those related risks. Hence, some degree of basic standardization for operational risk governance already exists in many organizations, and thoroughly documented and well-implemented risk management systems are available. One of the areas that is very important in the context of this book is safety management.

An SMS, as part of the risk management system, aims to ensure that the various safety risks posed by operating the facility are always below predefined and generally accepted company safety risk levels. Effective management procedures adopt a systematic and proactive approach to the evaluation and management of the plant, its products and its human resources.

To enhance safety for type II risks, the SMS considers safety features throughout scenario selection and process selection, inherent safety and process design, industrial activity realization, commissioning, beneficial production, and decommissioning. To enhance safety related to type I risks, both personal and group safety equipment are provided, training programs are installed and task capabilities are checked. Arrangements are made to guarantee that the means provided for safe operation of the industrial activity are properly designed, constructed, tested, operated, inspected and maintained and that persons working on the site (contractors included) are properly instructed.

Four indispensable features for establishing an organizational SMS are:
- the parties involved
- the policy objectives
- the list of actions to be taken
- implementation of the system

The essence of accident prevention practices consists of safety data, hazard reviews, operating procedures and training. These elements need to be integrated into a safety management document that is implemented in the organization on an ongoing basis. To enhance implementation efficiency, this can be divided into 11 subjects [6]:

1. **Safe work practices** – A system should be installed to guarantee that safe work practices are carried out in an organization through procedural and administrative control of work activities, critical operating steps and critical parameters, through pre-startup safety reviews for new and modified plant equipment and facilities, and through management of change procedures for plant equipment and processes.

2. **Safety training** – The necessity of periodically organizing training sessions emerges from the continuously changing environment of plants, installations and installation equipment. Employees and contractors at all levels should be equipped with the knowledge, skills and attitudes relating to the operation or maintenance of organizational tasks and processes so as to work in a safe and reliable manner. Safety training sessions should also lead to a more efficient handling of any incident or accident.

3. **Group meetings** – An organization should establish a safety group meeting for the purpose of improving, promoting and reviewing all matters related to the safety and health of employees. This way, communication and cooperation between management, employees and contractors is promoted, ensuring that safety issues are addressed and appropriate actions are taken to achieve and maintain a safe working environment.

4. **Pursuing in-house safety rules and complying with regulations** – A set of basic safety rules and regulations should be formulated in the organization to regulate safety and health behaviors. The rules and regulations should be documented and effectively communicated to all employees and contractors through promotion, training or other means and should be made readily available to all employees and contractors. They should be effectively implemented and enforced

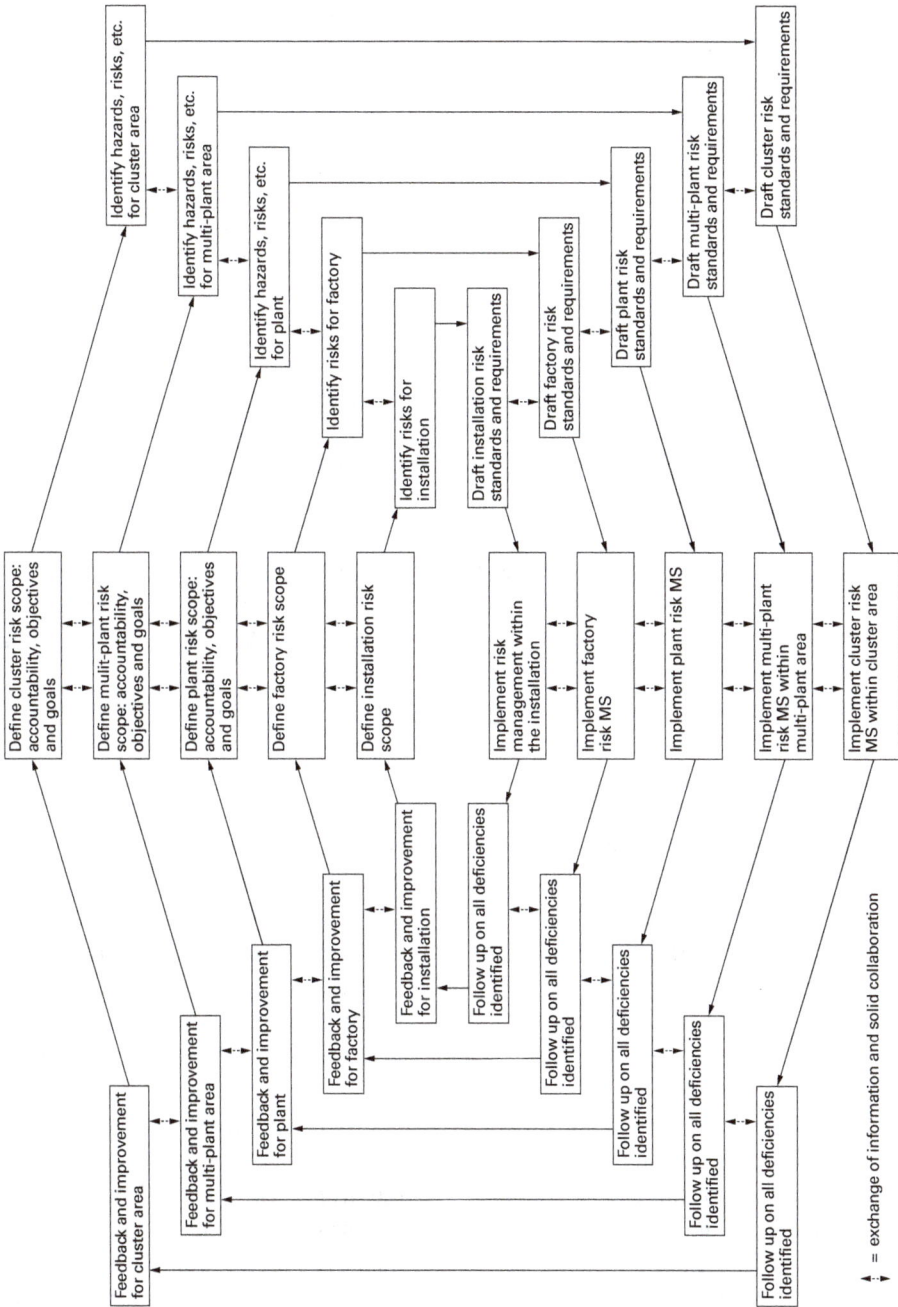

Fig. 10.2: Loop risk management structure at the installation, factory, plant, multiplant and cluster levels (inspired from Reniers [5]).

within the organization. The company rules should be in conformance with the legislative requirements, and rules that are non-statutory should conform to international standards and best practices.

5. **Safety promotion** – Promotional programs should be developed and conducted to demonstrate the organization's management commitment and leadership in promoting good safety and health behaviors and practices.

6. **Contractor and employee evaluation, selection and control** – The organization should establish and document a system for assessment and evaluation of contractors to guarantee that only competent and qualified contractors are selected and permitted to carry out contracted works. This way, personnel under external management but working within the organization are treated, evaluated and rewarded in the same manner (concerning safety issues) as internally managed personnel.

7. **Safety inspection, monitoring and auditing** – The organization needs to develop and implement a written program for formal and planned safety inspections to be carried out. The program should include safety inspections, plant and equipment inspections, any other inspections (including surprise inspections), and safety auditing. This way, a system is established to verify compliance with the relevant regulatory requirements, in-house safety rules and regulations, and safe work practices.

8. **Maintenance regimes** – A maintenance program needs to be established to ensure the mechanical integrity of critical plant equipment. In fact, all machinery and equipment used in the organization needs to be maintained at all times so as to prevent any failure of this equipment and to avoid unsafe situations.

9. **Hazard analysis and incident investigation and analysis** – All hazards in the organization need to be methodically identified, evaluated and controlled. The process of hazard analysis should be thoroughly documented. Written procedures should also be established to ensure that all incidents and accidents (including those by contractors) are reported and recorded properly. Furthermore, procedures for incident and accident investigation and analysis to identify root causes and to implement effective corrective measures or systems to prevent recurrence should be installed.

10. **Control of movement and use of dangerous goods** – A system should be established to identify and manage all dangerous goods through the provision of material safety data sheets and procedures for the proper use, storage, handling and movement of hazardous chemicals. To further ensure that all up-to-date information on the storage, use, handling and movement of dangerous goods in the organization reaches the prevention and risk management department, a continuously adjusted database with information should be established.

11. **Documentation control and records** – An organization should establish a central documentation control and record system to integrate all documentation requirements and to ensure that they are complied with.

These recommendations can be generalized to multiplant or cluster-related recommendations (please see [5]).

When drafting the standards and requirements, the Health and Safety Executive [7] indicates that it is essential to ensure standards are incorporated into business input activities through the following (see also [3]):

- Employees – such as defining physical, intellectual and mental abilities through job specifications based on risk assessment
- Design and selection of premises – such as consideration of the proposed and foreseeable uses, construction and contract specification
- Design and selection of plant – such as installation, operation, maintenance and decommissioning
- Use of hazardous substances – such as the incorporation of the principle of inherent safety and the selection of competent suppliers
- Use of contractors – such as selection procedures
- Acquisitions and divestitures – such as the identification of short-term and long-term safety risks associated with the organization's activities
- Information – such as maintaining an up-to-date system for relevant health and safety legislation, standards, and codes of practice

The Health and Safety Executive [7] (see also [3]) further lists some factors that should be included in an assessment of workplace safety standards:

- SMS – such as policy, organization, implementation, monitoring, audit and review
- Use of hazardous goods – such as the receipt, storage, use and transportation of chemicals
- Use of contractors – such as the provision of working documents and performance reviews.
- Emergency planning – such as the identification of emergency scenarios, liaison with the emergency services and the implementation of emergency planning exercises
- Disaster and contingency planning – such as the identification of disaster scenarios, preparation of contingency plans and the implementation of disaster planning exercises

Risk management systems are a must for organizations to handle risks at an operational level. Risks are diversified and omnipresent. For example, the chance that I will never finish typing this text exists, although its likelihood is rather very low. Several risks exist in this regard: the chair I am sitting in might break down, and I may fall, thereby crashing the laptop, or I may strike my head. My office may be set on fire. The ceiling may collapse upon me or my laptop. A plane might crash down on the building I am sitting in. A bomb may explode very nearby, etc. Risk management systems deal with assessing all these risks (estimating their consequences and likelihoods) and many others and treating them, i.e., trying to prevent them, and, in the case of an unfortunate event happening, despite all measures taken, trying to mitigate the consequences or transfer them.

10.3 A framework for risk and uncertainty governance

On the surface, risk management in an organization seems to be all about avoiding any type of unwanted event. All undesired happenings, either with large consequences or with a minor outcome, or with high or low likelihood, are treated by risk management within a company on the same psychological level. But as we have argued before in this book, this psychological level, and thus the management decisions taken, should be different for different types of risks.

It is obvious that by taking the positive side of risks into account, a decision-maker can be risk-taking in a case where a lot of historical data is available, as he has some knowledge from past events in this area, and the possible negative outcomes (which are never extremely severe in this case) can be predicted quite accurately by using scientific models, statistical methods, etc., because sufficient information is available. Hence, making profits in this type I area follows from taking positive risks and keeping the negative sides of these risks well under control. Profits are tangible and follow from investment choices, production decisions and risk management strategies (among others).

However, a decision-maker must be very careful to be risk-taking (for making profits) in the area of type II risks and uncertainties, as the possible negative consequences in these areas are possibly very high. Actually, making "profits" in the type II area follows from averting risks: profits in these areas are intangible and hypothetical. Non-occurring accidents (and their accompanying costs) (see also Chapter 9) resulting from risk-averting behavior of a decision-maker should be regarded as a true and large hypothetical benefit in these areas. In addition to classic industrial sectors, with the possibility of major accidents such as the chemical industry or the nuclear industry, the banking sector can also be used as an example. We should, e.g., be cautious in taking risks in an economic growth period for making huge profits (which is possible in the type II area), as in the event that a worldwide economic crisis occurs (as was the case in 2008), financial disaster may strike. Hence, non-occurring huge financial losses in a bank (which might be realized in the case of a sudden global economic crisis) as the result of careful risk management (aimed at making profits in the type I area) in the economic prosperous period, should be regarded as true and huge hypothetical gains in the type II area.

In summary, decisions to take risks, or indeed to avert them, depend on the character of the risks (i.e., the type of the risks) and their accompanying uncertainties and objectives. Moreover, hypothetical (short-term and long-term) benefits should be taken into account when taking decisions concerning type II uncertainties and risks.

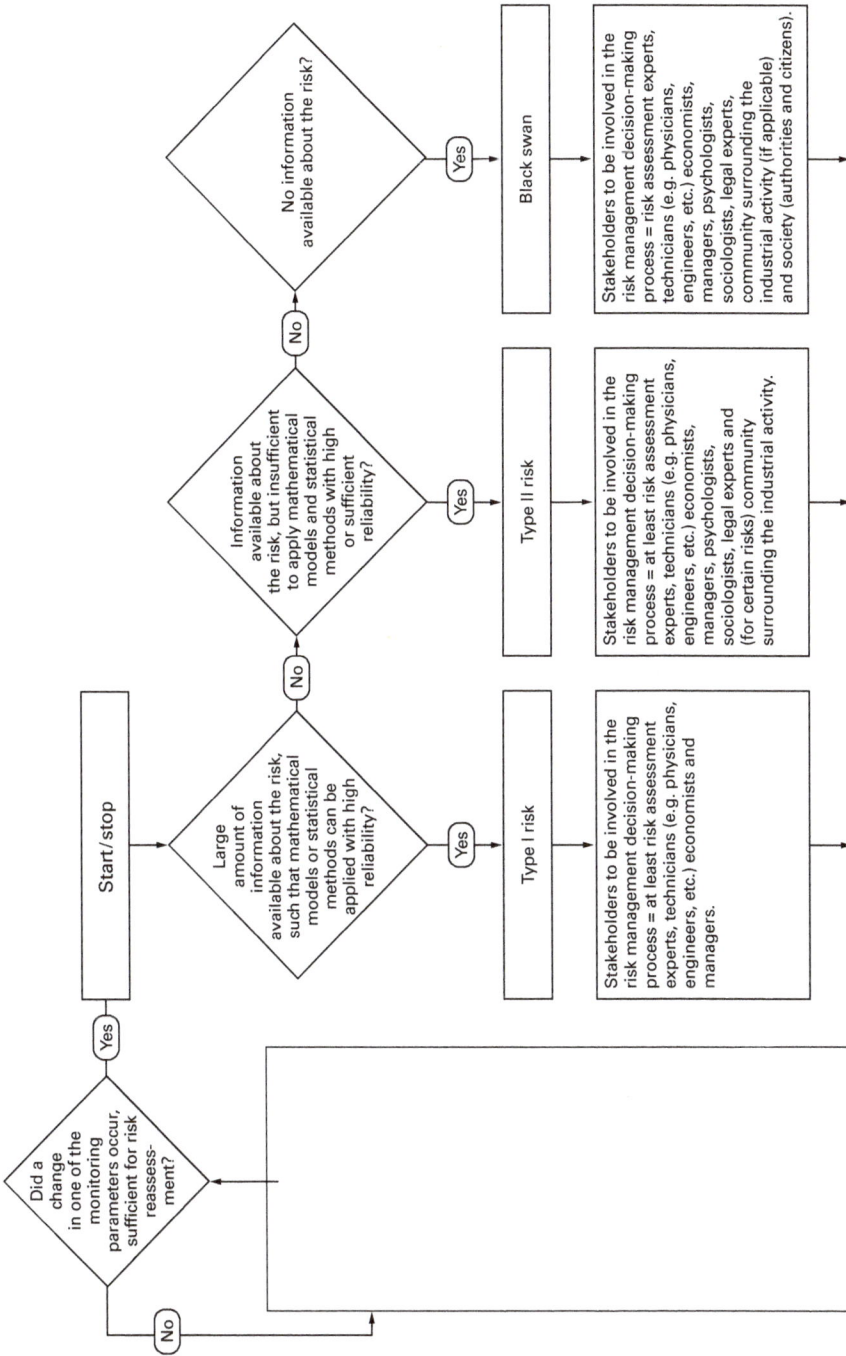

Fig. 10.3: Uncertainty/risk governance framework.

Determine risk tolerability criteria, risk acceptance criteria, ALARP region, risk targets, etc.

Carry out qualitative (e.g. checklist) or semi-quantitative (e.g. what-if analysis) risk assessments to determine risk treatment and mgt options for risk decision-making

Carry out risk assessments to determine risk treatment and mgt options for risk decision-making; risk assessment techniques may vary from very quantitative (e.g. QRA) to semi-quantitative (e.g. Hazop, SWIFT,...); risk assessment results should however always be interpreted with much caution!

Carry out qualitative (e.g. scenario building, future thinking,....) or semi-quantitative (e.g. what-if analysis) risk assessments to determine risk treatment and mgt options for risk decision-making; risk assessment results should always be interpreted with much caution

Risk treatment and management options are known; now consider economic issues related to safety (cost–benefit analysis, cost-effectiveness analysis, etc.) to evaluate options and adapt options in this sense

Risk treatment and management options are known; now use the precautionary principle to evaluate the options and take hypothetical benefits into consideration; adapt the options afterwards in this sense

Make sure that the risk treatment and management options take into account:
– short-term as well as long-term-view
– the possible existence of feedback loops and feedforward loops (hence, the possible systemic character of risks)
– possibly existing non-linearities

Risk managers take decisions, mainly based on technical assessment results

Risk managers, together with top management, take decisions, partly based on technical assessment results based on as much information as available, and partly based on qualitative judgments by a variety of disciplines and stakeholders

Monitoring/
use at least the following monitoring parameteres:
– new or changed working conditions concerning the risk?
– events, incidents, accidents happened concerning the risk?
– did legislation or regulations change concerning the risk?
– have external factors, influencing the risk, changed?
– were new insights gained by science concerning the risk?

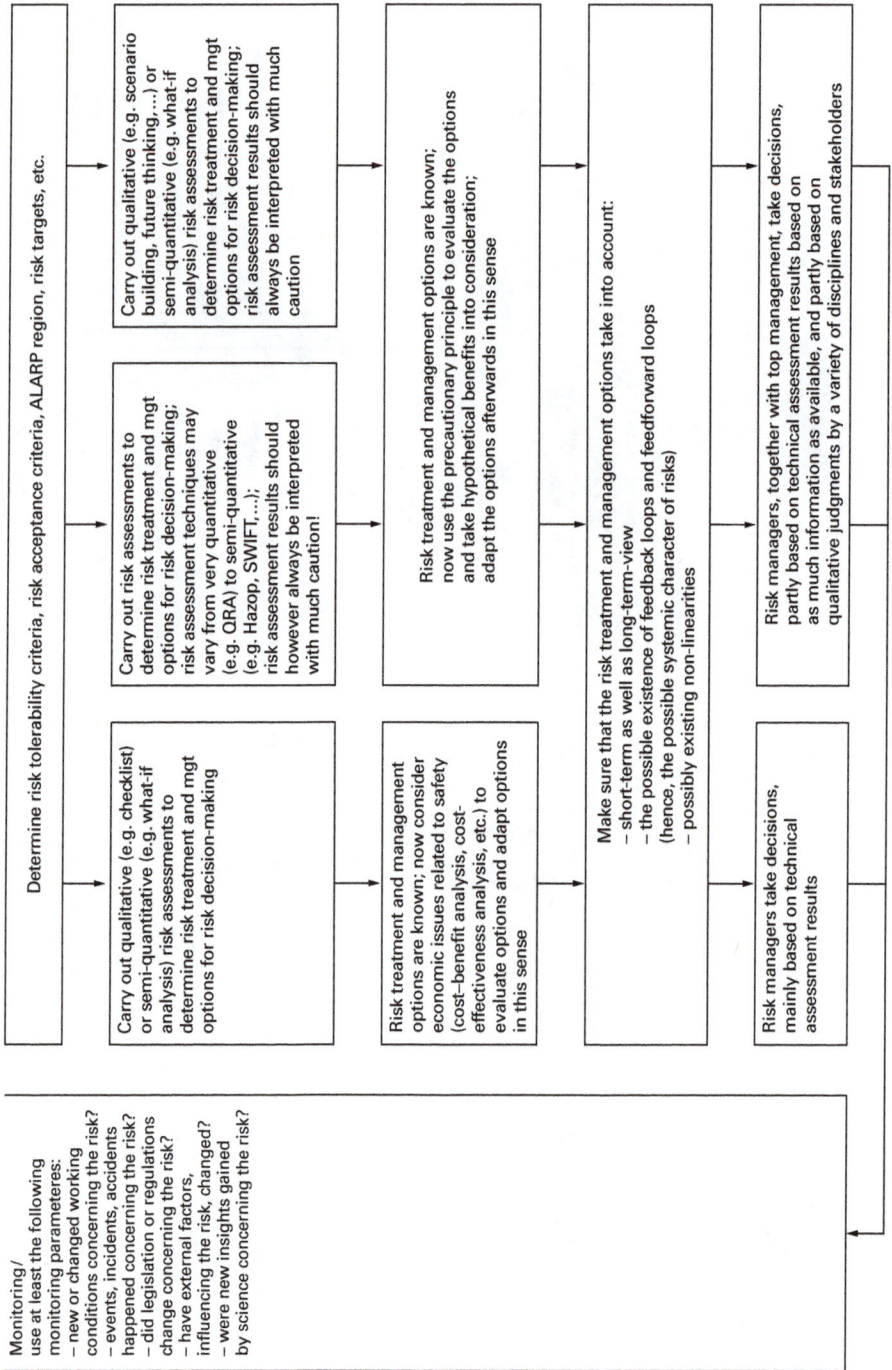

Fig. 10.3 (continued)

A framework for risk and uncertainty governance is suggested in Fig. 10.3 to meet all the aforementioned goals.

The framework from Fig. 10.3 first considers the different types of risk, depending on the amount of data and information that is available (see also [8]). For the different types of risks, different sets of stakeholders are involved. The more uncertainty there is about the risk, the more stakeholders need to be involved, as not only is the know-how to deal with risk assessment techniques required, but also open-mindedness, the willingness to collaborate viewpoints, know-hows and knowledge from other, non-purely technological disciplines, which can make the difference between good or poor risk decisions. In the next step of the framework, risk assessments are carried out. A risk assessment method that can, e.g., be used for all three types of risks, but especially for black swan risks, is scenario building. Scenario building is well known and much used by risk experts. Scenarios are drafted based on past events or by imagining future events using a variety of available techniques, such as classic risk analysis techniques and the Delphi method. Scenarios lead to a better understanding of possible futures, and what can be done to prevent some of the possible unwanted futures, and/or to enhance the emergence of possible wanted futures. It should be noted that the more information is available in case of type II risks, the less uncertainty there is, and the more quantitative character the used risk assessment technique can have when dealing with these type II risks. With no or extremely little data or information available, highly quantitative analyses should be avoided. At best, semiquantitative analyses might be employed for extremely high uncertainties, and the results should always be treated and interpreted with much caution.

When the decision options are known, some further principles, at least the precautionary principle, should be followed when taking decisions. The precautionary principle [9] states that it is not because a cause–consequence relationship between any two variables cannot be proven that such a relationship does not exist; in other words, unknown complicated relationships (characterizing any complex phenomenon) should not be disregarded, whether the consequences are desired or undesired.

In the last step, before carrying out the decision, a decision-maker should reflect on several important systemic principles. It is therefore recommended that every decision is approached by thinking in the long term (as well as the short term), thinking circularly (as well as being cause–consequence-minded) and thinking nonlinearly (as well as linearly) regarding potential results. Long-term-oriented thinking indicates that risk decisions should not only be taking the short and medium term into account, but also the (very) long-term, thus leading to more sustainable risk decisions. Circular thinking and nonlinear thinking are concepts used in systems thinking [10, 11]. As mentioned before, system thinkers see wholes that function rather than input–output transformers; they also see the parts of the whole and the relationships among them; they see patterns of change rather than static snapshots. Positive and negative feedback loops exist between events. These feedbacks are not necessarily linear, meaning that one should think in terms of "changes of A and B in the same or opposite direction," rather than "a predefined increase or decrease of B, caused by a specific increase or decrease of A." Such an approach guarantees more profound and well-considered decisions.

When decisions have been taken, the decision-maker should regularly monitor to check that no changes have occurred concerning the decision. If this was the case, the risk should be subjected to the risk governance framework again.

Summarizing, this risk governance framework aims to obtain objective and consistent results for the different existing types of risks and uncertainties. Assessments that are subject to small uncertainties should be treated differently from assessments subject to large uncertainties, in a way that under different assumptions and/or the use of other analysis, the decision options – and results concerning any uncertainty or risk – should not be different. Solutions and decisions should only depend – or partially depend – on probabilities and probability estimates under certain conditions and circumstances. As Aven [12] already suggests, risk assessments need to provide a much broader risk picture than is typically the case today. Separate "uncertainty analyses" should be carried out, extending the traditional probability-based analyses. The uncertainties accompanying the risks should be mentioned with every risk assessment result, and they should be used by the risk decision-makers. Only by this way, can more objective and higher-quality risk decisions be made. *Risk managers should be uncertainty experts!*

The framework is generic in its application, indicating that any uncertainty/risk can be tackled by it, and recognizing the negative as well as positive consequences of risks. In general terms, the negative side of risks should always be minimized, and the positive side should always be maximized. This framework recognizes that this is not always the best solution that leads to optimal decisions, and addresses the need to sometimes "balance" negative and positive consequences rather than opt for one or the other, to be able to ensure the long-term success of – and long-term profits within – any organization. By continuously monitoring and questioning its own decisions, the risk governance framework measures up to the science paradigm of always being critical and questioning any findings.

Company management should be aware that uncertainties and risks should not be rooted out, and that they should not be considered as some kind of evil that detracts from managers' abilities to manage with control in an organization. Uncertainty – and thus risk – should be recognized by company decision-makers as two-sided, creating obstacles for the organization in generating profits and ensuring consistent performance, as well as presenting opportunities for improvement and innovation. Moreover, different levels of an uncertainty require different stakeholders, different levels of quantitativeness of risk assessment techniques, different principles to be followed and so on. The risk governance framework takes this into account and offers decision-makers within organizations, the possibility to manage not only (negative) risks, but also uncertainties. Such an approach leads to better risk decision-making and long-term organizational success.

The perception of the situation, the circumstances, reality and so on are essential when assessing risks. Human perception is based on recognition and interpretation of patterns. All perceptions are therefore colored and influenced by experience and expertise, which – intuitively – anticipates everything we encounter, and makes an interpretation. This (scientifically called) "bias," taking the form of expectations, tendencies and premonitions, helped us to survive in the past. Indeed, making decisions based on very limited information was absolutely necessary (and a major advantage) to be able to survive. However, we do not live in a simple world anymore, dealing with simple tasks and simple processes, where only simple accidents can happen that lead to minor injuries or to single deaths at most. We are currently part of a very complex global society with complicated and complex industrial activities. The activities carried out in today's organizations may wipe out entire villages and even cities. Moreover, many people are connected via internet, social media and so on, and hence an incident and its potential consequences, or an accident and its real consequences, are swiftly shared by millions of people at the same time. This completely changed industrial environment requires an entirely different way of dealing with risks compared with any time in the last 50 years.

> It is indeed obvious: over the last 50 years, the time of simplicity and of simple single brain heuristics is definitely over. The post-atomic age requires us to find new ways of dealing with all the existing complexities and making the right decisions, despite these complexities. The answer to this, in the form of collaboration and perception improvement (often against intuition), is risk governance, and the risk governance framework helps to arrive at the right decisions.

10.4 The risk governance model

The most influential research into how people manage risk and uncertainty has been conducted by Kahneman and Tversky, two psychologists [13]. One of their most interesting findings is the asymmetry between the way people make decisions involving gains and decisions involving losses. This topic is also mentioned in Chapters 7 and 9, and explained in the light of what is discussed in each. The research results show that when the choice involves losses, people are risk-seekers, whereas when the choice involves gains, people are risk-averse. In other words, people tend to not gamble with certain gains, but they tend to gamble with uncertain losses. Other research results indicate inconsistency with the assumptions of rational behavior and people being loss-averse, rather than risk-averse. People are apparently perfectly willing to choose a gamble when they consider it appropriate, and they do not so much hate uncertainty as they hate losing. Losses are likely to provoke intense, irrational and abiding risk-aversion because people are much more sensitive to negative than to positive stimuli.

These results from human psychology, and human decision-making have important repercussions for risk decision-making in companies, especially concerning negative risks. Operational risk managers and middle or top management tend to gamble with uncertain losses, possibly leading to major accidents. They often fail to recognize the huge hypothetical benefits resulting from prevention.

Hence, because of the inconsistent understanding and meaning of the risk concept by people and the irrational risk decision-making, it is essential that "risks" should be regarded as "uncertainties" in companies. The less uncertainty there is concerning the possible consequences of a decision, the more it is possible to make adequate and good decisions. This is typically the case for type I risks and much less the case for type II risks.

But is there a model that can be used for risk governance? Indeed there is. Reniers [14] indicates that the use of a number of triangles allows any organization to optimize its risk decision and expertise process. In the model, 12 triangles are employed (see Fig. 10.4). Note that, to further generalize the model, a 13th triangle might be added, on the same level as the "risk" triangle and describing positive risk: opportunities–exposure–gains. We did not include this triangle in the figure, in order to keep the focus on negative risks, which are the main concern of this book.

Following the well-known PDCA loop of continuous improvement from quality management, a four-step plan (policy–decision–risk–culture or the PDRC loop of continuous improvement) is proposed as the basic structure to serve for the RGM. From a holistic viewpoint, risks, and their uncertainties, outcomes and management, should be the concern of organizations–authorities–academia; these three actors within society form the first triangle – "risk policy." This first triangle should be the cornerstone of solid and holistic risk management, creating the right circumstances and helping to induce collaboration between all parties involved. The second triangle, "decision," consists of information–options–preferences. It is obvious that decision-making always requires information, options and preferences, as without any one of these three, decisions can simply not be made. Each of the blocks of the subsequent triangle of the rad needs to be aware of the three blocks of the previous triangle of the rad: risk information has to be taken from academia, organizations and authorities, and the same holds for developing options and mapping or composing preferences. The third triangle, "risk," includes hazards, exposure and losses. For the different dimensions of the risk triangle, the decision triangle has to be considered. The fourth triangle, "organizational culture," includes people–procedures–technology. Each of these three domains, composing an organizational culture, has to take the risk triangle into consideration. In its turn, the organizational culture triangle serves as a guide for each of the domains within the risk policy triangle.

Hence, risk policy guidelines, rules and so on, made or investigated by authorities, organizations and academia, should ultimately lead to an efficient and effective organizational risk culture. This, in turn, should be used by risk policymakers as an input for risk policy guidelines, rules and so on, to continuously improve risk decision-making. People, procedures and technology, forming the backbone of this culture, should be continuously optimized through the PDRC loop displayed in Fig. 10.4.

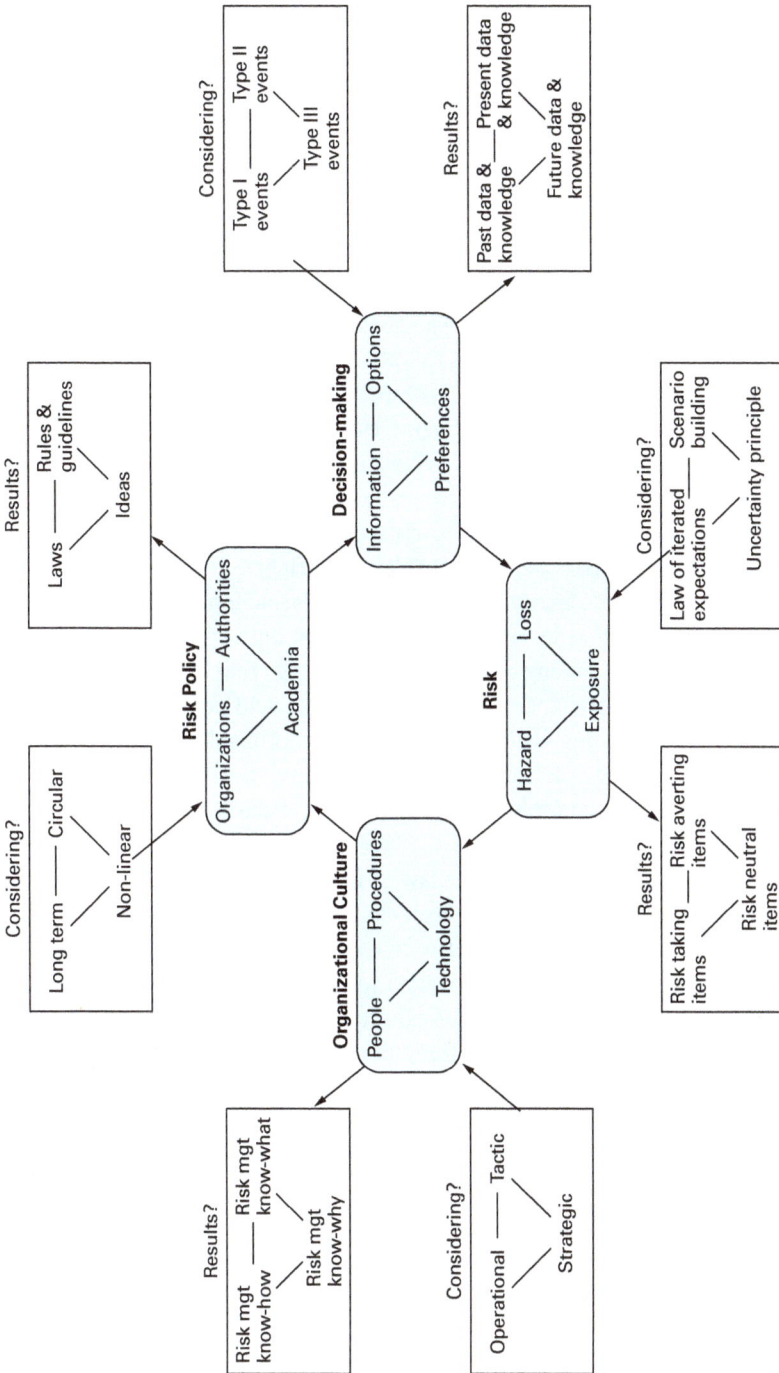

Fig. 10.4: Risk governance model.

An effective and efficient organizational culture implies that organizations are open-minded toward collaborating with other organizations and that they are prepared to search for the optimal way to decrease risks, also multiplant risks, and act accordingly (e.g., in their investment policies).

10.4.1 The "considering?" layer of the risk governance model

"Considering?" triangles can be used to gain a holistic indication of how to deal with the PDRC building blocks and what features should be taken into account when addressing the blocks belonging to the rad of the RGM. For example, it is recommended that academia–organizations–authorities within the risk policy triangle are all approached by thinking in the long term, thinking circularly and thinking nonlinearly. Hence, as far as risks and risk decisions and expertise are concerned, the way to think about people, procedures and technology by academia, organizations and authorities should be long-term-oriented, circular and nonlinear, on top of the regular thinking.

For the decision-making triangle, the "considering?" triangle relates to type I and II events. Thus, when exploring, identifying and mapping information, options and preferences from the viewpoints of academia, organizations and authorities focus should be on all three types of events. Among others, this implies, e.g., that external domino effects should be taken into account, and information, options and preferences should be developed regarding such events, even if they are highly improbable.

The way to deal with the risk triangle for gathering and drafting information, options and preferences on each of the blocks (i.e., possible hazards, possible losses and possible exposure) should be according to the law of iterated expectations, scenario building and the uncertainty principle. The law of iterated expectations simply states that if an event (e.g., an accident) can be expected somewhere in the future *with a certain probability*, then the event may also be expected at present. Following this principle implies that measures should be taken for preventing some events, among them some of the type II events, as if they might happen now. Of course, taking prevention measures depends, among other things, on the available budget and on the levels of uncertainty. Scenario building and the uncertainty principle (following the precautionary principle reasoning) were both explained earlier.

People, procedures and technology (forming the organizational culture triangle), each needs to be considered in an operational, a tactical and a strategic manner of thinking, whereby hazards, losses, exposure, opportunities and profits should be considered within every domain (people–procedures–technology).

10.4.2 The "results?" layer of the risk governance model

For each of the triangles of the loop, a "results?" triangle can be drafted. The risk policy triangle leads to concrete and abstract ideas, rules and guidelines and laws. The decision triangle leads to knowledge and information about the past, the present and the future. The risk triangle transpires into risk-averse, risk-taking and risk-neutral items of consideration. The organizational culture triangle leads to risk management know-how, risk management know-what, and risk management know-why in each of its domains. The loop of continuous improvement is closed by more concrete and abstract ideas, rules and guidelines and laws.

10.4.3 The risk governance model

If the triangles from the policy–decision–risk–culture rad, the "considering?" triangles and the "results?" triangles are displayed in an integrated way on the same Fig. 10.4, we create a model that can be used to continuously advance the optimization of risk decisions and risk expertise within and across any organization(s).

Similar to the fact that people do not want physicians only to be able to recognize well-known and "casual" diseases and not be able to detect a very rare disease, organizations should not be satisfied with risk experts and risk decisions tackling only well-known and "usual" (mostly occupational) risks and not considering out-of-the-ordinary-thinking risks (such as type II risks) or not using the proper method and the proper data and expertise to tackle certain risks or certain types of risks. In managing risks, to elaborate sustainable solutions, it is often much more important (and more difficult) to identify and define in detail the problem(s), than the solution(s). This can only be achieved by using a risk governance model, such as displayed in Fig. 10.4, integrating all possible viewpoints from diverse stakeholders, risk subjects, methodologies and methods, approaches, disciplines, etc. This model strives to make risk decisions truly more holistic, more systematic, more objective and more justified.

10.5 A risk governance PDCA

Risk governance should really be seen as a product of people. Risk decision results depend more than ever on the efforts made by management, stakeholders and employees. In order to make continuous progress, the commitment of these three clusters of people is indispensable. These lines of thought are the basic ingredients of a risk governance PDCA, which is visualized in Fig. 10.5. In agreement with traditional management systems, the idea of continuous improvement is depicted by the circular outline of the figure.

Figure 10.5 shows five inherent and characterizing components of a genuine risk governance policy connected with each other. The combined knowledge and commitment from central management and relevant stakeholders forms the foundation of a

Risk Governance Model (RGM)

Loop risk management

Risk Governance Framework

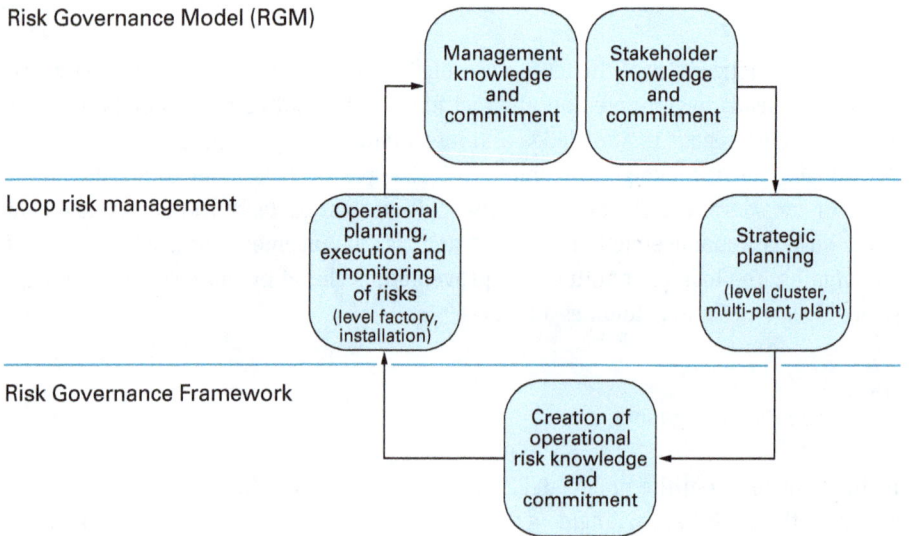

Fig. 10.5: Illustration of a risk governance PDCA (composed of "Mindset" (top of the figure), "Expertise and stakeholders" (middle of the figure), and "Knowledge and know-how" (bottom of the figure) from Fig. 2.2).

strategic and genuine risk governance policy. This should be evident for company employees. Only after this fulfilment is realized, an organization becomes able to concretize its strategy into operational actions and procedures. An accurate evaluation of these actions and their results should optimally result in more knowledge and a higher level of commitment on the part of management and stakeholders. All this can be achieved by using the RGM, as explained in the previous section. From this point of view, the upper sphere of knowledge and commitment can be interpreted as a reservoir that is being filled permanently with the essential fuel to undertake a new and more detailed run through the risk management system. Risk governance should, in this context, be interpreted as a gradual and endless process.

The huge contribution that is dedicated to the human influence in the conceptual and paradigmatically created risk governance cycle is a well-considered choice. Attitudes of people function as influencing factors within the cycle.

The interaction between attitudes of management and stakeholders should result in the choice of a limited number of strategic risk governance pillars at the cluster, multiplant and/or plant level, which need to form the foundations of the risk governance policy. Such pillars lead, in turn, to a risk governance framework that determines the limits of the risk governance policy, and that enables it to encompass all the risk management and risk governance efforts by the company.

Once the prospective risk governance policy is written, it has to be used to gain an understanding of the risks that have to be managed. Risks also need to be communicated to all the internal employees, as the understanding of risks, together with the commitment with respect to risk governance, can be interpreted as crucial factors toward an increased risk decision performance. On the one hand, this process should occur from a top-down approach. Already, in the selection process, management can decide to use the risk governance mentality of potential employees as a distinguishing parameter. Subsequently, after an employee's commencement of employment, he/she should still be informed regularly by the central management or by the risk management team about the risk governance course it takes, by means of frequent information sessions. On the other hand, a simultaneous bottom-up approach is required. Employees must feel that they possess a key that opens a gate toward an operational fulfilment of the chosen risk governance strategy. The design of some motivation programs or suggestion systems, e.g., on a central intranet, which enables employees to spread their ideas, is a concrete example of such a bottom-up approach.

Once employees are informed and convinced of the importance of a structured risk governance policy, top and middle management, together with the risk management team, can proceed to develop and implement operational planning of specific activities and instructions.

Risk governance can be described as a black box; even though it is fed with essential input by the management team and by risk managers, it basically depends on people's efforts. Every person should therefore evolve toward becoming a self-guided leader.

The application of consistent key performance indicators is a prerequisite that can be considered as an ever-returning issue within a long-term risk governance policy and which enables organizations to make their efforts measurable. An eventual underestimation of the importance of indicators within an organization makes it impossible to examine whether realized actions have led to significant progress. The operational information and scores on indicators that become available after the completion of the integral conceptual management system can be compared with the essential fuel to drive the internal motor of the risk management system. Realized scores on indicators should be seen as the material that drives an organization forward to new achievements and sharper objectives. In this context, corporate risk governance should really be seen as a gradual process that can never be ratified as being finished.

The completion of the integral conceptual management system may lead to the formal announcement of obtained results and desired actions for the future, based on information that can be filtered out of the circle. Management and the most prominent stakeholders can choose to publish this kind of information as documented proof of significant efforts and realized results.

10.6 Risk governance deficits

The IRGC defines risk governance deficits as "deficiencies or failures in the identification, assessment, management or communication of risks, which constrain the overall effectiveness of the risk governance process." Understanding how deficits arise, what their consequences can be and how their potential negative impact can be minimized is a useful starting point for dealing with emerging risks as well as for revising approaches to more familiar, persistent risks [15].

> **!** Risk governance deficits operate at various stages of the governance process, from the early warnings of possible risk to the formal stages of assessment, management and communication. Both underestimation and overestimation can be observed in risk assessment, which may lead to under-reaction or over-reaction in risk management. Even when risks are assessed in an adequate manner, managers may under- or over-react and, in situations of high uncertainty, this may become clear only after the fact. Human factors influence risk governance deficits through an individual's values (including appetite for risk), personal interests and beliefs, intellectual capabilities, the prevailing regulations or incentives, but also sometimes through irrational or ill-informed behavior [15].

IRGC defined two clusters identifying the causes of the most frequently occurring risk governance deficits:
- The assessment and understanding of risks (including early warning systems)
- The management of risks (including issues of conflict resolution)

For the first, they identified 10 deficits that can arise when there is a deficiency of either scientific knowledge or knowledge about the values, interests and perceptions of individuals and organizations.

1. The failure to detect early warnings of risk because of missing, ignoring or exaggerating early signals of risk.
2. The lack of adequate factual knowledge for robust risk assessment because of existing gaps in scientific knowledge, or failure to either source existing information or appreciate its associated uncertainty.
3. The omission of knowledge related to stakeholder risk perceptions and concerns.
4. The failure to consult the relevant stakeholders, as their involvement can improve the information input and the legitimacy of the risk assessment process (provided that interests and bias are carefully managed).
5. The failure to properly evaluate a risk as being acceptable or unacceptable to society, and the failure to consider variables that influence risk acceptance and risk appetite.
6. The misrepresentation of information about risk, whereby biased, selective or incomplete knowledge is used during, or communicated after, risk assessment, either intentionally or unintentionally.

7. The failure to understand how the components of a complex system interact or how the system behaves as a whole, thus a failure to assess the multiple dimensions of a risk and its potential consequences.
8. The failure to recognize fast or fundamental changes to a system, which can cause new risks to emerge or old ones to change.
9. The inappropriate use of formal models as a way to create and understand knowledge about complex systems (overreliance and underreliance on models can be equally problematic).
10. Failure to overcome cognitive barriers to imagining events outside of accepted paradigms ("black swans" or type III events). The acknowledgment that understanding and assessing risks is not a neat, controllable process that can be successfully completed by following a checklist.

For the second cluster (management of risks), they identified 13 deficits related to the role of organizations and people in managing risks, showing the need for adequate risk cultures, structures and processes [15]:
1. The failure to respond adequately to early warnings of risk, which could mean either underreacting or overreacting to warnings.
2. The failure to design effective risk management strategies that adequately balance alternatives.
3. The failure to consider all reasonable, available options before deciding how to proceed.
4. Inappropriate risk management occurs when benefits and costs are not balanced in an efficient and equitable manner.
5. The failure to implement risk management strategies or policies and to enforce them.
6. The failure to anticipate the consequences, particularly negative side effects, of a risk management decision, and to adequately monitor and react to the outcomes.
7. An inability to reconcile the time frame of the risk issue (which may have far-off consequences and require a long-term perspective) with decision-making pressures and incentives (which may prioritize visible, short-term results or cost reductions).
8. The failure to adequately balance transparency and confidentiality during the decision-making process, which can have implications for stakeholder trust or for security.
9. The lack of adequate organizational capacity (assets, skills and capabilities) and/ or of a suitable culture (one that recognizes the value of risk management) for ensuring managerial effectiveness when dealing with risks.
10. The failure of the multiple departments or organizations responsible for a risk's management to act individually but cohesively, or of one entity to deal with several risks.

11. The failure to deal with the complex nature of commons problems, resulting in inappropriate or inadequate decisions to mitigate commons-related risks (e.g., risks to the atmosphere or oceans).
12. The failure to resolve conflicts where different pathways to resolution may be required in consideration of the nature of the conflict and of different stakeholder interests and values.
13. Insufficient flexibility or capacity to respond adequately to unexpected events because of bad planning, inflexible mindsets and response structures, or an inability to think creatively and innovate when necessary.

Diagnosis and remedy of deficits is not a one-time event, but rather an on-going process of finding problems and fixing them. It relies on an interactive process between risk assessment and management, and between risk generators and those who are affected by risks.

10.7 Risk governance tips for improvement

Improving risk governance is essential for organizations to identify, assess, manage, and monitor risks effectively. They can not only safeguard their assets and reputation but also position themselves for sustained success in an increasingly unpredictable world. Some of the key reasons may be expressed as (non-exhaustive list): Loss mitigation, informed decision, regulatory compliance, operational efficiency, resilience, reputation protection, growth and opportunity, and competitive advantage.

Below are listed ten practical tips to strengthen a risk governance framework:
1. Establish a clear governance structure
   - Define roles and responsibilities, create a dedicated risk committee and align with organizational goals.
2. Enhance risk awareness and culture
   - Promote a risk-aware culture, offer training and education, and ask leadership to set example.
3. Develop a comprehensive risk management framework
   - Adopt an industry-standard framework (ISO, COSO, NIST), document policies and procedures and identify all risk categories.
4. Integrate technology and analytics
   - Leverage risk management software, predictive analytics and cybersecurity focus.
5. Conduct regular risk assessments
   - Conduct periodic reviews, scenario planning and stakeholder input.
6. Improve reporting and communication
   - Maintain transparent reporting, key risk indicators (KRIs) and real-time updates.

7. Strengthen risk mitigation and controls
   - Embed controls in processes, risk response plans and third-party risk management.
8. Monitor and review continuously
   - Make dynamic adjustments, conduct independent audits and build feedback loops (based on lessons learned).
9. Engage stakeholders
   - Involve the board, engage employees and collaborate with external people.
10. Align risk governance with ESG goals
    - Incorporate sustainability risks and stakeholder expectations.

Implementing these tips requires consistent effort and commitment from leadership. Over time, these measures will build resilience and strengthen the organization's ability to manage risks effectively. They foster stability, sustainability and stakeholder confidence in a volatile world.

10.8 Conclusions

Risk governance cannot take place in isolation, as argued by Renn [1]. Risk governance cannot be applied in a standard way at all locations, and in all political cultures, organizations and risk situations. It should be open to flexibility and adaptation in order to reflect the specific context of risks. In modern societies, a myriad of risk-influencing factors come into play when considering the wider environment of risk governing. To build an adequate capacity for successful risk governance at all levels, an organization may employ the risk management system, including standards and guidelines (operational), the risk governance framework (tactic) and the RGM (strategic), as elaborated and explained in this chapter.

References

[1]   Renn, O. (2008). Risk Governance. Coping with Uncertainty in a Complex World. London, UK: Earthscan Publishers.
[2]   Hilb, M. (2006). New Corporate Governance. Berlin, Germany: Springer.
[3]   Fuller, C.W., Vassie, L.H. (2004). Health and Safety Management. Principles and Best Practice. Essex, UK: Prentice Hall.
[4]   International Risk Governance Council. (2006). White Paper on Risk Governance Towards an Integrative Approach. Geneva, Switzerland: IRGC.
[5]   Reniers, G.L.L. (2010). Multi-plant Safety and Security Management in the Chemical and Process Industries. Weinheim, Germany: Wiley-VCH.
[6]   Oil and Petrochemical Industry Technical and Safety Committee. (2001). Code of Practice on Safety Management Systems for the Chemical Industry. Singapore: Ministry of Manpower.

[7] Health and Safety Executive. (1993). Successful Health and Safety Management, HS(G)65. Sudbury, UK: HSE Books.

[8] Reniers, G. (2012). From risk management towards uncertainty management. In: Risk Assessment and Management. Zhiyong, Editor. Cheyenne, WY: Academy Publishing.

[9] O'Riordan, T., Cameron, J. (1996). Interpreting the Precautionary Principle. London, UK: Earthscan Publishers.

[10] Senge, P. (1992). De Vijfde Discipline, De Kunst En De Praktijk Van De Lerende Organizatie. Schiedam, The Netherlands: Scriptum Management.

[11] Bryan, B., Goodman, M., Schaveling, J. (2006). Systeemdenken, Ontdekken Van Onze Organizatiepatronen. Den Haag, The Netherlands: Sdu Uitgevers.

[12] Aven, T. (2010). Misconceptions of Risk. Chichester, UK: John Wiley & Sons.

[13] Tversky, A., Kahneman, D. (2004). Loss aversion in riskless choice: A reference-dependent model. In: Preference, Belief, and Similarity: Selected Writings. Shafir, Editor. Cambridge, MA: MIT Press.

[14] Reniers, G. (2012). Integrating risk and sustainability: A holistic and integrated framework for optimizing the risk decision and expertise rad (ORDER). Disaster Adv. 5: 25–32.

[15] International Risk Governance Council. (2009). Risk Governance Deficits: An Analysis and Illustration of the Most Common Deficits in Risk Governance. Geneva, Switzerland: IRGC.

11 Examples of practical implementation of risk management

This chapter illustrates, through several examples, how to put engineering risk management into practice. We will focus on "research and teaching" topics as it is more complex and more challenging to implement the "engineering risk management" principles and methods discussed in this book, in such research environments, where a continuous evolution and constant changes of the risks and working environment and of the production processes (e.g., lab syntheses) are taking place. As already indicated in earlier chapters, safety management (as part of risk management) is a quality system used to encompass all aspects of safety throughout an organization. It provides a systematic way to identify hazards and control risks while maintaining assurance that the risk controls are effective. It is a global challenge for each organization to establish a safety management program and plan.

Research and teaching within certain fields of knowledge and science have an array of unique hazards that reflect both the variety and the continuous evolution of their operations. These hazards include chemical, physical, biological and/or technical facets. For example, there is an increasing awareness of reactive chemistry hazards. While controlling these hazards is frequently accomplished through engineering approaches such as ventilation and procedures, the long history of repeated incidents suggests that a more formal approach to hazard recognition and management is required.

Academia is composed of many different actors such as scientific staff, researchers, teachers, technicians, students, apprentices, administrative staff, short-term visitors and external stakeholders. Those people have different skills, education and knowledge. Hence, an overall safety management approach should address the different requirements needed by the diverse population.

Research activities have become more complex over the last decades with more interrelationships and interdependencies. Moreover, new technologies and innovations in developing new materials introduce new risks. This complexity, combined with increasing multifunctional use of space and increasing population densities with high turnover, creates larger risks to society (and to the research community), while at the same time their acceptance is decreasing. Public perception is generally years behind current practices and reality.

Many risk analysis techniques and risk management emerged in the industry from the 1960s onward. This can be regarded as a reaction to some major accidents, as well as the desire to achieve higher performance and improve production, quality and workers' health. Often regarded as centers of conceptualization and theoretical modeling, high schools and universities – the academia/research in a broad sense – are hardly comparable to the industry regarding safety management. The academic world remains also the headquarters of experiment validation associated with a concept of free re-

https://doi.org/10.1515/9783111493633-011

search. This makes it an environment particularly prone to risk. Indeed, experiments have not always been carried out without incidents or accidents. Several accidents are reported in academia, of which some remarkable ones can be presented as:

- 2006, Mulhouse (France): Explosion (followed by a fire) in the university's chemistry building. As consequence, one dead person and several injured.
- 2007, Taipei City (Taiwan): Blindness after a chemical experiment at University of Technology.
- 2008, Delft (the Netherlands): Fire due to a short circuit at the Technical University causing considerable financial losses.
- 2009, UCLA, Los Angeles (USA): Explosion (followed by a fire) in the University's chemistry building. As a consequence, one person died.
- 2010, Texas Tech University (USA). A student received severe burns and lacerations to his face and hands when a mixture of nickel hydrazine perchlorate exploded in a chemistry department laboratory.
- 2011, Yale, New Haven (USA): A student killed in a chemistry lab by being pulled into a piece of machine-shop equipment.
- 2012, Princeton, New Jersey (USA): Three people sent to hospital; 300 evacuated due to a wrong mix of nitric acid and solvents.
- 2012, Shanghai (China): Graduate student at university opens gas cylinder and dies from inhaling of the gas.
- 2013, Colorado Springs, Colorado (USA): A chemical incident in a student lab at Colorado College sent 13 people to the hospital. The group was exposed to titanium tetrachloride.
- 2013, Middleburg, Eastern Cape, South Africa (RSA): Six people died in an explosion at the Rolfe Pharmaceutical Laboratory.
- 2014, Minneapolis, Minnesota (USA): An explosion in a chemistry lab at the University of Minnesota injured a graduate student. The student was making trimethylsilyl azide.
- 2014, San Antonio, TX (USA): A lab technician at Southwest Research Institute (SwRI) was killed after a fatal accident in one of their labs (he was struck by an object from a machine he was operating).
- 2015, Tsinghua University in Beijing (China): A researcher died after a hydrogen storage cylinder unexpectedly exploded.
- 2016, Hawaii university (USA): Postdoctoral researcher lost her arm and sustained burns to her face and temporary loss of hearing due to hydrogen/oxygen explosion.
- 2017, Bristol (UK): A student at the University of Bristol unintentionally made an explosive, prompting a building evacuation.
- 2017, Harare (Zimbabwe): A student at the University of Zimbabwe died from severe burns he suffered when performing an experiment.
- 2018, Nashville, Tennessee (USA): 17 people were injured when a classroom science experiment caused a flash fire.

- 2018, Beijing (China): A chemical explosion on campus at Beijing Jiaotong University killed three students (working on a wastewater treatment experiment in a science laboratory full of flammable materials, which exploded upon contact with air).
- 2019, UCLA, Los Angeles (USA): One person was injured in an explosion involving acetone and occurring in a lab fume hood.
- 2019, Haifa (Israel): Professor Emeritus at Technion – Israel Institute of Technology died in an explosion involving hydrogen research at his lab at the Department for Materials Science and Engineering.
- 2020, Schenectady, New York (USA): A tank used to treat avocados exploded at a lab at Innovative Test Solutions. Kapp, a former mayor, later died from his injuries.
- 2021, Gubbio, Perugia (Italy): An explosion at a Green Genetics cannabis lab killed a 52-year-old worker.
- 2021, Beijing (China): A graduate student was killed in a laboratory blast at the Institute of Chemistry of the Chinese Academy of Sciences.
- 2022, Multan (Pakistan): A lab technician died as a result of a chemical explosion at the Government Shahbaz Sharif Hospital.
- 2023, Visakhapatnam (India): A pipeline carrying ethanol exploded at GMFC Labs due to a generation of static energy; one dead, three injured.
- 2024, Chennai (India): A student was killed in an explosion while carrying out an experiment with some chemicals.

However only very few are reported and presented in the open literature:
- 2006, Mulhouse (France): Explosion (followed by a fire) in the university's chemistry building. As a consequence, there was one fatality and several were injured [1].
- 2008, Delft (the Netherlands) Fire caused by a short circuit at the Technical University causing considerable financial losses [2].
- 2009, UCLA, Los Angeles (USA): Explosion (followed by a fire) in the university's chemistry building. As a consequence, there was one fatality [3].
- 2010, Texas Tech University (USA): A student received severe burns and lacerations to his face and hands when a mixture of nickel hydrazine perchlorate exploded in a chemistry department laboratory [4].
- 2011, Yale, New Haven (USA): A student was killed in a chemistry lab by being pulled into a piece of machine-shop equipment [5].

One of the most important factors is that "engineering risk assessment" should be built into scientists' routines. As an illustration, each chemical substance used comes with a list of potential risks and appropriate safety precautions through the material safety data sheet (MSDS), although unpredicted toxicity can affect even the most careful chemist. According to Peplow and Marris [1], it seems clear that academic labs are more dangerous than those in industry, because of their more relaxed approach toward safety.

Despite awareness about the growing risks in the academic/research world, risk management in this environment is even more complex compared to industry because of certain inherent specificities. Moreover, management of change, as expressed by Langerman [6], is even more critical as research laboratories are undergoing continuous and rapid changes. Furthermore, teaching laboratories are occupied by inexperienced operators who are being exposed to new situations. Existing methodologies for risk assessment are hardly directly applicable.

> **!** Implementing a risk management concept in complex systems requires a deep understanding of the possible risk interactions. Complex systems reside not only in a multifaceted combination but also with the difficulty that they evolve in the uncertainty region. This means that we have to deal with scarce or missing information and nonetheless evaluate the risk that we might face.

11.1 The MICE concept

Langerman [7] discussed the "lab process safety management" (PSM) approach for chemical labs, which was designed to help define when changes need to be handled in a coordinated and structured manner. This methodology is mainly process-oriented and does not satisfy the global/overall approach of how global safety management processes should be implemented in a research or teaching dedicated environment. Eguna et al. [8] presented some comments about the management of chemical laboratories in developing countries. An initial safety audit revealed plenty of room for improvement. Therefore an eight-session workshop was conducted for the laboratory personnel over a period of 8 weeks.

These two examples, among others, indicate that there is plenty of room for the implementation of global safety and risk management in research and teaching institutions. One solution is the implementation of a safety management program called MICE (management, information, control and emergency) based on solid education adapted to the target audience. This program, comprising four levels, is similar to the Deming wheel process or the improved plan-do-check-act as described by Platje and Wadman [9]. The four components of the MICE concept are [10, 11]:

- **M** – The management step
- **I** – The information and education step
- **C** – The control step
- **E** – The emergency step

11.1.1 The management step

The management step concerns different topics such as:
- The welcoming and training of new collaborators (every collaborator, independently from their activity, or student going to practical labs, should have a course introducing them to the basics of safety, fire-fighting training and first aid)
- The decentralized safety management and organization where each research and teaching unit has a safety delegate or coordinator (acting as a first-line safety actor)
- Lab-door panels (including information on present hazards, responsible and contact persons, prohibitions and requirements, safety classification, cleaning issues, etc.) on every research and teaching lab
- The hazard mapping of all research/teaching labs and offices allowing identifying laboratories with a high-level of danger or cumulative hazards
- Near-miss, incident and accident web-based interface and database allowing for analyzing and implementing adequate corrective measures in order to avoid the event's repetition

11.1.2 The information and education step

The information part of the MICE program is mainly related to targeted education or workshops for students (bachelor, master or PhD students), coworkers, researchers, technicians, teachers, administrative and technical staff as well as to external contractors. Websites especially dedicated to safety should be developed including a comprehensive online safety manual, tutorials on different hazards that collaborators could face in their activities, training videos on how to behave in case of emergency, how to deal with special hazards or how to safely operate in chemical labs and where someone could find help from a safety specialist. Emergency equipment and their use should also be depicted with training videos, operating manuals and directives.

Newsletters, information panels or paper information could be used as extra communication tools; however, we should not forget that nowadays everyone is submerged in papers and emails. A proactive response should be implemented instead of passive communication means.

11.1.3 The control step

Every management process needs a control step. This could be realized by safety audits of each research and teaching lab in order to ensure that the minimal safety requirements are satisfied. This ensures that the management of any change is always covered and mastered. Effectively, as process and procedures are rapidly evolving in research, we have to make sure that adapted safety management is equally reactive and proactive. These audits also have an educational issue as they should be realized in the presence of the individual unit safety coordinator explaining the observed deviation and remediation to be implemented. It allows for rapidly accessing the involved risks and implementing the adequate corrective measures in terms of prevention and protection.

11.1.4 The emergency step

The final step is the emergency step. Despite what one could imagine, emergency is not entirely related with professional intervention squads such as firemen, first-aid and technicians, but also with education on how to behave correctly in case of accident (call the center, evacuation drills, behavior, rules, first intervention, training, etc.) and also on how to act after intervention squads have left. In our opinion, emergency is also concerned with remediation and how to recover from physical or material damage.

> The MICE concept allows for the implementation of a safety management system covering the management step, the information and education, the control part and finally the emergency. It is comparable to quality management where all the facets have to be covered in order to implement an appropriate process.

11.2 Application to chemistry research and chemical hazards

Research and teaching labs can be too crowded, and such overcrowding raises the risk of spills. Waste disposal becomes a major issue. Most chemistry labs have open bottles where solvents are dumped along with the black gunk left from failed reactions. Wrong manipulation, storage or disposal of chemicals can cause great damage whether it occurs on industrial plants, in academia or at home. Among the numerous reasons, lack of knowledge and haste are the most common ones. Except for a few substances subject to international agreements, the academic world benefits from great latitude in the use of chemicals.

Chemical management is a crucial step in order to ensure safety [12]. Obviously, many chemicals should not be manipulated without any confirmation that the chosen equipment is adapted for the purpose and that workers are correctly trained, especially regarding the manipulation of carcinogenic, mutagenic, reprotoxic substances (CMR), highly toxic compounds or substances with high energetic reactivity. In order to validate the safety measures at the workplace, safety and health professionals must have access to the information about the substances used throughout the laboratories. Management of chemicals must find a process that ensures staff have a safe work environment without impairing their innovative thinking. To address this issue, imagine a comprehensive chemical management flowchart starting from the ordering of chemicals and ending with waste disposal (see Fig. 11.1).

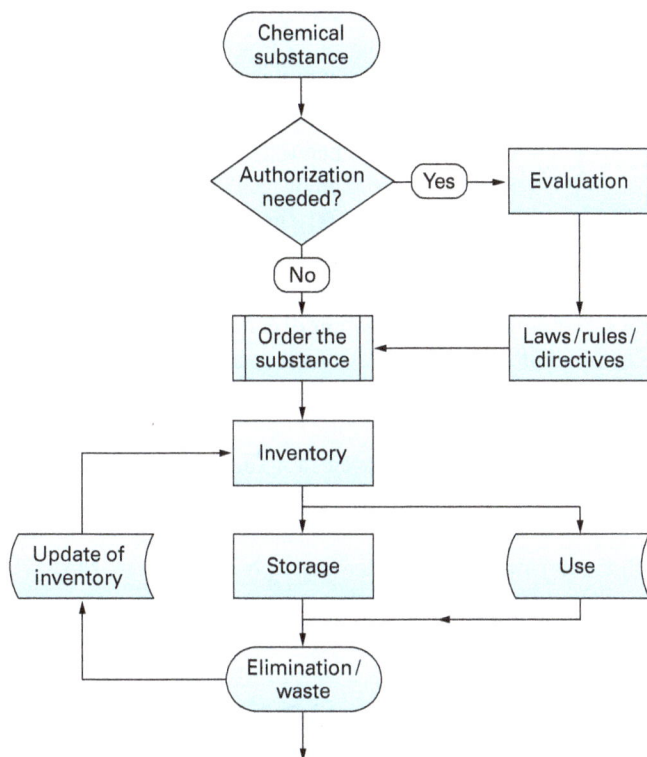

Chemical management flowchart.

1. Ordering chemicals and substances subject to authorization
   - Chemical management must start at the ordering stage. All substances must be bought through chemical stores responsible for general negotiations with suppliers, for checking compliance between ordering and shipping as well as for reporting every chemical into the inventory database (see below). Special treatment should be devised for substances that lead to serious health concerns, in particular, class 1 CMRs or highly toxic substances. Coworkers must obtain authorization to use these chemicals based on a comprehensive work conditions analysis. This quality process allows them to get support from specialists regarding the possibility of replacing all applicable substances with less problematic ones to verify if safety measures and operating procedures are sufficient, adequate and adapted for the planned project and eventually to determine the need of monitoring measures.

2. Inventory and storage
   - All chemicals have to be inventoried independently of their physical state (solid, liquid or gas) in a dynamic central database (intranet interface), taking into account chemical information (quantity, purity, MSDS), as well as logistical information (storage place, owner, date). Furthermore, the inventory allows the users to know if a substance is already present and could be borrowed for an initial test. Moreover, a computerized inventory is a powerful tool, enabling safety and health specialists to search for all chemicals with a uniform specific hazard statement or to prevent aging degradation by checking storage duration.
   - Storage must be considered as a first-line safety measure to prevent undesired events. According to CLP regulation (EC implementation of GHS, EC regulation [13]) substances must be stored in an appropriate manner, taking into account chemical compatibility and segregation depending on their intrinsic reactivity. This part of quality management should also be verified by workplace audit controls.
3. Waste management
   - At the end of the process, wastes are treated in a similar manner as pure compounds. They are separated by chemical compatibilities and properties. Self-reactive mixtures have to be rendered inert, as reactive mixtures should not be mixed with incompatibles ones. Once correctly conditioned, they are collected by disposal contractors.

> **!** Chemical management is a major issue due to the fact that a chemical substance does not cease to be active throughout its entire life. Wastes, at the end of their life, are still active and thus representing even more hazards compared to pure compounds. In fact, wastes turn into mixtures of unknowns, due to continuous composition evolution by successive adding. One should not forget that until their destruction (by incineration or physical treatments), chemicals remain active; they are designed for and used because of their activity.

11.3 Application to physics research and physics hazards

There are many hazards related to physics: high magnetic fields, ionizing and nonionizing rays, cryogenics, lasers, noise, work in hot or cold environment, engineered nano objects, electricity, etc. It would be illusory to draw a comprehensive list of those activities as they are as diverse as imagination could bring. Moreover, often these technologies are combined or linked together, leading to even more complex risk management [14]. It is therefore important to have a rapid hazard assessment method that could be incorporated into a risk management process.

Let us illustrate these concepts by applying a risk management process to the use of cryogenics fluids. Cryogenic liquids are liquefied gases that are kept in their liquid state at very low temperatures. They have boiling points below −150 °C (e.g., boiling point of helium and nitrogen are, respectively, −269 and −196 °C) and are gases at normal temperatures and pressures. These gases must be cooled below room temperature before an increase in pressure can liquefy them. Different cryogens become liquids under different conditions of temperature and pressure, but all have two properties in common: they are extremely cold, and small amounts of liquid can expand into very large volumes of gas (e.g., the ratio volume of gas to volume of liquid at 1 bar and 15 °C is 738 for helium and 1,417 for neon). However, some of them are flammable (e.g., hydrogen, methane, ethylene and ethane) and/or toxic (e.g., ozone, carbon monoxide and fluorine), adding another dimension to the abovementioned risks.

The vapors and gases released from cryogenic liquids also remain very cold. They often condense the moisture in air, creating a highly visible fog. In poorly insulated containers, some cryogenic liquids actually condense the surrounding air, forming a liquid–air mixture. With the exception of liquid oxygen, which is light blue, all the liquid cryogens are colorless. The properties of many cryogens of being colorless and odorless make them impossible to detect and discriminate by eye or by the sense of smell.

Everyone who works with cryogenic liquids (also known as cryogens) must be aware of their hazards and should know how to work safely with them. Before applying any strategies of risk management we must understand which hazards we might face.

11.3.1 Hazards of liquid cryogens

Cryogen hazards can be summarized in two categories: health hazards and hazards related to material properties, refrigerant properties and condensation mechanisms. Often those hazards cannot easily be seen and need to be supported by external means like advanced modeling and visualization techniques [15].

The health hazards can be listed as:

- Cold burns and frostbite – Exposure of the skin to a cryogenic liquid or its cold vapor/gas can produce skin burns similar to heat burns. Frostbite is caused by prolonged exposure of unprotected skin to cold vapors or gases. Once tissues are frozen, no pain is felt and the skin appears waxy and of yellowish color.
- Contact with cold surfaces – If unprotected skin comes into contact with cold surfaces, like uninsulated pipes or vessels, the skin may stick and flesh may be torn off on removal.
- Effect of cold on lungs – Patients suffering from bronchial asthma or chronic obstructive lung diseases often experience aggravation of bronchospasm on exposure to cold environment. Inhalation of cold mist, gases or vapors from the evaporation of cryogenic liquids worsens the degree of airway obstruction in sensitive patients. Short exposure creates discomfort even in normal subjects and could damage the lungs in case of prolonged exposure.
- Hypothermia – Exposure to low air temperatures can cause hypothermia. Hypothermia is a condition associated with the decrease of body temperature below

35 °C. The susceptibility of a person to hypothermia depends upon the temperature, the exposure time and the individual concerned (older people are more likely to succumb). When the body temperature goes below 33 °C, the victim can fall unconscious and after some time could fall into coma.

- Asphyxiation – As mentioned above, cryogenic liquids have a very large expansion rate. The liquid evaporates in such a large volume of gas that air displacement can considerably reduce the amount of oxygen available. Neon, hydrogen and argon have the highest expansion rates. Argon and nitrogen are heavier than air and tend to accumulate in low-lying areas such as pits and trenches. They can also seep through porous materials, fissures in the soil, cracks in concrete, drains and ducts. The cold vapors may collect and persist in confined spaces creating an oxygen deficiency hazard.

- Toxicity – Acute or chronic exposure of the considered substance will induce health effects depending on the concentration and the exposure duration. This effect is not specific to cryogens and more related to chemical specificities.

- Thermal radiation burns – Exposure to thermal radiation caused by the combustion of flammable cryogenic gases can produce first-, second- or third-degree burns. The degree of severity of a burn depends on the combustion temperature of the gas–air mixture, the distance of the victim from the heat source and the time of exposure.

- Blast/explosion injuries – A blast is a consequence of a confined combustion of a flammable gas. A leak of a flammable cryogenic fluid in a confined space is extremely dangerous because the flammable gas–air mixture would tend to cumulate in the area. If the gas mixture is ignited, a large overpressure is produced with dramatic consequences.

Hazards related to material properties, refrigerant properties and condensation mechanisms could be listed as:

- Brittle fractures – Some properties of materials can change significantly with temperature. Special care should be taken in selecting the material for equipment employed at cryogenic temperatures because of the risk of brittle fractures. A ductile material if placed in tension will stretch. For low values of tension, the elongation is proportional to the tension, and as the stress is removed the material returns to its original length (elastic behavior). For higher values of tension, the elongation is permanent. Eventually the material will break at a maximum tension characteristic of the material, called "tensile stress" or "ultimate stress." A brittle material does not exhibit this behavior; it rather suddenly breaks at high enough tension with no permanent deformation observed prior to rupture. As temperature is lowered, some materials undergo a change from ductile to brittle behavior. Fractures and cracks cause spills and leakages of the cryogenic liquid and gas.

- Thermal contraction leakage – Most materials have a positive thermal expansion coefficient meaning that when they warm up they expand and when cooled down they contract. Pipes, vessels and joints employed with cryogens must be of materials carefully selected according to their thermal expansion coefficient. Cooling down a material from room temperature to cryogenic temperatures causes a significant thermal contraction. For example, when going from ambient to cryogenic temperatures iron-based alloys contract about 0.3%, while many plastics contract well over 1%. Great stress is produced at the joints of rods or pipes of a cryogenic system if these are not allowed to contract freely when cooled down. Such stresses might result in broken joints or cracks along the pipe that would produce gas or liquid leakages.

– Overpressure – As mentioned above, cryogenic liquids vaporize into large volumes of gas. The normal heat inlet through the insulated walls and pipes of the storage vessel raises the temperature of the cryogenic liquid. For a steady heat flow the liquid boil-off rate of the vessel could be determined by the amount of liquid that slowly vaporizes as temperature rises. As the liquid vaporizes, the pressure inside the vessel increases. In this situation, overpressure could be caused by failure of the protection systems such as relief valves or rupture disks or the loss of thermal insulation. If the thermal insulation is damaged, for example if the vacuum in the vacuum jacket of the vessel is compromised, this could produce a large boil-off rate that the relief valves could not manage. Pressure can rise extremely high when cold liquid and vapor are stored in pressurized vessels that are not adequately vented and when refrigeration is not maintained. A pressure buildup may produce a burst with the sudden release of large quantities of cold liquid and gas as well as projection of mechanical parts.

– Combustion-induced flammable cryogens – Deflagration occurs when a portion of a combustible gas–air mixture is heated to its ignition temperature. As the combustion starts, the heat released is sufficient to ignite the adjacent gas, producing the propagation of the flame front through the combustible mixture. For example, the deflagration velocity of hydrogen is 2.7 m/s.

– Combustion caused by an oxygen-enriched atmosphere – Special care should be taken when using liquid oxygen as coolant. Oxygen is not flammable by itself but it supports combustion. Oxygen is heavier than air and tends to accumulate in low-lying areas such as pits and trenches and can also seep through fissures in the soil, cracks in concrete, drains and ducts. The cold vapors may collect and persist in these confined spaces, creating an oxygen-enriched atmosphere. Personnel should not enter areas where the atmosphere is rich in oxygen (>22%). Hair, clothing and porous substances may become saturated in oxygen and burn violently if ignited. Clothes contaminated with oxygen should be kept for at least 15 min in open air. Combustible material in the presence of an oxygen-enriched atmosphere ignites more easily, burns much more vigorously and can react explosively. Moreover, materials normally noncombustible in air such as stainless steels, mild steel, cast iron and cast steel, aluminum and zinc become combustible in an oxygen-enriched atmosphere.

– Condensation hazards – Because of their very low temperature, almost any cryogen can solidify water or carbon dioxide. The presence of solid particles within a fluid system can cause damage to the system and hazardous situations such as overpressure and leakages. The solid particles can erode valves and gaskets producing leakages. If a large amount of solid particles cumulate inside pipes or in proximity to relief valves, these could be blocked and prevent gas from being released. If gas is trapped inside the system, pressure could increase to dangerous levels. Another issue related to the condensation of air is the phenomena of oxygen-enrichment. The development of an oxygen-enriched atmosphere is a fire and explosion hazard. Air is composed of 21% oxygen and 78% nitrogen. Liquid air produced by condensation on cold surfaces does not have the same composition of the vapor being condensed. Because of the higher boiling point of oxygen (90.2 K) than nitrogen (77.4 K), oxygen condensates preferentially to nitrogen. It will produce liquid air with an oxygen concentration that could reach 50%. Liquid air is therefore very rich in oxygen, and an explosive hazard is more present.

11.3.2 Asphyxiation

We do not intend to discuss all the hazards linked with the use of cryogens, for such a discussion would fall out of the scope of this book. We will concentrate on asphyxia, caused by the oxygen content lowering, and discover how to apply the risk management process to this particular situation.

Ventilation is a key issue when handling and storing cryogenic liquids. Large quantities of cryogenic liquids must be stored in open air or in well-ventilated areas. Small, unventilated rooms should be avoided to prevent buildup of the gas as the cryogen evaporates. A well-ventilated laboratory has between five and ten air renewals per hour. But what will happen if some liters of a cryogen leaks into the room? Is there a health problem? What could be the consequences?

In order to answer those questions, trials were held on a real scale where several liters of liquid nitrogen were poured on the floor, and the evolution of the oxygen content was measured at different locations and heights in the room. Results indicate that there were huge differences depending on the ventilation efficiency. This led to the conclusion that an oxygen detection system should be installed if the quantity of stored cryogen in the room is above 0.4 L/m^3 of space when correctly ventilated and above 0.3 L/m^3 for unventilated rooms. These values correspond to the observed limits where the oxygen content was below 19% for more than 3 min at a height of 1.60 m (being the height-average of the human face). An oxygen detector should be installed depending on the quantity of liquid cryogen stored in the room. The detector is equipped with a visual alarm that becomes active if the oxygen level goes below 19%. If the oxygen level drops below 18% an acoustic alarm is also activated indicating the evacuation of the room. The instrument should undergo regular calibration checks and routine maintenance to ensure reliable performance. Moreover, a clear labeling of the room door should advise people entering that the room may have potential asphyxiation risks when the alarm is on (see Fig. 11.2).

Fig. 11.2: Alarm warning and door panel indication.

This example shows that even in the absence of regulation it is possible to implement a risk management strategy by replacing rules by experiments and interpreting them to define internal regulation that could be applied in similar situations.

Most of the physical hazards have in common that they are not correctly evaluated by human senses. Technical measures should inform our senses and mind that we are facing a hazard and that a risk might be present. Often, their actions on human bodies are not directly connected to their use; i.e., cryogens are used to attain low temperatures, but one of their drawbacks is the possibility of death by asphyxiation in case of a large release in the environment.

11.4 Application to emerging technologies

Emerging technologies have in common that they are evolving in higher levels of uncertainty and complexity, accelerating speed and competency-destroying change. These are among the characteristics that make managing emerging technology distinct from managing established technology. We could then define emerging risks (ERs) as new or already known risks that are difficult to assess ("known unknown risks," see Chapter 2). The IRGC goes a bit further by defining three categories of ERs [16]:

1. Uncertain impacts: Uncertainty resulting from advancing science and technological innovation
2. Systemic impacts: Technological systems with multiple interactions and systemic dependencies
3. Unexpected impacts: Established technologies in evolving environments or contexts

The OSHA [17] definition of ER stipulates that any risk that is new and/or increasing is emerging. By "new" they define that:

- the risk did not previously exist and is caused by a new process, new technologies, new types of workplace or social or organizational change or
- a long-standing issue is newly considered as a risk because of a change in social or public perception or
- new scientific knowledge allows a long-standing issue to be identified as a risk.

The risk is "increasing" if:
- the number of hazards leading to the risk is growing, or
- the likelihood of exposure to the hazard leading to the risk is increasing (exposure level and/or the extent of human values exposed), or
- the effect of the hazard is getting worse (severity and consequences and/or the extent of human values affected).

SwissRe, a reinsurance company, defines ERs as newly developing or changing risks which are difficult to quantify. The loss potential of these risks is currently difficult to estimate, but they may have a major business impact on the insurance industry [18]. They provide insights into four environments: "societal, political, technological and natural and competitive and business" and the respective macrotrends in ERs as depicted in Fig. 11.3.

All these definitions have in common that often their acute and chronic impacts are not well known. This constant transformation requires a dynamic approach. We need to scan, monitor and respond to technologies and strategies that are constantly in motion. The main challenge for the future resides in dealing with multiple interacting risks. Most of them will be not sufficiently known or even unknown, leading them to remain in the uncertainty zone.

The main question to answer is: "How do we protect against something where we have insufficient information about its consequences?" Knowing what to implement and when to make the change can make anyone wander around in a haze of confusion, but with a few process steps, the management can become much simpler. However, given the risky and often unproven nature of these technologies, considerable confusion exists on how to manage their implementation [19]. The key issue is the ability to adequately learn about new advances, boiling down to a three-step phase:
1. Keep your eyes open for new technologies that might assist you. Read articles and scour technology sites where emerging technologies often are introduced (see "Mindset," Fig. 2.1).
2. Evaluate the technology against your strategy. Will this new technology increase efficiency, allow entering new markets, expediting your development or reduce costs? (See "Stakeholders and expertise," Chapter 2).
3. Ask several questions and make yourself the knowledge expert if you think the technology might be useful. Reach out to those who wrote articles on the subject. Attend specialized conferences in order to make your network broader to learn faster (cf. "Knowledge and know-how," Chapter 2).

Political and economic environment

- Macroeconomic fragility
- Challenged globalisation
- Geopolitical & economic instability
- Rising interest rates and risk of persistent inflation
- Infrastructure funding needs

Competitive and business environment

- Re/insurance value chain disaggregation and rise of alternative re/insurance providers
- Consolidation of platforms as a business model through strategic partnerships
- Regional champions going global
- Increasing digital customer interaction
- Increasingly litigious environment
- Rising importance of Environmental, Social and Governance (ESG)

Demographic and social environment

- Shifting demographics and global aging
- Growing middle class in high growth markets
- Longevity & radical medical innovation
- Prevalence of mental health issues
- Mass migration & urbanisation
- Changing workplace and talent gaps
- Rising social inequality & unrest

Technological and natural environment

- Addressing physical climate change risks
- Rising importance of biodiversity and ecosystem services
- Transition to a low carbon economy
- Expansion of digital & cyber risk
- Data as an asset
- Impact of generative AI
- Digital products and processes
- Disruptive digital technologies
- Autonomous transportation & robotics

Fig. 11.3: Screening of interdependent macrotrends [18].

The second issue is the implementation of these new advances:

1. Outline the benefits and risks of the change. Prepare a cost–benefit analysis that lists what you hope to gain with the change and any risks that might be associated with that change. Sometimes just looking at this list can help make the decision to move forward or to drop the project.

2. We often hear that small is beautiful, so start small when possible. Begin with a limited test of the technology. A mini-implementation can help you evaluate new technologies within your own products, processes and services much better than any literature.

3. Establish and document a communication plan. This should include project communications, such as status, timetables, phases, issue resolution and cost. It also should include how you will communicate with employees or externals assisting with the implementation.

4. Plan a fallback path. Newer technologies may have "bugs" or not work as promised. When implementing, be on guard for the unknown. This way, if something comes up that makes the implementation a bad idea, you have a way to scale back to what you had before.

5. With unknowns we must be proactive. It is better to act instead of react, especially when we are driven by the need to find solutions to short-term challenges. Given the current economic climate, the quality of foresight and planning based on longer-term planning are not really in vogue.

Bhattacherjee [20] discussed key organizational factors affecting the implementation of emerging technologies. He concluded that adequate efforts and resources must be devoted to understanding and managing these challenges. IRGC developed four risk governance dimensions, including 11 themes for improving the management of ERs. The grouping begins with risk governance (a concern of strategy, top management and organizational design), then moves to an organization's risk culture, to training and capacity building and, finally, to adaptive planning and management [21]:

I. Risk governance: strategy, management and organizational matters:
   1. Set ER management strategy as part of the overall strategy and organizational decision-making.
   2. Clarify roles and responsibilities.
II. Risk culture:
   1. Set explicit surveillance incentives and rewards.
   2. Remove perverse incentives to not engage in surveillance.
   3. Encourage contrarian views.
III. Training and capacity building:
   1. Build capacity for surveillance and foresight activities.
   2. Build capacity for communicating about emerging issues and dialoguing with key stakeholders.

3. Build capacity for working with others to improve the understanding of, and response to, ERs.
IV. Adaptive planning and management:
   1. Anticipate and prepare for adverse outcomes.
   2. Evaluate and prioritize options; be prepared to revise decisions.
   3. Develop strategies for robustness and resilience.

Improving the management of ER requires improvement in communication to identify and characterize such risks. Transparency is a precondition for having the research and innovation perceived as balanced, fair and beneficial for the society. What is beneficial for the society is, however, not precisely defined and it can be a topic of major differences in opinion, depending on the stakeholder groups in society. In particular, the question, "Is a particular innovation or new technology beneficial for the society?" (e.g., engineered nanotechnologies) often cannot be answered in a straightforward way. This mainly implies that the question should be posed in the reversed way: "How can we be sure that the innovation will not involve risks that we do not want to accept?" This is the question that leads to the precautionary principle (better safe than sorry) [21, 22]. It states that "in the absence of suitable hazard data, a precautionary approach may need to be adopted." But the practical implementation of the general principles poses a lot of challenges and leads to different solutions. When dealing with the uncertainty zone, we do not have strict answers or even questions. Any discussion about ER may start with the question: "While it is emerging, it is not yet a risk, when it is a risk, it is no longer emerging." The concern about ER is magnified by the fact that our knowledge about the phenomenon is incomplete and we are not sure what exactly we are taking about. But should this indicate that nothing has to be done? The answer is clearly "no." The route map when dealing with ER could be listed as [23]:
– earlier recognition of ER;
– more systematic recognition of ER by evaluating precursors (on web, papers, conferences, debates, etc.) and monitoring their development, identifying similarities with known risk and their precursors (find analogies);
– better identification of critical ER;
– recognition of interdependencies and relations among all risks;
– improving knowledge on triggers, drivers and factors of ER;
– setting up a monitoring process and follow-up; and
– systematic interlinking among hazards, vulnerabilities and stakeholders.

Emerging technologies have in common that they are evolving in higher levels of uncertainty and complexity, accelerating speed and competency-destroying change. ERs can be defined as new or already-known risks that are difficult to assess such as: uncertain impact, systemic impacts or unexpected impacts. Mastering those risks requires some prerequisites such as keeping eyes open for new technologies, evaluating them and asking as many questions as needed to gain knowledge.

11.4.1 Nanotechnologies as illustrative example

As this field is rapidly evolving according to the progress being made worldwide, this section should be taken as a snapshot of what is the current state of the art.

How do you protect people from materials with properties that are unknown? The approach is similar to training firemen to fight fires from unknown material.

The properties of manufactured nanomaterials (materials made of nanoparticles smaller than 100 nm produced intentionally by humans, ENP) are paving the way for a wide variety of promising technological developments. However, due to the many uncertainties, conventional assessment of the associated hazards and exposure based on quantitative, measurable criteria is difficult. These uncertainties will only be removed as scientific understanding of the properties of nanomaterials advances.

The handling of nanomaterials is a challenge because of the unknowns involved. If there is a known impact, does it arise from only one part of the material distribution? Doing nothing is not acceptable; this indicates that education guidance and handling procedures must be developed. Key elements must be largely disseminated.

Furthermore, given the current state of knowledge on manufactured nanomaterials, it is highly likely that many years will pass before we know precisely which types of nanomaterials and associated doses represent a real danger to humans and their environment. Indeed, the assessment of potential health effects following exposure to a chemical must consider the extent and duration of exposure, the biopersistence and interindividual variability, all subjects on which we have practically no knowledge for the field of nanomaterials [24].

It is therefore extremely difficult to conduct a quantitative risk assessment in most work situations involving nanomaterials with the currently available methods and techniques. It will be challenging, at best.

With unknowns, we must be proactive. In many cases, we do not know what we are looking for. So the question is: "How do we proceed?" While we do not have all the answers, we can take a number of precautions. The focus of our actions must be to:
- Keep ourselves safe.
- Keep our colleagues safe.
- Keep the general population safe.
- Keep the facilities safe.
- Keep the environment safe.

Nanosafety is a growing concern that many institutions are working on because nowadays. Today's environment requires that people and organizations are responsible for their actions. As the number of engineered nanomaterials (ENMs) used in research increases with an incredible speed, health and safety specialists are continuously faced with the challenge of evaluating the risks involved with these materials. Nowadays there is not enough information about their toxicology and new materials are continuously being developed. Preliminary scientific results indicate that ENM might have a damaging impact on human health, which makes it even more important to have the right mitigation measures in place [25].

We will not discuss all the available information from the literature but we will focus on the management aspects of several different safety methodologies or procedures. We have chosen to compare results using three different methodologies: the tree decision [26], the control banding NanoTool 2.0 [27] and the ANSES method [24]. For supplementary methods, we refer to [28–31].

The decision trees developed by Groso et al. [26] and Buitrago et al. [33] are, based on new information concerning the hazards of ENMs, to improve a previously developed risk assessment tool [32] by following a simple scheme to gain in efficiency. In the first step, using a logical decision tree, one of the three hazard levels, from H1 to H3, is assigned to the nanomaterial. Using a combination of decision trees and matrices, the second step links the hazard with the emission and exposure potential to assign one of the three nanorisk levels (Nano 3 highest risk; Nano 1 lowest risk) to the activity. These operations are repeated at each process step, leading to the laboratory classification. The third step provides detailed preventive and protective measures for the determined level of nanorisk. The methodology provides a list of required risk mitigation measures [technical (T), organizational (or procedural, P) and personal (P)]. This tool is intended for researchers to self-quantify their hazard and risk level. Depending on the nano-hazard classification, several measures are defined. It has to be noted that it is the sole method of defining and proposing measures at the strategic, technical, organizational and personal levels. Moreover, they also define adequate measures for visitors, technical and maintenance staff, intervention squads, pregnant women and medical survey. Schwab et al. [34] added the urgent need for actions in managing nanowaste as depicted in their paper in *Nature nanotechnology*.

NanoTool [27] was developed to support first-line occupational health professionals and researchers in evaluating the potential risks related to production and downstream use of nanomaterials in a research work environment. In the NanoTool, CMR toxicity, general toxicity and dermal toxicity of parent material are used as distinct parameters with assigned severity points. In the CB NanoTool, parameters related to emission potential are dustiness/mistiness (substance emission potential) and amount of ENP handled (activity emission potential). The CB NanoTool links the hazard and exposure bands, which have the same ranges of scores, into four risk levels and consequently to control bands linked to the risk levels.

ANSES [24] is intended to be used by those adequately qualified in chemical risk prevention. It uses the classification of either the bulk material or an analogous substance as the starting point for the hazard-banding process if the ENP is not a biopersistent fiber. Hazard parameters such as dissolution time and reactivity may increase the hazard band. ANSES covers emission potential by initial banding based on the physical state of the material, ranging from solid (exposure band 1) to aerosol (exposure band 4). Further modification of the bands (increment) is possible either due to the substance emission potential or the process operations (activity emission potential). In ANSES, the five hazard and four exposure (emission potential) bands are directly linked into five control bands. The hazard band dominates the allocation of the control band (or risk level) because the highest hazard band, e.g., in case of persistent fibers or lack of information, requires the highest control band independent of the exposure band.

Let us discover, using a very simple example, the outcomes using the three different methods. An illustrative example is the preparation of a wafer on which will be deposited a black ink containing carbon black ENPs. The process could be described as:

1. Carbon black particles (FW200), $d = 13$ nm with a specific surface area of 550 m^2/g, are received in 500 g containers.
2. Carbon black is weighted in order to distribute the powder into smaller containers.
3. About 30 g of carbon black is weighted from a small container.
4. Surfaces surrounding the container and outer walls of container are cleaned.
5. Weighted carbon black is added to the previously prepared liquid resin.
6. The prepared mixture is stirred in a closed flask to obtain the desired ink.
7. The ink is deposited by a pipette on the wafer.
8. The wafer is baked in a hermetically closed oven.

We will concentrate on step 3 as it is the most hazardous in relation to exposure. Using the three methods, we end up with the results expressed in Tab. 11.1.

Tab. 11.1: Results of the evaluation using three different evaluation methods.

	Tree method	CB NanoTool	ANSES
Hazard level	Level 3 on the scale of 3	Level 3 on the scale of 4	Level 5 on the scale of 5
Control measures	Technical, organizational and personal measures for Nano 3	Containment of the process	Full containment and review by a specialist

We can see that the classification is rather similar for the three methods, being in the highest hazard zone. However, they are largely different when looking at the measures that should be followed. The tree method [26] is the most comprehensive, indicating in detail what should be applied, then the NanoTool indicates that the process should be confined and ANSES adds to this that a review should be made by a specialist.

We may then raise the following questions:
– Who is the specialist when a lot of unknowns are predominant and when we act in the uncertainty zone?
– How should we apply the precautionary principle?
– How should we implement safety and risk management in such conditions?

Answering those questions requires applying risk management principles expressed in the preceding chapters, taking into account the uncertainty zone. It is hard to find one "expert" who is capable on his/her own of addressing complex systems; it should be a multidisciplinary team effort.

When using any kind of methodology, we should not forget that the ultimate goal is to safely operate a hazardous process. The more information we have, the better will be the outcome.

11.5 Tips for implementing risk management in practice

Traditional risk management methods are often fragmented, isolating risks into separate silos. These approaches typically concentrate on mitigating uncertainties related to physical and financial assets, with an emphasis on loss prevention rather than enhancing enterprise value. As a result, they fall short of providing the comprehensive framework organizations need to redefine the value of risk management in a rapidly evolving landscape. To address this gap, organizations must adopt modern risk management practices that are proactive, forward-looking, and better equipped to evaluate and manage uncertainties while creating sustainable value for stakeholders.. Some tips could be expressed as follows:

1. Conduct the enterprise risk assessment: This assessment identifies and prioritizes the organization's risks using the business strategy as context. It provides data for the formulation of appropriate responses to risks, including information on the current state of risk management capabilities to prioritize risks.
2. Articulate the ERM vision using the gaps around priority risks: This step alienates the economic rationale for moving forward. The vision is a shared understanding of the role of risk management in the organization and the capabilities needed to manage its key risks. A working group of senior managers should be empowered to articulate the role of risk management in the organization and define the relevant goals and objectives for the organization.
3. Advance the organization's risk management capabilities for one or two selected priority risks: The organization should focus on improving its risk management capabilities in an area where management knows that improvements are needed. Like any other initiative, ERM has to start somewhere. Since there are many possible starting points, the beginning and end of the process must be well-defined since the beginning.
4. Assess the capacity of the existing infrastructure and develop a strategy to advance it: Oversight, control and discipline are required to advance critical risk management capabilities. The policies, processes, organization and reporting that provide this oversight, control and discipline are referred to as "ERM infrastructure." Its purpose is to eliminate significant gaps between the current and desired state of the organization's capabilities to manage its key risks.

5. Advance risk management capabilities for other key risks: Once the first four steps have been completed, it will often be necessary to update the assessment to reflect changes. Once the priority risks have been redefined based on the updated assessment, management needs to determine the current state of risk management capabilities for each risk and then assess the desired state.

11.6 Conclusions

If there are so many good models on risk management (see Chapter 3), then why is there still such a high failure rate of projects? It is very easy to acknowledge that there are large risks present when undertaking development projects. It can therefore be tempting for development managers to completely ignore engineering risk management approaches; risk management may imply that the failure of the project is almost a certainty. The key aspects are:

– Focus on sustainability: Leverage existing practices for risk management purposes.
– Be pragmatic: Customized strategies should be supported by simple and efficient methods that meet the needs of, and add value to, managers.
– Take a balanced approach: Balance between level of investment, value expected from investment and the capacity of the organization.
– Be realistic: The sophistication of the risk management regime must be in step with the maturity of the other management processes of the organization.
– Provide leadership: Establish champions across the organization, with clear accountabilities.

When you have effective risk management in place, you can focus your planning on avoiding future problems rather than solving current ones. You can routinely apply lessons learned in order to avoid crises in the future rather than fixing blame. You can evaluate activities in work plans for their effect on overall project risk, as well as on schedule and cost. You can structure important meeting agendas to discuss risks and their effects before discussing the specifics of technical approach and current status.

Above all else, you can achieve a free flow of information at and between all program levels, coordinated by a centralized system to capture the risks identified and the information about how they are analyzed, planned, tracked and controlled. You can achieve this when risk is no longer treated as a four-letter word, but rather is used in your organization as a rallying perspective to arouse creative efforts.

With effective risk management, people recognize and deal with potential problems daily, before they occur, and produce the finest product they can within budget and schedule constraints. People, work groups and projects throughout the program understand that they are building just one end product and have a shared vision of a successful outcome.

Engineering risk management should:
- create value,
- be an integral part of organizational processes,
- be part of decision-making,
- explicitly address uncertainty,
- be systematic and structured,
- be based on the best available information,
- be tailored,
- take into account human factors,
- be transparent and inclusive,
- be dynamic, iterative and responsive to change and
- be capable of continual improvement and enhancement, etc.

Effective risk management requires the same steps as the decisions encountered at a stoplight. Objectives (getting to our destination) must be clear, and attributes of achievement (we must get there before a certain time) must be included. We assess the compliance of others (did everyone stop for the red?) and manage uncertainty based on our risk tolerance (is there time for us to cross on the yellow light?).

Our action/inaction is guided by our analysis and our risk tolerance. Common carriers (railway, bus and air) are very aware that their customers have delegated risk management to them and generally operate very conservatively (by regulation and by choice), so as to ensure that they are not more risk-tolerant than their most risk-intolerant client.

References

[1]   Peplow, M., Marris, E. (2006). How dangerous is chemistry. Nature. 441: 560–561.
[2]   Meacham, B., Park, H., Engelhardt, M., Kirk, A., Kodur, V., Van Straalen, I., et al. (2010) Fire and Collapse, faculty of architecture building, delft university of technology: Data collection and preliminary analysis. Proceedings, 8th International Conference on Performance-Based Codes and Fire Safety Design Methods, Lund University, Sweden.
[3]   Kemsley, J. (2009). Learning from UCLA. Chem. Eng. News. 87: 29–34.
[4]   Johnson, J. (2010). School labs go under microscope. Chem. Eng. News. 88: 25–26.
[5]   Foderaro, L.W. (2011). Yale Student Killed as Hair Gets Caught in Lathe. New York Times April 13th.
[6]   Langerman, N. (2008). Management of change for laboratories and pilot plants. Org. Proc. Res. Dev. 12: 1305–1306.
[7]   Langerman, N. (2009). Lab-scale process safety management. Chem. Health Saf. 6: 22–28.
[8]   Eguna, M.T., Suico, M.L.S., Lim, P.J. (2011). Learning to be safe: Chemical laboratory management in a developing country. Chem. Health Saf. 6: 5–7.
[9]   Platje, A., Wadman, S. (1998). From Plan-Do-Check-Action to PIDCAM: The further evolution of the Deming-wheel. Int. J. Proj. Man. 16: 201–208.

[10] Meyer, Th. (2012). How about safety and risk management in research and education?. Proc. Eng. 42: 934–945.

[11] Marendaz, J.L., Friedrich, K., Meyer, Th. (2011). Safety management and risk assessment in chemical laboratories. Chimia. 65: 734–737.

[12] Brückner, S., Marendaz, J.-L-, Meyer, Th. (2016). Using very toxic or especially hazardous chemical substances in a research and teaching institution. Safety Sci. 88: 1–15.

[13] European Council. (2008). Regulation on classification, labeling and packaging of substances and mixtures. Commission Regulation, (EC) No 1271/2008, Official Journal L. 338: 005–0052.

[14] Marendaz, J.L., Suard, J.C., Meyer, Th. (2013). A systematic tool for Assessment and Classification of Hazards in Laboratories (ACHiL). Safety Sci. 53: 168–176.

[15] Iannarelli, R., Novello, A. M., Stricker, D., Cisternino, M., Gallizio, F., Telib, H., Meyer, Th. (2019). Safety in research institutions: How to better communicate the risks using numerical simulations. Chem. Eng. Trans. 77: 871–876.

[16] International Risk Governance Council. (2011). Improving the Management of Emerging Risks. Lausanne, Switzerland: International Risk Governance Council.

[17] European Agency for Safety and Health at Work. (2010). European Risk Observatory Report – European Survey of Enterprises on New and Emerging Risks. Managing Safety and Health at Work. Brussels, Belgium: Publication Office of the European Union.

[18] SwissRe Institute. (2024). Swiss Re SONAR: New emerging risk insights. June, 1–52.

[19] Day, G.S., Schoemaker, P.J.H., Gunther, R.E. (2000). Wharton on Managing Emerging Technologies. 1st edn. New York: John Wiley & Sons, Inc.

[20] Bhattacherjee, A. (1998). Management of emerging technologies: Experiences and lessons learned at US West. Inform. Manag. 33: 263–272.

[21] European Union. (1992). Treaty of Maastricht on European Union. Off. J. C. 191.

[22] Commission of the European Communities (2000) Communication on the precautionary principle, COM 1.

[23] iNTeg-Risk FP7 (European Framework Program) Project (2012) 4th iNTeg-Risk Conference, "Managing Early Warnings – What and how to look for?" Stuttgart. http://www.integrisk.eu-vri.eu.

[24] ANSES (2010) Development of a Specific Control Banding Tool for Nanomaterials. Request No.2008-SA-0407.

[25] Novello, A.M., Buitrago, E., Groso, A., Meyer, Th. (2020). Efficient management of nanomaterial hazards in a large number of research laboratories in an academic environment. Safety Sci. 121: 158–164.

[26] Groso, A., Petri-Fink, A., Rothen-Rutishauser, B., Hofmann, H., Meyer, Th. (2016). Engineered nanomaterials: Toward effective safety management in research laboratories. J. Nanobiotechnol. 14.

[27] Zalk, D.M., Paik, S.Y., Swuste, P. (2009). Evaluating the control banding nanotool: A qualitative risk assessment method for controlling nanoparticle exposures. J. Nanopart. Res. 11: 1685–1704.

[28] Van Duuren-Stuurman, B., Vink, S.R., Verbist, K.J., Heussen, H.G., Brouwer, D.H., Kroese, D.E. et al. (2012). Stoffenmanager Nano version 1.0: A web-based tool for risk prioritization of airborne manufactured nano objects. Ann. Occup. Hyg. 56: 525–541.

[29] California Nanosafety Consortium of Higher Education. (2012). Nano toolkit. In Working Safely with Engineered Nanomaterials in Academic Research Settings. De La Rosa Ducut, J., Editor. Riverside, CA: University of California.

[30] Schulte, P., Geraci, C., Zumwalde, R., Hoover, M., Kuempel, E. (2008). Occupational risk management of engineered nanoparticles. J. Occup. Environ. Hyg. 5: 239–249.

[31] Federal Office for Public Health. Precautionary matrix for synthetic nanomaterials. (accessed in 2012) http://www.bag.admin.ch/nanotechnologie

[32] Groso, A., Petri-Fink, A., Magrez, A., Riediker, M., Meyer, Th. (2010). Management of nanomaterials safety in research environment. Part. Fibre Toxicol. 7: 40.

[33] Buitrago E., Novello A.M., Fink A., Riediker. M., Rothen-Rutishauser B., and Meyer T. (2021). NanoSafe III: A User Friendly Safety Management System for Nanomaterials in Laboratories and Small Facilities. Nanomaterials. 11(10): 2768.

[34] Schwab F., Rothen-Rutishauser B., Scherz A., Meyer T. Begüm B. Karakoçak, Fink A. (2023). The need for awareness and action in managing nanowaste. Nat. Nanotechnol. Doi: 10.1038/s41565-023-01331-4.

12 Concluding remarks

In any organization, practices regarding engineering risk management are, without a doubt, what make the difference in safety performance, and, by that extent, in business performance. When it comes to making a difference, there is no goal more important than ensuring that risks (positive and negative) are well managed and that every employee goes home safe (and happy) every day. The managers of long-term successful organizations are indeed well aware that decreasing negative uncertainties and risks and guaranteeing safety for their employees makes the difference between their companies and less successful organizations. Hence, engineering risk management is very important.

Risk is understood as an uncertain consequence of an event or an activity with respect to something that human's value. In its simplest form, risks refer to a combination of two components: the likelihood or chance of potential consequences and the severity of consequences of human activities, natural events or a combination of both. Such consequences can be positive or negative, depending on the values that people associate with them. This book therefore also looks at the risk concept from an ISO 31000 perspective, that is, risk is "the effect of uncertainty on objectives." For truly understanding risks equals truly understanding the possible effects, refer the accompanying uncertainties and the related objectives attached to future scenarios.

Often, engineers concentrate their efforts on (predominantly negatively evaluated) risks that lead to physical consequences in terms of human life, health and the natural and built environment. Risks also address impacts on financial assets, economic investments, social institutions, cultural heritage or psychological well-being, as long as these impacts are associated with the physical consequences. In addition to the strength and likelihood of these consequences, this book emphasizes the distribution of risks within and across organizations and, over time, space and people. In particular, the timescale of appearance of adverse effects is very important and links risk management to sustainable development (delayed effects).

We distinguish risks from hazards. Hazards describe the potential for harm or other consequences of interest. These potentials may never even materialize if, for example, people are not exposed to the hazards or if the targets are made resilient against the hazardous effect (such as immunization). In conceptual terms, hazards characterize the inherent properties of the risk agent and related processes, whereas risks describe the potential effects that these hazards are likely to cause on specific targets such as buildings, ecosystems or human organisms and their related probabilities.

There are a lot of books and other literature that discuss how to conduct nonfinancial risk management. However, these works usually only focus on the purely technical aspects of what risk management is all about, which is only a part of the

https://doi.org/10.1515/9783111493633-012

story. Or, they discuss it from a non-technological viewpoint and they only very briefly discuss technical matters of risk management. This book tries to provide an inclusive overview and treats risk management from an engineering standpoint, thereby including technological ("engineering") topics (such as what risk assessment techniques are available and how to carry them out, how to conduct an event analysis, etc.), as well as non-technological ("management") topics (risk management concepts, crisis management, risk governance, economic issues related to risks, etc.). There is a reason why it is so difficult to present and write an inclusive book on risk management: risk management is confronted with three major challenges that can be best described using the terms "complexity," "uncertainty" and "ambiguity." These three challenges are not related to the intrinsic characteristics of hazards or risks themselves but to the state and quality of knowledge available about both hazards and risks:

–   Complexity refers to the difficulty of identifying and quantifying causal links between a multitude of potential causal agents and specific observed effects.
–   Uncertainty is different from complexity, but often results from an incomplete or inadequate reduction of complexity in modeling cause–effect chains. It might be defined as a state of knowledge in which, although the factors influencing the issues are identified, the likelihood of any adverse effect or the effects themselves cannot be precisely described.
–   Ambiguity – Whereas uncertainty refers to a lack of clarity over the scientific or technical basis for decision-making, ambiguity is a result of divergent or contested perspectives on the justification, severity or wider "meanings" associated with a given threat.

In economics, the "Trias Economica" exists, implying that a healthy economy needs three partners: financial institutions (banks), insurance companies and all other companies. If one of these partners falls away or does not perform its function in the economy, the economy will heavily suffer. In risk management, a parallel can be drawn: the "Trias Risico" (risk trias) is proposed, indicating that, to exist, a risk needs three factors: one or more hazards, exposure to the hazard and possible loss. If one of these factors is eliminated, the risk ceases to exist. Risk management therefore is concerned with influencing/decreasing the importance, in any way (this can be extremely technical to procedural, to purely human factor, communication, etc.), of one or a combination of these three factors. This seems to be very simple, but it is not. Every individual factor and all its influencing parameters can be very complicated or complex to deal with, showing high levels of uncertainty, high levels of ambiguity, high requirements of knowledge and collaboration, (hidden) relationships between the parameters, etc. Because of this complexity, a systemic engineering approach is absolutely needed, besides an analytic engineering approach.

In the past, "risk management" was sometimes a synonym for "insurance management." Luckily, this is not at all the case anymore. Nowadays, organizations are well aware that nonfinancial risk management is a very important domain that indirectly leads to a large amount of financial gains. However, the domain is so wide that there seems to be no consensus about what are the best approaches and models to use, which are the most efficient analysis techniques, etc. This book therefore fills a gap in presenting in an easy and legible way, the knowledge and know-how about risk management in order to be successful. Indeed, applying the various ideas, concepts and techniques provided and elaborated in this book leads to more effective and more efficient risk decision-making and to better or to optimal results.

Nonetheless, we have to accept that the most debated part of handling risks refers to the process of delineating and justifying a judgment about the tolerance or acceptability of a given risk. The term "tolerable" refers to an activity that is seen as worth pursuing (for the benefit it carries), yet it requires additional efforts for risk reduction, within reasonable limits. Whereas, the term "acceptable" refers to an activity where the remaining risks are so low that additional efforts for risk reduction are not seen as necessary.

In the end, which risk can be accepted, tolerated or just taken will largely be driven by other factors than purely technical ones. Society, culture, ethics, economy, regulations and business appetite will largely influence the decision-makers. However, to take the best decision, it should be based on solid analysis and evaluation in order that the risk taken – because "zero risk" or "absolute safety" is a myth – is decreased as much as sustainably possible to a level that not only the business is taking, but also the community as a whole. A major challenge for risk managers and engineers resides in the best supply of information, and, in a broad sense, communication, to decision-makers or policymakers in order to achieve the best risk governance for a better world.

In popular perception, rocket science is the totemic example of complex and/or complicated science. The two terms are frequently mixed up, but they are fundamentally different in the following ways:
- Complicated refers to a situation, system, or problem with many interconnected parts or variables, but one that follows predictable patterns and rules.
- Complex, instead, refers to a situation, system, or problem with many interconnected and interdependent elements that interact unpredictably and dynamically.

Risk management is indeed no rocket science. It is much more complex than rocket science! After all, there are not ten possible approaches to send a rocket to the moon or to Mars. There are, nonetheless, tens of thousands of possible risks, and easily hundreds of possible approaches to manage them, leading to as many different possible outcomes and realities. Risk management is far more complex than it is merely complicated.

Einstein mentioned that if he had 1 h to solve a problem, he would spend 45 min in understanding and analyzing the problem, 10 min to perform a critical review and finally use the last 5 min to solve it. Transposing this to risk management leads to exactly the same concept, except that the last 5 min will be taken for the decision-making process.

To conclude, we, the authors of this book, are well aware that in today's world, managers know that *safety risks* (not necessarily *security risks* though) are a top priority. That is reality. Managers don't want to see someone working under their supervision get hurt. Nowadays, good safety performance is a requirement for a successful career. Moreover, safety is no longer just a "priority." It has become an organizational "value," listed in many mission statements. Safety *is* really that important. Case closed. Or is it?

Well, if every manager truly understood how essential safety is, most workfloors would look more than a little bit different. There wouldn't be a shortage of staff to perform operational tasks, operating equipment would be very well and timely maintained, unsafe shortcuts to get the work done would be nonexistent, no safety problems would be ignored for reasons of time and/or money, safety trainings would be extensive, competences would be regularly checked and improved, there would be large health and safety departments staffed by employees with a variety of knowledge disciplines to solve multi- and transdisciplinary problems, etc., to name a few. Reality is different, as we all know: that is simply an observation, not an indictment. Managers and leaders should understand that safety, along with productivity and innovativeness, are actually the most important business objective, they have to be long-term profitable and for their organizations to be sustainable and to be around, long after they are gone.

Index

https://doi.org/10.1515/9783111493633-013

www.ingramcontent.com/pod-product-compliance
Lightning Source LLC
Chambersburg PA
CBHW080656220326
41598CB00033B/5222